MANUAL OF METEOROLOGY

MANUAL OF METEOROLOGY

VOLUME IV

METEOROLOGICAL CALCULUS: PRESSURE AND WIND

(A REVISED EDITION OF PART IV, 1919)

BY

SIR NAPIER SHAW, LL.D., Sc.D., F.R.S.

*Late Professor of Meteorology in the Imperial College of Science and
Technology and Reader in Meteorology in the University of London;
Honorary Fellow of Emmanuel College, Cambridge; sometime
Director of the Meteorological Office, London, and President
of the International Meteorological Committee*

WITH THE ASSISTANCE OF

ELAINE AUSTIN, M.A.

*of the Meteorological Office
formerly of Newnham
College, Cambridge*

CAMBRIDGE
AT THE UNIVERSITY PRESS
MCMXLII

CAMBRIDGE
UNIVERSITY PRESS

University Printing House, Cambridge CB2 8BS, United Kingdom

Cambridge University Press is part of the University of Cambridge.

It furthers the University's mission by disseminating knowledge in the pursuit of education, learning and research at the highest international levels of excellence.

www.cambridge.org
Information on this title: www.cambridge.org/9781107475496

© Cambridge University Press 1931

First edition 1919
Second edition 1931
First published 1931
Reprinted 1942
First paperback edition 2014

A catalogue record for this publication is available from the British Library

ISBN 978-1-107-47549-6 Paperback

PREFACE TO VOLUME IV

THE volume now presented is a revised edition of Part IV which was issued in advance in 1919. There are besides two chapters introductory to the theory of motion under balanced forces which had been adopted in Part IV as the basis of the relation between pressure and wind, and two chapters supplementary, summing up the position with regard to the theory of the general circulation of the atmosphere and its local disturbances. Here we need only account for the limitations of the work now ended.

Its purpose was to give the reader an idea of the general problem of the meteorology of the globe and of the material and methods which are available for its solution.

More narrowly stated the problem is to provide a rational quantitative explanation of the sequence of the phenomena of weather in any locality—its first stage to trace the phenomena to their natural causes and identify the physical and dynamical processes by which the sequence is controlled.

In *Nature* the account which has been given of the physical processes of weather in Vol. III has been called discursive, and with at least equal justice the epithet may be applied to the account of the dynamical processes in this last volume; and there is no reason for resentment on that account. Fifty years ago similar remarks were often made about Maxwell's *Theory of Heat*, "an extraordinary textbook," the perusal of which may even now be recommended to students of meteorology as a good resolution for every New Year, though neither author nor editor nor publisher (up to 1891) had found himself able to make an index to it. Nature is herself very discursive in the demands which she makes upon her various sciences for providing the sequence of our weather. The book does not aspire to be regarded as an abstract even of meteorological literature, still less of dynamical or physical. It adheres as strenuously as circumstances permit to the idea of weather as a manifestation of the transformations of energy taking place in the atmosphere. Anything which helps the comprehension of the actual structure and processes as affecting the transformation of energy would be helpful to the book; but the accretions around different sections of the subject, instrumental, observational, methodical, theoretical or literary, even if the point of attachment is physical or dynamical, are deliberately left for the reader's enjoyment of private excursions with occasional finger-posts to remind him of their existence.

Discursiveness is shown not merely in the sections of the subjects treated but also in the mode of treatment itself. Results have been quoted without exhibiting the algebraical reasoning employed in the original papers. Algebra has powerful attractions for its familiars; but it is not everybody's hobby. We shall have satisfied our aspiration if those who seek can find. There is no part of the subject in which a knowledgeable friend is not welcome. Our own obligation to D. Brunt on that score alone is always in mind.

The plan of the complete work as modified by the experience gained in its construction stands as follows:

Vol. I, 1926, Meteorology in History leading up to modern instruments and methods of investigation.

Vol. II, 1928, Comparative Meteorology setting out the accumulated data in maps, diagrams and tables.

Vol. III, 1930, The Physical Processes of Weather with examples of their application.

Vol. IV, 1931, Meteorological Calculus for expressing the relation of pressure and wind.

A general summary of the contents has been added to exhibit the manner in which the material has been arranged.

Additional acknowledgments for the material of this volume are indicated in the Table of Contents, p. xiii, and List of Illustrations, p. xvii. We can only reiterate our grateful thanks to the Meteorological Committee of the Air Ministry and to the Director of the Meteorological Office for their continued support, to Mr R. G. K. Lempfert, C.B.E., Mr D. Brunt and Commander L. G. Garbett, R.N., for assistance with the proofs, to the Cambridge University Printer and his staff of compositors and readers for the form in which the work is presented.

Since Part IV was written originally with the needs of aviation in view the attention of the world's meteorologists has been concentrated upon that part of the general subject, and the information available in 1919 will seem very small compared with what has been published since that date. But still it provides material for the portals of the subject, and though perhaps perspective makes the portals of 1919 seem insignificant compared with the masses of material now available for those who look through them, the alignment may be useful in the development of the structure.

The body of the volume, chapters numbered III to X in place of I to XI, remains devoted to the study of the kinematics of the atmospheric structure and is here reprinted with necessary alterations and a few additions. The two new chapters at the beginning introduce the classical methods of dealing with the dynamical aspects of the problem of atmospheric motion and incidentally illustrate the importance of the idea of stratification by entropy. The other two at the end summarise the view attained when the situations expressed in the three previous volumes, historical, geographical and physical, are taken into account.

No general solution of the dynamical problem of the atmospheric circulation is offered but some suggestions of a connected system are afforded in the final summary by harmonies in the correlation of some well-recognised parts.

In some respects the attitude adopted is one for which it would be difficult to find approval in the meteorological canon. The following statements are designed to express the more pertinent innovations.

1. Records of horizontal wind by pressure-tube anemometers are put forward as a concise statement of the dynamical problem of the atmosphere and a guide to the order of consideration of the elements of structure as set out in chapters III to X.

2. In the free air, motion is the controlling dynamic feature of the circulation. Pressure-gradient, stabilised by the persistence of horizontal motion on the rotating earth, appears as a static index of the motion. It is not regarded as a source of energy apart from that indicated by horizontal differences of entropy.

3. The balance of horizontal wind, pressure-gradient and centrifugal force is disturbed by turbulence due to rigidity and friction at the surface and also by convection which operates wherever there is a difference of entropy between masses of air in juxtaposition, most notably at points along the "polar front."

4. Convection is regarded as the effect of gravity redressing in any horizontal surface differences of entropy that may be developed by conduction of heat to or from the ground, by the emission or absorption of radiation or by the condensation or evaporation of water.

5. Apart from the disturbing effects of conduction and turbulence at the surface and of local convection, the relation of
horizontal wind to pressure-gradient, which was and still is the
fundamental principle of the volume, is further developed and confirmed; but motion has taken the place of pressure as the "independent variable" in the gradient equation, that is to say the
pressure-distribution is regarded as the "banking" required for the
maintenance of the air-currents.

6. Motion is derived ultimately from convection with the aid of
conservation of momentum, but the derivation is intricate and only
partially explored.

7. Since entropy-difference acting through gravity is the cause
of convection, it follows that the elements by which the whole
drama of the atmosphere is staged are two, viz. entropy-difference
and wind-velocity. Entropy-difference is contingent upon the
condition of the air in respect of water-vapour.

Entropy (with water-vapour as an accomplice) and air-motion
are the joint rulers of the atmosphere; gravity, pressure, the centrifugal force of rotation whether of the earth or of a local mass (and
incidentally the author) are their obedient servants.

NAPIER SHAW

31 *March* 1931

PREFACE TO PART IV 1919

WITHIN the past four years urgent questions have been addressed to the Meteorological Office from many quarters about the winds. Some of them refer to the winds at the surface, a subject of immemorial antiquity which has only within the last score of years been subjected to comparatively accurate measurement; and others refer to the winds of the free atmosphere beyond the reach of the highest point upon which an anemometer could be fixed.

The work of the Meteorological Office aided by the contributions of data from various departments of the naval, military and air-services has provided the material for answers to these questions, but the first conclusion derived from the study of the material is that the answers to all the questions cannot be treated separately. They all form part of a description of the structure of the atmosphere, a structure so complicated, even so perverse, as to make the attempt to describe it intelligibly without some guiding principle, or to deal with it piecemeal, hopeless.

We have found a guiding principle of great practical utility in the relation of the wind to the distribution of pressure which can be deduced from the assumption that as a general rule the motion of air in the free atmosphere follows very closely the laws of motion under balanced forces depending on the spin of the earth and the spin in a "small circle" on the earth. And therefore, in order to provide the best available answers to the questions put to us, we have studied the relation of the winds to the distribution of pressure at the surface and in the free atmosphere.

This study has led to setting out what amounts almost to a general meteorological theory. It forms Part IV of this manual and includes within its scope the best answer which we are able to give to the general questions put to us.

Behind it lies the vast accumulation of facts obtained by the industry and perseverance of meteorological observers all over the world which are represented compendiously for practical meteorologists by a series of normal values of meteorological elements of every kind. These are summarised in Part I (Vol. II) as a general survey of the globe and its atmosphere which is based upon the great but unfortunately still incomplete work of Hildebrandsson and Teisserenc de Bort, *Les Bases de la Météorologie dynamique*.

But anyone who takes an intelligent interest in the structure of the atmosphere, whether he regards it in detail or thinks only of its more general features, must have a working knowledge of the physical properties of air, and that is no slight matter. Maxwell's wonderful *Theory of Heat* with all its digressive chapters might have been written for the purpose, for all that it contains is extraordinarily appropriate. Only one additional chapter, on the difficult subject of the thermodynamical properties of moist air, is required.

The physical properties of air form the subject of Part II (Vol. III).

Part III (Vol. IV) contains the formal setting out of the dynamical and thermal principles upon which theoretical meteorology depends, and which find their application in Part IV. It is necessarily technical but again its main outlines are sketched by the hand of a perfect master of the art in Maxwell's *Matter and Motion*.

The whole is preceded by a historical introduction and a statement of the position of the general meteorological problem at the present day (Vol. I); because the history of the study of weather forms a striking example of the interaction of the progress of science and the creation of the instruments which it uses.

Part IV is issued in advance because what is contained therein has not hitherto been presented in a collected form. It represents the progress made chiefly by those who have been associated in the work of the Meteorological Office in the past twenty years. We owe our success to the fortunate circumstances of our meteorological *terrain*.

For the other parts of the subject all meteorologists have the same sources of reference open to them if they care to use them.

Our concern in this work is to present a summary of them in the most handy form for conveying an idea of the information which is available. For the survey of the meteorology of the globe Bartholomew's *Atlas of Meteorology* by Buchan and Herbertson is in itself an admirable compendium.

The special climatological atlases, of Russia by General Rykatchef, of India by Sir John Eliot, of Canada by Sir Frederic Stupart, the great work on the climatology of the United States of America by A. J. Henry and the less complete but still notable work by C. M. Delgado de Carvalho on the Meteorology of the United States of Brazil, invite contributions to the common stock of knowledge on the part of other countries so that the student of meteorology may not continue to be dependent upon the data in their original form, which are only contained in a few of the libraries of any country.

The physical and dynamical principles upon which the processes of weather depend are the common property of all students of physics. If those to whose care the progress of physics is entrusted had taken the physical problems of the atmosphere under their charge as their predecessors did before the advent of the electrical era one half at least of this book might have been more effectively dealt with by other hands.

A work of this kind necessarily depends in very large measure upon illustrations which often represent, in the most succinct manner, results of observations which cannot be transcribed in words or formulae. The original drawings are the only satisfactory evidence for writer or reader because in the gradual development of the science what at one stage of our knowledge appears to be a superfluous accident may become the starting point of a new advance. The author and the Meteorological Committee, at whose instance this work was undertaken, desire here to place on record their acknowledgment to the Controller of H.M. Stationery Office, the Board of Trade, the Ordnance Survey and the Advisory Committee for Aeronautics for permission to use illustrations which have appeared in the publications of H.M. Government.

And similar acknowledgment is due to the Royal Society, the Royal Meteorological Society, the Carnegie Institution of Washington, Professor McAdie of Harvard University and particularly to C. J. P. Cave of Stoner Hill.

Particulars of the extent to which illustrations have been borrowed are set out in the list contained in pp. xvii to xx.

The author desires also to express his thanks to an old friend and colleague, Mr J. B. Peace, Printer to the University of Cambridge, for the care which he has given, in the difficult circumstances of the later stages of a great war, to the arrangement of the book and the form of the illustrations.

NAPIER SHAW

METEOROLOGICAL OFFICE, LONDON
    9 *December* 1918

A preliminary issue of a limited number of copies of this part of the work for official use has enabled me to obtain from colleagues and friends some corrections and suggestions which have now been incorporated in the text or added as notes.

N. S.

17 *May* 1919

TABLE OF CONTENTS

VOLUME IV. METEOROLOGICAL CALCULUS: PRESSURE AND WIND

LIST OF ILLUSTRATIONS

NOTE. Many of the diagrams are copied, with modifications in some instances but more generally in the original form, from the publications to which reference is made. Those marked * are from Official Publications of H.M. Government and permission for their reproduction has been obtained from the Controller of H.M. Stationery Office. The mark † indicates original records of the Meteorological Office.

Thanks are also due to the Royal Meteorological Society for fig. 45 and to the Royal Aeronautical Society for fig. 1.

FIGURE NUMBERS

In this revised edition the figures have been re-numbered consecutively throughout the volume in accordance with the practice in the other volumes. In the original edition the figures of each chapter were numbered separately. For the illustrations of the original edition the original figure-number is inserted after the number on the revised plan.

The relation between the current numbers and the original chapter-numbers of Part IV is indicated in the following table:

Part IV fig.	1	2	3	4	5	6	7	8	9	10
				Figure numbers in new volume						
Chap. II	17	18	19	20	.	.	.	.	.	.
„ IV	23	24	25	.	.	.	.	.	.	.
„ V	28	29	30	31	32	33	34	.	.	.
„ VI	35	36	37	38	39	40	41	42	43	44
„ VII	49	50	.	.	.	.	.	.	.	.
„ VIII	22	21	.	.	.	.	.	.	.	.
„ IX	52	51	.	.	.	.	.	.	.	.
„ X	59	60	61	63	64	62	65	66	.	.
„ XI	67	71	72	68	69	70	.	.	.	.

The Frontispiece of Part IV has become fig. 74.

CHAPTER I

THE LAWS OF ATMOSPHERIC MOTION

Another storm a'brewing—I hear it sing i' the wind
Tempest, Act II, Sc. 2.

NEWTON'S LAWS OF MOTION

THE problem of dynamics as applied to the atmosphere is to trace the relation between the motion of the air as observed in the winds at the surface by anemometers, in the upper air by aeroplanes, or as inferred from the motion of clouds or balloons, and the "forces" which are assumed to be operative in producing or in changing the motion. For the solution to be accepted the relation established must be quantitative as well as qualitative, that is to say the forces which are involved must be shown to be adequate to give not merely a general idea of the phenomena which have to be explained but also an accurate measure of the motion which has been observed.

When a satisfactory relation of that kind has been established between the observed motion of the air and the forces available to produce it and maintain it, the same process can be applied to changes in the present state of motion, which will be produced by the operation of the available forces, provided these can be expressed numerically. The prediction of weather will then become a "mathematical certainty."

The classical example which mathematicians have in mind when they endeavour to trace a quantitative or numerical relation between the forces operative in the atmosphere and the motion which results therefrom is the extraordinarily accurate solution of the problem of the dynamics of the solar system as expounded by Sir Isaac Newton in his *Philosophiae Naturalis Principia Mathematica* which contains also applications of the same principles to other dynamical problems.

For more than two centuries that exposition was regarded as the outstanding example of the ingenuity of the human mind—*genus humanum ingenio superavit*—and may still be so regarded unless the modification introduced in the present century by Prof. A. Einstein on the new principle of relativity of motion be considered a still more impressive example of human genius.

The mysterious influence concealed under the name of gravity has become a spatial contortion, and we have found the new point of view useful for weather-study in relation to the other mysterious influence "entropy."

The efforts which have been made hitherto by mathematicians to solve the problem of the dynamics of the earth's atmosphere have been guided by the Principia of Newton, and based upon three laws or axioms, which express in singularly well-chosen words the general ideas of force in relation to motion. They were evolved in the seventeenth century from the observations of astronomers and the experiments of natural philosophers. In the original Latin, which hardly needs translation, they are as follows:

Axiomata, sive Leges motus

Lex I. Corpus omne perseverare in statu quo quiescendi vel movendi uniformiter in directum, nisi quatenus illud a viribus impressis cogitur statum suum mutare.

Lex II. Mutationem motus proportionalem esse vi motrici impressae, & fieri secundum lineam rectam qua vis illa imprimitur.

Lex III. Actioni contrariam semper & aequalem esse reactionem: sive corporum duorum actiones in se mutuo semper esse aequales & in partes contrarias dirigi.

It is one of the great achievements of the enunciation of the laws of motion that they embody a conception of force which can be employed in mathematical argument, namely that the measure of force, whatever name be given to it, is the rate at which the momentum of the moving body is changing. Momentum must be understood to mean a new physical quantity expressed numerically as the product of the mass of the moving body and its velocity. If the rate of change is uniform, as it is assumed to be in many illustrative examples, the force is expressed by the actual change of momentum in the unit of time—in other words the mass multiplied by the acceleration.

The conception of force

The reader will notice that he is assumed to be acquainted with what is to be understood by *vis* (force), *vires impressae* (impressed forces), *vis motrix impressa* (impressed motive force) and *actio, reactio*, for which the terms action and reaction are used in English. And indeed the conception of force, which is of course fundamental if forces are to be employed to calculate motion, is one which no ordinary person makes any difficulty about understanding. And yet, when the basis of the understanding comes to be examined, it is difficult to realise that in the specification of any force anything more is meant than that directly or indirectly, somehow and somewhen, the force has shown a capacity for resisting or balancing the attractive force of gravity.

It was doubtless the *brut* realism of this statement compared with the idealism of measuring force by momentum, which is never or hardly ever exactly possible, that led some distinguished engineers and mathematicians to insist upon measuring forces in terms of that of gravity upon a pound, and to turn a scornful lip towards the poundal and the foot-poundal for measuring force and work, which were offered by the advocates of systematic international measurement as a concession to British prejudice in such matters.

Let us note that, in order to be consistent in estimating force, the motion must be referred to the proper centre, otherwise we may arrive at paradoxical conclusions. For example, consider two independent planets, masses M and m, acting upon one another with some force like gravity, and consider the motion of each as observed from the other. Dealing with a period of time so short that the variation in the intensity of the force is insignificant the changes in their velocity in unit time are V and v. The force on the one is MV and the force on the other mv. But since velocity is relative the approach of M to m is the same as the approach of m to M; hence v and V are equal; but by Law III the action MV is equal to the reaction mv. Hence the two masses must be equal—which does not appear in the original hypothesis. The velocity measured ought not to be that of one body relative to the other but of each with regard to the "centre of mass" of the two.

The principles of Newton in the hands of mathematicians have been found serviceable in the solution of many dynamical problems relating to the heavenly bodies, the earth and its figure, the sea and its tides, the air and its pressure; in particle dynamics, rigid dynamics and hydrodynamics; and the feeling is quite general among mathematicians that the line of approach to the solution of the general meteorological problem is by the mathematical evaluation of the effect of known forces upon the state of the atmosphere which can be regarded as initial. In the case of the atmosphere the question that arises is whether we really know all the forces which are operating to produce and maintain the motion which may be observed in a sample of air.

It is this practice of assuming a knowledge of the forces in order to compute the motion instead of using the observed motion in order to infer the forces that constitutes the difference between the deductive, or mathematical, and the inductive, or observational method of treating the subject, either of which may furnish appropriate material for this volume.

The recognised forces

What then, let us ask, are the forces which meteorologists can offer to mathematicians for their enterprise? Clearly gravity, the great Newtonian force, is one, of which the character and magnitude are quite well understood even if its origin has still to be accounted for.

The latest view of mathematical physicists, as expressed by Einstein, seems to be that the force of attraction, or the change of momentum which is its equivalent, is an affection of space in the neighbourhood of material objects such as the bodies of the solar system; and it may be worth the reader's while to think of that mode of explanation in relation to the restriction of the motion of air to an isentropic surface which we have set out in chap. VI of vol. III.

Secondly there is centrifugal force which is perhaps not really a Newtonian force at all, and yet comes naturally to the mind when one thinks of a heavy bob, as that of a pendulum, whirled round one's head at the end of a string. Something is wanted to account for the pull of the bob on the string, and if we say that the pull, which requires the tension of the string to balance it, is centrifugal force we can urge in justification that on occasion it may be more than the string can bear and it is centrifugal force which produces disruption. If not, what does? In the atmosphere a centrifugal force which is never absent except at the very poles is that due to the rotation of the earth. It takes a hand in every meteorological phenomenon by pushing sideways any air that moves.

Next pressure, the statistical expression of momentum transferred by the impact of the molecules of a parcel of air upon its boundaries, in practice only a special manifestation of the force of gravity. If we consider the motion of the air we regard the bombardment of any part of its boundary by the molecules of its environment as producing a force normal to the surface bombarded, measured by the product of the pressure and the area of the surface, and balancing the pressure of the interior.

Fourthly—flotation, or, as commonly used in meteorology, convection, on the Archimedian principle of an upward force, equal to the weight of fluid displaced, acting upon any body immersed in a fluid, this also is another effect of gravity.

Fifthly—surface tension, the expression of what is called capillarity, which moulds water-drops into spherical shape as we have seen in vol. III. The peculiarities of this form of force are held to be responsible for the dynamical manifestations of thunderstorms, but it operates only in the surfaces of liquids, unless the colloidal properties of mixtures of foreign bodies with air may be in some way associated with it.

Sixthly—the force of turbulence, the tendency of moving air when it is passing other air with sufficient rapidity to roll itself up into something suggestive of a whirl or vortex. We shall have something to say later about the vitality of this particular mode of motion.

Seventhly—friction, a mysterious force which intrudes itself whenever and wherever the surfaces of bodies in contact slide, or even tend to slide, one past the other; and which, so far as the atmosphere is concerned, has something to do with viscosity—of which later.

There are other forces which have sometimes to be taken into account by those who would refer to Newton's laws of motion as embodying the principles upon which an answer must be found to the oft-repeated question, "Will it rain to-morrow?" But the seven examples which we have cited are sufficient to indicate that the problem is liable to a good deal of complication.

The classical mathematical method of dealing with any dynamical problem is to write down equations representing the balance between changes of motion (in terms of momentum), regarded as unknown, and impressed forces, regarded as known when normal information is available.

Natural philosophers have not always waited for that stage; they have on occasions proposed instead such approximations to a solution as can be obtained by general reasoning from their knowledge of the nature and distribution of the operative forces.

As we have already seen (vol. I, p. 288), that kind of reasoning provided an explanation of trade-winds and monsoons, of land- and sea-breezes, based upon general considerations of thermal convection and the rotation of the earth. Such explanations appeared for many generations in the text-books of physical geography; but there was no attempt to obtain numerical values either of the direction or of the velocity of the winds or their variation from season to season. In such questions however it is unsafe to disregard any influence or assume a complete knowledge of the causes, and so long as there is any uncertainty about the causes it is fair to say that for nearly all purposes an accurate description of the phenomena is the best substitute for an explanation.

Let us accordingly pass in review the steps which have been taken towards the numerical expression of the forces which we have enumerated.

1. **Gravity**—strictly speaking the force of attraction between all material substances in the universe, which in the case under our consideration would add to any mass, free to move, a velocity of approximately 981 cm/sec in every second, directed towards the centre of gravity of the earth. In practice however the gravity denoted by the letter g takes account also of the earth's rotation and is the acceleration of any body, free to move, along the "vertical," that is to say at right angles to the "horizontal" or "level" surface which forms the conventional boundary of a fluid earth. It points to the centre of gravity of the earth only at the equator and at the pole; elsewhere in the northern hemisphere it points south of the centre of gravity, in the southern hemisphere to the north of it.

Its numerical expression for a point on the earth's surface is given by the equation:

$g = 980 \cdot 617 \, (1 - \cdot 00259 \cos 2\phi) \, (1 - 5z/4\epsilon)$ c, g, s units, where ϕ is the latitude, z the height above sea-level, and ϵ the earth's radius

$= 980 \cdot 617 \, (1 - \cdot 00259 \cos 2\phi) \, (1 - 1 \cdot 96 \times 10^{-7} z)$, where z is the height in metres.

[ϵ is chosen as a symbol for the earth's radius because it comes near to being a semicircle with a radius from its middle point.]

This formula takes into account the additional attraction of the high ground and supposes the mean density of the elevated area to be equal to one-half of the mean density of the earth.

980·617 is the value of the gravitational acceleration in c, g, s units at sea-level in latitude 45°. For the determination of gravity at points above the earth's surface, the factor $1/(1 + z/\epsilon)^2$, which equals approximately $(1 - 2z/\epsilon)$, replaces $(1 - 5z/4\epsilon)$.

(*Computer's Handbook, Introduction*, M.O. 223, 1921, p. 9.)

2. **Centrifugal force.** The numerical expression for the effect produced at a point on the earth's surface by the rotation of the earth is represented by an acceleration $\omega^2 \epsilon \cos \phi$ outwards, perpendicular to the axis of rotation, where ω is the angular velocity of the earth and $\epsilon \cos \phi$ the distance of the point affected from the axis of rotation. We have seen that the value of g takes account of the earth's rotation. The vertical component gives a force $\epsilon \omega^2 \cos^2 \phi$ in diminution of g. The horizontal component gives a force $\epsilon \omega^2 \cos \phi \sin \phi$ directed towards the equator. This is balanced by the inclination of the earth's surface to the surface of a true sphere of the same mass and volume as the earth. The geometrical slope towards the pole which is represented by the excess of the equatorial radius (6377 km) over the polar radius (6356 km), and which therefore gives a greater distance from the earth's centre at the equator than at the poles, is known as the geoidal slope (vol. III, p. 296). It does not mean that a body on a perfectly smooth sea actually drifts from the equator to the pole, but it would so drift (and the water too) if the earth's rotation should by any chance ease off.

3. **Pressure.** We have so often used the expression of pressure in previous volumes that here we need only remind the reader that in computing pressure we assume a quiescent atmosphere (vol. III, p. 215) and express the pressure numerically with the aid of Laplace's equation $dp = - g\rho dz$. In considering the motion of an element of fluid it is the difference of pressure on two opposite sides that counts as the *vis impressa* upon the element.

4. **Convection.** This likewise has been in familiar use all through the work of the three previous volumes. Chapters VIII and X in vol. III have already been devoted to it. The numerical expression for the force of flotation of a volume v of air of density ρ in an environment of density ρ' is $\iiint g\,(\rho' - \rho)\,dx\,dy\,dz$ or, if local variation of density is negligible, $g\,(\rho' - \rho)\,v$.

5. **Capillarity.** We have already explained that capillarity only comes in when water-drops or other fluid bodies are under consideration. We may refer to the chapter on the subject in Maxwell's *Theory of Heat*, or any other text-book of physics, for the details of the numerical expression of the force. The effect of capillarity is expressed as a tension in the surface as though it were made of flexible material. Its numerical value is expressed by the force per unit of length of a line drawn in the surface. The tension of the surface between water and air is 74 dynes per cm, and of that between mercury at 290·5tt (17·5° C) and air 547 dynes per cm.

6. **Diffusive forces: viscosity and turbulence.** These forces cannot be evaluated like the five which we have considered. Their numerical expression is a statistical one and is derived from a consideration of experiments on diffusion in which the progressive distribution over one part of space, of matter or energy drawn from an adjacent part of space, is watched.

7. **Friction.** Here we have to distinguish between viscosity which gives the equivalent of a frictional or tangential force between two parts of a fluid moving one past the other and the friction between a fluid and a solid surface over which it is moving. The former may be regarded as subject to a general law for which a coefficient of viscosity is appropriate, the latter depends not only on the viscosity of the fluid but also on the nature of the solid surface.

CONSERVATION OF MASS AND ENERGY

The numerical expression of the diffusive and frictional forces of the atmosphere requires a more formal introduction than the familiar forces of gravity, centrifugal action, pressure and convective force; but before dealing with that part of the subject we may remind the reader that the whole of the calculus of weather which we have in view accepts as primary conditions the laws of conservation of mass and conservation of energy as explained in the introductory paragraphs of the chapter on "Air as Worker" (chap. VI, vol. III). We must include these conditions among the laws which govern the movements of the atmosphere. Let us therefore state them.

Law IV. Conservation of mass. In the computation of any atmospheric movement the expression of the distribution of the mass of the moving parts must account for any changes which may take place in the boundaries of a selected parcel and the mass contained within them on the understanding that the total mass of the whole system, viz the moving air and its environment, is unalterable.

The expression of this principle of the conservation of mass is referred to the coordinates in which the position of any parcel is expressed and will be included as the "equation of continuity" in our setting out of the general equations of motion of a parcel of air in the free atmosphere.

In considering the behaviour of "Air as Worker" in chap. VI of vol. III we have not paid any special attention to the particular forms of "work" which the air may perform, whether the transference of energy expressed by work results in the kinetic energy of moving mass, or any other of the forms which we have enumerated; but in the dynamics of the atmosphere these differences in the forms of energy are precisely the subjects of study. It will be sufficient here if we quote the statement of the principle as expressed by Maxwell.

Law V. Conservation of energy. The total energy of any material system is a quantity which can neither be increased nor diminished by any action between the parts of the system, though it may be transformed into any of the forms of which energy is susceptible.

This principle indeed is so vital to the study of the dynamics of a material system that it can be used as the starting-point for the expression of the equations of motion following the method of Lagrange to which reference will be made in chap. II.

What exactly is the material system under which the energy of the movements of the atmosphere can be studied requires a little consideration because it must take account of all the operating forms of energy of which gravity and solar radiation are the most important. But perhaps it will be sufficient if we regard the rotating earth with its atmosphere as the material system, allowing for radiation as energy supplied from space or lost thereto.

CONSERVATION OF MOMENTUM, LINEAR AND ANGULAR

While we are dealing with questions of conservation we must remember certain conditions relating to momentum.

A particular form of conservation is implied in Newton's third law. The equality of action and reaction between two bodies requires that if the action and reaction are measured by change of momentum, in any case of the influence of one body on the other the gain of momentum by the one body corresponds with the loss of momentum in the same direction by the other. Hence the momentum and indeed the component of momentum in any given direction is conserved during the dynamical operation of one body on another.

We must understand that the measure of the motion is duly taken with regard to the common centre of gravity of the two, otherwise we get into the difficulty suggested on p. 2.

This form of conservation is a notable matter because the two bodies regarded as a system may lose energy while momentum is conserved.

For example, two bodies of equal mass impinging one on the other with equal velocities with reference to their common centre will have the kinetic energy of both annihilated during the transference of momentum on impact, and lost unless the heat equivalent of the energy is brought into account. For this form of energy the laws of motion make no allowance.

There is another form of conservation of momentum which can be deduced from the laws of motion, namely that of angular momentum (see p. 45). That also has interesting aspects from the point of view of energy.

The angular momentum or moment of momentum of a moving mass with regard to a point in the plane of its motion is measured by the product of the momentum *mv* and the distance *s* of the line of momentum from the point. It is therefore represented algebraically by *mvs* and geometrically by twice the area of the triangle formed by joining to the point of reference the extremities of a line representing the momentum.

If the force under the action of which the mass is moving is a central force, that is if it always passes through the point of reference, it can never produce any acceleration at right angles to itself, and therefore though the velocity may change the moment of momentum is not affected. It remains constant throughout the motion. The motion will be such that the area of the triangle formed by joining the point of reference to the extremities of the line representing the velocity will be constant.

This is the expression of the law of equal areas, known as Kepler's first law of the motion of the planets with reference to the sun.

What is true of a planet revolving round the sun is equally true of a ring of particles rotating about a centre, and therefore true for a ring of air, or part of a ring, rotating about the earth's axis.

It is a property of great importance in the study of atmospheric motion, and we therefore enunciate:

Law VI. *Conservation of angular momentum.* Any portion of a ring of air rotating about the earth's axis under the influence of forces which are directed to or from the axis will conserve its moment of momentum or angular momentum.

An experiment in illustration of this law is described by Aitken (see p. 256).

THE LAW OF DIFFUSION

Let us now consider in greater detail the nature of the diffusive forces. In the illustration of the superior mirage on p. 61 of vol. III we have referred to gradations of density of a solution of sugar as produced by the gradual diffusion of sugar from the bottom of the vessel upwards through the water of the layers above it. The gradation may be expressed by the strength of the sugar solution (the amount of sugar per unit of volume) at different heights above the layer at the bottom. The process is expressed by an equation of the type $\dfrac{d\theta}{dt} = \mu \dfrac{d^2\theta}{dx^2}$, derived by J. B. J. Fourier for the diffusion of heat by conduction in a bar of metal and known by some as Fourier's equation and by others as Fick's equation.

The diffusion may be the diffusion of heat (conduction), the diffusion of a salt through water, or of water-vapour through air, or the diffusion of momentum between two streams of air with different velocities (viscosity), or the diffusion of potential temperature (entropy) by turbulence. All these processes are reduced to one form of expression by the consideration that the quantity of the element which diffuses across any area is proportional to the change in the strength of the element along its path. The coefficient of proportionality is known as the coefficient of diffusion. Hence at any point of its progress the rate of change (dm/dt) in the strength of the diffusing element is proportional to the space rate of change of the gradient of the element along the line of travel. In algebraical form the law is expressed as $\dfrac{dm}{dt} = \mu \, \dfrac{d^2m}{dx^2}$, where μ is a constant different for each of the different diffusive processes.

Thus we may enunciate as a law which takes its part in the control of atmospheric motion:

Law VII. In an atmosphere stratified in layers, when the rate of change of an element with respect to time is proportional to the space rate of change of the gradient of the element in the direction of flow, the element is said to be diffusing.

The simplest case is that of conduction of heat through a body of homogeneous material in which the diffusing element may be taken to be the energy expressed by temperature (vol. III, p. 223), and we can consider the flow of heat from one side of a conducting plate to the other.

Viscosity[1]

In the same way we can treat viscosity which is used in the expression of the force at a surface of separation in a stream of air the layers of which are in relative motion. The average momentum of the molecules of air is greater in a layer of greater velocity and the force arises, as we have said in chap. VIII of vol. III, from the exchange of mass between two layers in contact in consequence of the inherent velocity of the movement of the molecules of which the gas is composed. The mean square of the molecular velocity of a gas is $\overline{V}^2 = 3p/\rho$, where p is the pressure, ρ the density (which must be expressed in terms of the fundamental units). Hence regarding air as a "homogeneous" gas with a density of ·001161 at 300tt, and pressure 10^6 c, g, s units, we get $\overline{V}^2 = (3 \times 10^6 \times 10^3)/1\cdot161$ cm^2/sec^2 = $2\cdot583 \times 10^9$; $\overline{V} = 5\cdot08 \times 10^4 = 50800$ cm/sec as the velocity of mean square at that temperature.

The exchange that takes place in consequence of the lively bombardment of one layer by its neighbour carries fast-moving air downwards and slow-moving air upwards and tends to equalise the momentum much in the same way as, on a larger scale, the turbulence of the flowing air produces a diurnal variation of wind-velocity as explained by Espy and Köppen, see p. 96.

The effect of viscosity in a stream of air, the consecutive layers of which show velocity increasing at the rate dV/dz per unit of distance z across the stream, is a retarding force F opposite to V acting upon each unit area of the faster moving layer such that $F = -\mu dV/dz$, where μ is called the coefficient of viscosity.

The numerical value of μ for air at $0°$ C in c, g, s units is 0·000168. Hence in a horizontal air-current which increases in the vertical at 10 m/sec per km of height, or ·01 cm/sec per cm, the retarding force upon any square centimetre of any layer is ·01 × ·000168 dynes, or $1\cdot68 \times 10^{-2}$ dynes per square metre; the rate of loss of momentum across that area of the upper layer is $1\cdot68 \times 10^{-2}$ g cm/sec^2.

[1] See a lecture on 'Turbulence,' by G. I. Taylor, *Q. J. Roy. Meteor. Soc.*, vol. LIII, 1927, p. 201.

The dimensions of F are M/LT^2 and in consequence the dimensions of μ are M/LT, that is mass per unit of length time.

So far we have supposed the *direction* of the stream V to be the same at all levels, differing only in speed in consequence of the viscous friction. That might be so in a laboratory experiment, but in the free air the earth's rotation always gives a force varying with the speed and at right angles to the motion.

To allow for that in the frictional force we must take that force as equivalent to the rate of change of momentum in the stream. F will be equal and opposite to dM/dz and therefore proportional to dV/dz.

If we can disregard variations in the density of the viscous substance and concern ourselves only with the relative motion of successive layers, it may be convenient to base the calculations on a volume-unit instead of the ordinary mass-unit. Thus we can choose the mass of unit volume of the fluid as mass-unit instead of a gramme, understanding, of course, in that case that all transference of momentum is by change of velocity without any change of density, a condition not strictly satisfied in the case of air or any other gas, but sufficiently nearly so for most practical purposes. When this mass-unit is employed, the coefficient of viscosity is called "kinematic" as distinguished from the original dynamic coefficient and is denoted by ν. The dimensions of ν are L^2/T.

To keep the viscosity equation numerically true the transfer of momentum per unit of area, which is expressed in ordinary c,g,s units as $F = -\mu dV/dz$, becomes $F' = -\nu dV/dz$, as expressed in "kinematic units" where $\nu = \mu/\rho$. In these kinematic units the unit of mass is the mass in grammes of a cubic centimetre of the air.

With viscosity measured in this kinematic fashion, force is the rate of change of the momentum of a cubic centimetre. The loss of momentum per second across a square metre of surface in the case quoted above for air of density ·00125 g/cc is $800 \times 1.68 \times 10^{-2}$ or 13·44 cm²/sec. This expresses the transfer of momentum from an upper surface to a lower one across a layer in which the rate of change of velocity and consequently the transfer (or conduction) of momentum is uniform throughout the layer. What the upper surface loses the lower surface gains: each of the intermediate surfaces receives the same amount from the one next above and transmits the same to the one next below.

In order to study the changes in the distribution of velocity in the intervening layer, taking into account variations of velocity in the horizontal as well as in the vertical, the unit of volume is convenient.

Consider an element δx, δy, δz, with velocity V at the base and $V + \dfrac{\partial V}{\partial z}\delta z$ at the upper surface. The force in c, g, s units on the lower surface is $F = -\mu\,\delta x\,\delta y\,\partial V/\partial z$ and on the upper surface being opposite in direction to F is

$$-\left(F + \frac{\partial F}{\partial z}\delta z\right) = \mu\,\delta x\,.\,\delta y\,.\,\partial V/\partial z + \mu\,\delta x\,\delta y\,\frac{\partial^2 V}{\partial z^2}\,\delta z.$$

Hence $-\dfrac{\partial F}{\partial z}\delta z = \mu\dfrac{\partial^2 V}{\partial z^2}\,\delta x\,.\,\delta y\,.\,\delta z$, or if R is the force in dynamic units on the element per unit of volume $R = \mu\,\partial^2 V/\partial z^2$.

In kinematic units the accelerating force per unit of volume $R' = \nu\partial^2 V/\partial z^2$.
In the latter case the numerical value of the density does not appear in the equation. We can use the same equation for fluids of different density and compare the behaviour of the volume-elements of each by comparing their kinematic viscosity. But force must be understood to be expressed in terms of the mass of unit volume multiplied by the rate of change of velocity.

Turbulence, eddies, vortices and stream-line motion

Besides the exchange of molecules which is treated as viscosity there is another action between two streams of air in relative motion on which much more energy is spent but of which the mode of generation is still obscure; that is the interaction of the streams which causes "turbulence." Turbulence is also caused in a stream which passes solid obstacles, and when a stream is turbulent, for either cause, the motion of a particle is far from being the shortest distance between two points; it takes part in the circulation of eddies while it is making its way with reduced speed down stream.

The curious thing about turbulence is that it is only excited when the relative motion between the stream and the obstacles, or between an upper fluid and a lower, exceeds a certain limit. It is therefore something dependent upon the mutual relations of parts of a fluid structure. So long as the motion is slow (the actual velocity to be exceeded depends on the viscosity, being indeed greater the greater the viscosity) the fluid creeps past obstacles or allows its neighbour to creep past itself, without any confusion; but once the limiting velocity is passed the motion becomes confused, full of eddies with greater or less regularity of occurrence, and mixing takes place in a few seconds which it would take many days for viscosity to accomplish.

The whole study of vortex motion belongs to the department of turbulence; but if we regard a vortex as consisting of particles which are describing circles about a recognisable axis, a stream may exhibit remarkable turbulence without the formation of any definite vortex.

Possibly the action may be associated with wave-motion. We have explained already that when air passes over water, or light air passes over heavy air, wave-motion is set up in either medium, or in both; and in consequence the surface, originally smooth like the undisturbed sea, becomes rough; friction of a different kind arises involving vertical components of force and vertical motion of the fluids; the action is in consequence intensified and what starts as a scarcely noticeable deformation of the surface may become a succession of breaking waves which do in fact represent a very effective form of turbulence.

To pursue the suggestion further we require to know the law of wave-length in relation to density of such disturbances as those of V. Bjerknes in chap. I of vol. III; and if it should turn out that the wave-length increases as the difference of density of the two fluids diminishes (and it should do so according to Helmholtz's theory of dynamical similarity), the turbulence or relative motion in the upper air would be on a very large scale and might explain some features of atmospheric motion.

The study of turbulence is largely the creation of Osborne Reynolds[1]. He explained that turbulence tended to occur in the region of contact of two flowing fluids in the following circumstances.

1. If there is a particular variation of velocity across the stream, as when a stream flows through still water.
2. When the fluids are bounded by solid walls.
3. When the solid boundaries are divergent.
4. When the curvature is such that the velocity is greatest on the inside.

On the contrary, turbulence was less marked:

1. When there is viscosity, or fluid friction, which continually destroys disturbances.
2. When the fluids have a free surface.
3. When the solid boundaries are convergent.
4. When the curvature is such that the velocity is greatest on the outside.

We shall have to notice the difference in behaviour when we come to consider in subsequent chapters the vortical motion of fluid which has a core revolving like a solid, i.e. with no relative motion, and is surrounded by a "simple vortex" in which velocity is inversely proportional to the distance from the axis and is therefore liable to be turbulent near the core where V is great.

Osborne Reynolds illustrated his suggestions by attractive experiments with coloured bands introduced into water, which made visible the unseen and unsuspected disturbance of the flow.

When there is no turbulence, or the amount of turbulence is negligible, the flow of the fluid is said to be "stream-line" motion; it follows the outline of a solid, or the boundary between two liquids or two gases, without any eddies, and at each point of its track the motion is tangential to the "stream line" which marks the actual track of successive particles during the flow unless the character of the motion is changed by the variation of the dynamical condition.

The transition from stream-line motion to turbulent motion was shown, by Reynolds's experiments on the flow of water through tubes, to occur when the velocity V is so great that Vl/ν passes a critical value determined for the special case, ν being the kinematical viscosity of the fluid and l the linear dimension of the system. The critical value Vl/ν is of such importance that it carries the name of its discoverer as the Reynolds number.

Stream-line motion lends itself readily to the illustration of hydrodynamical equations, turbulent motion has to be treated statistically like the motion of the molecules of gases.

Turbulence is due to the action and reaction between the different parts of the fluid in which it is formed, and in that respect may be compared or contrasted with the motion of a rigid body and therefore approximately with the motion of a real solid which is never perfectly rigid.

We shall see that the equations of motion of a material particle may be extended to a rigid body with only the complication of the rotation of the mass about some axis through its centre of gravity, which is manageable.

[1] *Phil. Trans. Roy. Soc.*, 1883; *Proc. Roy. Inst.*, 1884; *Papers on Mechanical and Physical Subjects*, vol. II, 1881–1900, Cambridge University Press, 1901.

In the case of a fluid no such simplicity is possible and yet the motion of the fluid does entail action between the several parts and, in so far as the mutual action tends towards the properties which are represented by solidity, or in a perfect form by rigidity, the fluid will begin to imitate the solid; but its effort at imitation of rotation about an axis is generally very imperfect and incalculable. Still there is an element of rotation in it which forms the starting-point for the development of vorticity and vortex motion. Some forms of vortex may be said to have a "solid core" as explained by Oberbeck and represented in chap. IX, and the motion of a quasi-solid core of that character has an element of permanence peculiar to itself.

Turbulence occurs whenever a sample of air travels brusquely past another; a single pulse of motion produced by a flapper can be recognised easily, travelling across a large room with a velocity which is not that of sound but is suggestive of that of a vortex-ring. Indeed, the brusque motion of air relative to its environment after passing an obstacle produces the transient semblance of a vortex, and if the obstacle is a plate with a suitable hole in it a vortex-ring is produced which lasts until the motion is frittered away. There is however no dynamical measure of turbulence; all that can be done to characterise the condition of the atmosphere in respect of turbulence is to register the mixing which the turbulence produces, the mixing, for example, of cold air below with warmer air above by observing the temperature of the structure; the mixing of moist air with dry by observing the humidity. Regarded in that way turbulence follows the same law as diffusion but with a constant 100,000 times, or in extreme cases 1,000,000 times, as great as the ordinary coefficient of diffusion. As turbulence finds its visible expression in vortices and eddies the result of its action is sometimes called eddy-diffusion, and its constant is called the coefficient of eddy-diffusion.

Turbulence is hindered by viscosity and its influence is also less active where an upper layer moves over a lower one of markedly lower density. Air relatively warm may travel over a layer of cold fog for a whole day without seriously disturbing it; but when the densities of the two layers become equalised, and still more when the upper layer has less entropy than the lower, the turbulence is very vigorous.

THE LAW OF DYNAMICAL SIMILARITY

We have noticed that the effects of all the different examples of diffusion are covered by a common algebraical formula. This leads us to refer to a general principle with regard to the intrinsic similarity of dynamical operations. It is based upon dimensional equations derived from the numerical relations of physical quantities expressed in terms of systematic units (see p. 17). It was developed by Gauss, Helmholtz, Maxwell and Rayleigh[1], and more recently by other exponents of aeronautical science and has indeed been found of great practical service in aeronautical research because it enables the behaviour of a system in natural conditions to be inferred from the behaviour of a model on a small scale under artificial conditions.

[1] Advisory Committee for Aeronautics, *Report*, 1909–10, p. 38; 1910–11, p. 26.

The principle is known as the law of dynamical similarity: we may call it the law of the working model and express it as follows:

Law VIII. If for any dynamical system a numerical relation has been expressed between the elements of the system (e.g. the extent of the material affected, the density of the working substance, the velocities of the moving parts and the influences which cause or control the motion), and if the scale of one of the elements be changed the scales of the other elements must also be changed in the ratios which are necessary in order to maintain the relation between the numerical dimensions of the system. The ratios of change can therefore be ascertained by an examination of the dimensional equation derived from the numerical formula.

In Volume III there is displayed an obvious geometrical similarity between a flowing river and a lightning-flash. So that in one sense at least there is an analogy between the flow of water in a river and the flow of electricity in a lightning-flash. Whether the analogy has any physical significance, at the moment we are not prepared to say; but in the same volume we have also noticed another kind of similarity between different physical processes. Wave-motion, for example, is shown in the sea, in the air, and in the hypothetical medium which carries waves of light or heat. All these waves have certain analogous characteristics—wave-length, period of vibration and velocity of travel—and we may grant that these quantities, or elements, are dependent upon the properties of the medium concerned, its density, elasticity or gravity, by equations of similar form, differing only in the numerical values which are related if they satisfy the equation.

For example we have the universal relation $\lambda = V\tau$, where λ is the wave-length, V the velocity of travel of the wave, and τ the period of an oscillation of one of the particles which is affected by the wave. So we may conclude that if in one medium a wave has a length of 1 metre when the period is 1/1000 second, in another medium in which the period is the same and the velocity 10 per cent less the wave-length will also be 10 per cent less.

So also with longitudinal vibrations by which sound is transmitted; we know that in air the velocity is equal to the square root of the ratio of the elasticity to the density. A similar formula must hold for water of which the density is about 800 times that of air. The elasticity in c, g, s units is about 14000 times that of air, so the velocity of sound in water is to that in air as $\sqrt{14000}$ is to $\sqrt{800}$, or just over four times that in air. The leading change here is from a small scale density to a large scale density.

Fujiwhara (*loc. cit.* on p. 256) compares water-vortices and air-vortices and finds that a water-vortex of ·5 cm to 1 m in diameter corresponds with an air-vortex of 380 m to 4000 km and time-scale between 77 times and 320,000 times.

In like manner whenever we have a formula which connects the elements of a dynamical system we can infer the conditions on the large scale from those on a small scale if we are allowed to make the necessary adjustments in the scale of the auxiliary elements.

As examples of apparent similarity which is not dynamical we may contrast the velocity of propagation of the waves described in chap. I of vol. III and the ripples on the surface of water or mercury.

Example from gravity-waves and ripples

The velocity of travel of a gravity-wave in a medium of density ρ must depend upon the wave-length, the acceleration of gravity (which will be independent of the density) and the density. Hence we may write $v = k\lambda^x g^y \rho^z$,

or in dimensional form $\qquad LT^{-1} = L^x L^y T^{-2y} M^z L^{-3z}$,

and by equating the indices of the terms in L, M and T we obtain $z = 0$, $y = \frac{1}{2}$, $x = \frac{1}{2}$ and therefore $v^2 = k^2 \lambda g$, or for gravity-waves the velocity of travel is proportional to the square root of the wave-length.

In the case of ripples the velocity depends on surface-tension, s, density and wave-length. Surface-tension has dimensions M/T^2.

Hence, writing $v = k s^x \lambda^y \rho^z$,

$$LT^{-1} = M^x T^{-2x} L^y M^z L^{-3z};$$

hence $x = \frac{1}{2}$, $y = -\frac{1}{2}$, $z = -\frac{1}{2}$ and $v^2 = k^2 s/\lambda\rho$, or for capillary ripples the velocity is inversely proportional to the square root of the wave-length.

Working models of the atmosphere

In relation to the law of dynamical similarity as providing a criterion of the reality or aptitude of a working model in the representation of natural phenomena, let us give our attention for a few moments to various working models of the development of a vortex with a vertical axis as representing a cyclonic depression, a tropical revolving storm or a tornado, three recognisable features or elements of atmospheric structure which are certainly suggestive of vortical motion and differ one from the other in size and intensity of motion. The cyclonic depression may be of 2000 to 4000 kilometres diameter with winds of the order of 20 metres per second, the tropical revolving storm of 750 kilometres diameter and winds of 50 metres per second, and the tornado with a diameter of 1 kilometre and winds of 100 metres per second.

The models of vortical motion which have been suggested in the volumes of this work as representing possible modes of development of vortical motion from a quiescent atmosphere, start from (1) the formation of a vortex with a vertical axis in the water of a basin which has an outflow at the bottom and the variations of that kind of experiment described by Aitken and referred to on p. 256. (2) W. H. Dines's tornado model in the Science Museum in London in which a narrow vortical column about 1 metre in height is formed in a glass cube 1 metre each way with a louvred vertical opening in each side. The effective cause of the rotation is the removal of air from the middle of the top by a small fan-wheel. (3) A model described as a cyclone-gatherer in *The Air and its Ways*, in which the vertical height is reduced to a few centimetres, the boundary is cylindrical with louvred openings and the motive power is a vertical current of air blown upwards through circular openings in glass plates forming the top and bottom and representing the horizontal stratification of the natural atmosphere.

We will add (4) a wire model of a cyclonic depression by J. Bjerknes, shown in fig. 213 of vol. II, to represent his experience of the air-motion; and (5) another wire model, shown in fig. 220 of the same volume, which is practically of the same form with the addition of a spiral core, and was designed to associate the air-motion with the distribution of rainfall in the depression of November 11–13, 1901; (6) Exner's[1] experimental arrangement for observing the effect of a flow of cold air from the polar regions, and (7) Schmidt's[2] illustration of the advance of a front of heavy water.

Others are described in Weickmann's article on 'Mechanik und Thermo-dynamik der Atmosphäre,' Kapitel 80, in Gutenberg's *Lehrbuch der Geophysik*, 1929.

A complete equation necessary for the application of the law of dynamical similarity to the model could perhaps be made by a formula for the energy of rotational velocity of the air in terms of the rate of removal of air from the core, which would involve the dimensions of the air-space, and the power required to remove it. These are not available. The only characterisation that we have is the linear dimension and we notice that in Mr Dines's model the vertical dimension is about twenty times the linear diameter of the vortical column. In the cyclone-gatherer the vertical dimension is about one-twentieth of the horizontal. In the wire models the vertical and horizontal dimensions are approximately equal.

None of these proportions is really natural; we must look elsewhere for a working model of the cyclone, tropical revolving storm or tornado.

The construction of a working model which with the aid of the principle of dynamical similarity can be employed as an accurate representation of atmospheric motion has been discussed by C. G. Rossby in the *Monthly Weather Review*[3]. The consideration is important because the chief advantage of the principle of similarity is that if all conditions are satisfied a working model becomes in effect a physical integration of equations which cannot be solved directly on the natural scale.

Starting from the hydrodynamical equations of motion of a fluid with a known coefficient of viscosity which we shall give in the next chapter, Rossby shows that one condition for dynamical similarity is that the Reynolds number (Vl/v) must be the same for the model as in the natural atmosphere. From this an important inference can be drawn, namely, that if the linear dimension l of a system is increased in any ratio and the viscosity is not increased the effect will be *ceteris paribus* the same as if the viscosity were proportionally diminished in the enlarged system so that, as we shall see later (chap. II, p. 58), although its indirect effect in producing turbulence may be immense, on the vast scale of the atmosphere viscosity can be of very little direct importance in modifying the motion.

Gravitational acceleration is the same in the original and the model; consequently one condition of similarity is expressed by the dimensional equation $L/T^2 = 1$ and leads directly to the inference that if the linear dimension of a model is one nth of the original the time-scale of the motion is $1/\sqrt{n}$ of the

---

[1] *Dynamische Meteorologie*, Wien, 1925, p. 341.

[2] *Meteorologische Zeitschrift*, vol. XXVIII, 1911, p. 355; the diagram is reproduced in *Forecasting Weather*, 2nd ed. 1923, p. 340.

[3] June, 1926, vol. LIV, pp. 237–40.

original. Hence with a model in which 1 metre corresponds with 2000 km of nature, a day's atmospheric operations would be accomplished in a single minute and a year's operations in about six hours.

Further difficulties arise on account of the natural effect of the earth's rotation which would have to be represented in a model by an artificial rotation, and that implies an artificial direction for the resultant "vertical" force, and consequently an artificial surface (a shallow paraboloidal dish of which the edge for a 2-metre model is 2 mm above the centre of rotation) to form the equivalent of the earth's horizontal surface.

And that is not the end of the difficulty: the hydrodynamical equations make no provision for thermal effects; and on the small scale there would be nothing to correspond with the horizontal or isentropic stratification of the real atmosphere; so the atmosphere must be regarded as incompressible and isentropic. Moreover, the equations in three dimensions are unworkable and must therefore be reduced to two dimensions by disregarding any vertical velocity or its changes, which indeed amounts to disregarding the most visibly impressive of the phenomena of weather. On the other hand, the initial disadvantage of the small vertical thickness of the atmosphere compared with its horizontal extension is overcome by Rossby by a separate ratio of the change of vertical dimensions.

The whole discussion makes it clear that it is scarcely possible to hope for a dynamically similar model of the working of the atmosphere over a great area. For the present we may have to follow the example of aeronautical research and confine ourselves to problems of such small scale as are not obviously influenced by the earth's rotation. Such experimental models may suggest general principles the expression of which on the natural scale may be sought for in a closer investigation of local atmospheric structure. From this point of view a linear dimension of 100 kilometres would be already too large. So it would seem that in the line of approach to representation by a working model, the tornado comes before the great cyclonic depression, and for that we require some estimate of the energy involved and the forces by which it is expressed.

NOTE ON DIMENSIONAL EQUATIONS

On p. 14 we have introduced as law VIII the principle of dynamic similarity which, in aerodynamics, is the foundation of the method of study of actual conditions in air-currents by the examination of scale models in a wind-channel, and in doing so we have referred to a dimensional equation. The use and meaning of dimensional equations are mainly electromagnetic accomplishments and can hardly be reckoned as part of the ordinary equipment of students of weather; but they are not without interest or importance in the study. To interpolate an explanation where the word is mentioned would break the thread of the narrative and exaggerate the author's reputation for discursiveness. In order, therefore, not to leave the curious reader without assistance we add here a note on dimensional equations and change of units extracted from a back number of the *Proceedings of the Cambridge Philosophical Society*. We have amplified it by an assertion of the absolute claims of temperature to the status of a physical quantity with dynamic dimensions which have hitherto been neglected by both student and professor.

Since the introduction of methods of measuring electrical and magnetic quantities in absolute measure considerable attention has necessarily been turned to the question of the dimensions of units and dimensional equations. Maxwell, as is well known, has in various places discussed such questions, and they naturally form an important part of Everett's "Units and Physical Constants." But I do not recollect having anywhere seen any precise statement of the manner in which dimensional equations arise and what their actual significance is. I therefore venture to suggest the following exposition of the method of deducing dimensional equations, and I do so with more confidence as there seems a general tendency to attribute to the well-known symbol in square brackets more of the attributes of an actual concrete quantity than it is justly entitled to.

We may accept in the first place, as usual, that the complete expression of a physical quantity consists of two parts and may be represented by the symbol q [Q], where q represents the numerical part of the expression and [Q] *the concrete unit of its own kind* which has been selected for the measurement of the quantity.

The unit [Q] is initially arbitrary for every kind of quantity. There exist however certain quantitative physical laws which really express by means of variation equations relations between the numerical measures of quantities. We may take for instance the following to be the expression of Ohm's law: "The numerical measure of the current in an elongated conductor varies directly as the electromotive force between the ends of the conductor." Or Oersted's discovery may be summed up as follows: "The numerical value of the force upon a magnetic pole placed at the centre of a circular arc of wire conveying a current, varies directly as the strength of the pole, as the length of the wire, as the strength of the current, and inversely as the square of the radius of the arc." A very large number of similar instances might be given.

We may thus take as the expression of a physical law the general form

$$q \propto x^\alpha y^\beta z^\gamma \ldots,$$

where $q, x, y, z \ldots$ are the *numerical measures* of the different quantities concerned in the relation.

We may of course express the variation equation in the form

$$q = k\, x^\alpha y^\beta z^\gamma \qquad\qquad \ldots\ldots(\text{1}),$$

where k is some constant whose value in general alters, if we alter the units in which the different quantities are measured, for by so doing we alter in the inverse ratio the numerical values $x, y, z \ldots$.

We may adopt one of two courses with respect to the quantity k.

(1) If all the possible variables have not been accounted for we may regard k as a fresh variable. This has been done in the instance first quoted, viz. in that of Ohm's law, where k depends on the nature of the conductor. Thus the reciprocal of k in that instance is now generally known as the "resistance" of the conductor, and we re-state the law thus: "The current in the elongated conductor varies directly as the electromotive force between the ends and inversely as the resistance of the conductor," and the expression of the law becomes $c = k'e/r$. So that we are still left with an equation of similar form, and hence may regard the equation (1) as the final general form of the expression of any physical law.

(2) We have already mentioned that the numerical value of k depends upon the units [Q], [X], [Y], [Z] ... employed to measure the different quantities. We may therefore assign to k any value we please by a suitable choice of any one of the units [Q], [X], [Y] ....

For many reasons it is convenient that k should be unity, and therefore the most usual assumption is that the unit of $[Q]$ should be so chosen that k shall be unity.

In the same manner x, y and z can generally be connected by physical laws with the three units of mass, space and time, and we may thus obtain $k = 1$ for a large number of physical equations, provided the whole series of units are chosen on the principle here indicated. We thus see that systems of units can be formed based on three fundamental units, such that a whole series of physical laws, expressing relations between the quantities measured, can be represented by ordinary equations with constant unity, instead of by variation equations. We thus arrive at systems of units founded on this principle, and a unit belonging to such a system is called an absolute unit. For such a unit the right of arbitrary choice has been given up, and it is agreed that the choice shall be directed by a consideration that the quantity k in certain equations shall be made equal to unity.

It follows from this that when the three fundamental units are selected the rest of the units belonging to the system are thereby defined, and that if the fundamental units are altered, corresponding alterations must take place in the whole system based upon the three fundamental units, in order that the k's may be still maintained equal to unity.

Let us consider the change from one system of absolute units to another, both founded upon the same principle, that is to say, both agreeing that the same k's shall be unity.

The equation between the *numerical measures* of q, x, y, z ... thus becomes for *both* systems

$$q = x^\alpha y^\beta z^\gamma....$$

Let the unit of x be changed from $[X]$ to $[X']$, that of y from $[Y]$ to $[Y']$, and of z from $[Z]$ to $[Z']$. In consequence that of q changes from $[Q]$ to $[Q']$. Then if q', x', y', z' be the new numerical measures of the same actual quantities measured, we have

$$q\,[Q] \equiv q'\,[Q'], \quad x\,[X] \equiv x'\,[X'], \quad y\,[Y] \equiv y'\,[Y'], \quad z\,[Z] \equiv z'\,[Z'] \quad ......(2).$$

And since the equation between the numerical measures is by agreement the same as before, since both systems of units are absolute,

$$q' = x'^\alpha y'^\beta z'^\gamma...;$$

$$\therefore \frac{q'}{q} = \left(\frac{x'}{x}\right)^\alpha \left(\frac{y'}{y}\right)^\beta \left(\frac{z'}{z}\right)^\gamma...,$$

and hence

$$\frac{[Q']}{[Q]} = \left(\frac{[X']}{[X]}\right)^\alpha \left(\frac{[Y']}{[Y]}\right)^\beta \left(\frac{[Z']}{[Z]}\right)^\gamma....$$

Thus if the fundamental units X, Y, Z be changed in the ratios

$$\xi : 1, \quad \eta : 1, \quad \zeta : 1,$$

and the derived unit in the ratio $\rho : 1$, then

$$\rho = \xi^\alpha \eta^\beta \zeta^\gamma.$$

This statement may be evidently expressed by the relation

$$Q = X^\alpha Y^\beta Z^\gamma \qquad ......(3),$$

where, now, Q, X, Y, Z *no longer represent concrete units but the ratios in which the derived unit* $[Q]$ *and the fundamental units* $[X]$, $[Y]$, $[Z]$ *respectively are to be changed*, it being understood that the same method of defining the absolute system is to be adopted throughout.

The equation (3) giving the ratio in which a derived unit is changed when the fundamental units are changed in any given ratios is called a dimensional equation, and is very convenient for determining the factor of conversion for any unit, from one absolute system to another governed by the same principles.

Generally speaking a dimensional equation is used to show the relation of each side of an equation to the fundamental units with the understanding that the index of the power of each fundamental unit must be the same on each side. Thus, in the ordinary dynamical equation $s = \frac{1}{2}gt^2$, if we assume that the dimensions of g are $L^x M^y T^z$, disregarding the numerical constant, we may write the dimensional equation

$$L = L^x M^y T^{z+2},$$

and equating the dimensional index of each fundamental unit on either side

$$1 = x, \quad 0 = y, \quad 0 = z + 2.$$

In like manner we might tackle the adiabatic equation $pv^\gamma = $ constant; but in that case we should find that the constant was not independent of the change of units and that the relation was not a simple one.

The following rule for calculating the factor of conversion when the dimensional equation is given is easily remembered. If in the dimensional equation we substitute for the symbols of the fundamental units the value of each old unit in terms of the corresponding new one, the result gives the factor for converting the numerical measure of a quantity from the old system to the new. And if, on the other hand, for the symbols of the fundamental units there be substituted the new units in terms of the old, the result gives the factor by which the old derived unit must be multiplied to give the new derived unit.

We may under certain circumstances work backwards from the dimensional equation to the physical law, in case we know the dimensional equations from other sources. The problem in this case is practically knowing the dimensional equation for q to determine α, β, γ.

This may be applied for instance to prove the equation for the velocity of sound, viz. $v \propto \sqrt{p/\rho}$, because that equation equalises dimensions on either side. But it involves the assumption that the velocity depends upon no other quantities than p and ρ, and thus we get no indication of the dependence of the velocity upon either temperature or the ratio of the specific heats because neither of these quantities is measured in units which vary when the fundamental units of length, mass, and time vary.

The method of proving a physical law by means of a dimensional equation may thus be sometimes misleading. Any dimensional equation may be expressed as the product of two others, one of which may be a dimensional equation of recognisable form, but it does not necessarily follow that there is any physical interpretation corresponding to it.

The dimensional equations for electrical quantities on the electromagnetic system may be deduced from those on the electrostatic system by multiplying or dividing by a dimensional equation representing some power of a velocity. In this case a physical meaning can be assigned to the velocity, namely that of propagation of electrical disturbance.

The dimension of temperature

Let us pursue the considerations set out in this note with a little more precision with regard to temperature.

The note regards temperature as having no dimensions; but let us recognise that for air p/ρ is constant and proportional to the temperature on the absolute scale; thus for absolute measure a, we get $a \propto l^2/t^2$ and absolute temperature assumes the dimensions of geopotential, namely energy per unit mass. Incidentally this provides an answer to those critics who regard the expression of temperature on the absolute scale as something of the same order as its expression in the arbitrary scales of Fahrenheit or Celsius.

Let us take the physical equation $pv = Ra$, where v is the volume of unit mass of air, p its pressure and a its absolute temperature. R is a constant equal to $2 \cdot 870 \times 10^6$ for dry air when c, g, s units are used, or $2 \cdot 876$ for air which contains water-vapour enough to saturate it at the normal freezing point of water.

If we extend our vision beyond the range of practical meteorology we may have to recognise R as not absolutely conditioned by temperature measured on a conventional scale. The same may similarly be said about the ordinary statement of Ohm's law. Within the meteorological range we are justified in working as though R were constant. In that case $pv = Ra$ is one of the equations which can be converted into $pv = a$ by a suitable choice of units. We know that pv expresses energy per unit mass and will be given in ergs if p and v are given in c, g, s units.

Taking then $R = 2 \cdot 876 \times 10^6$ and $a = 273$ we get pv equal to $7 \cdot 851 \times 10^8$ ergs; and the same gas would have energy of 10^9 ergs at 348; 8×10^8 ergs at 278; 9×10^8 ergs at 313; $1 \cdot 072 \times 10^9$ at 373.

Hence, if temperature were expressed in a unit of 10^9 ergs per gramme, the equation $pv = a$ would hold and a would equal 1 at 348, $\cdot 8$ at 278, $\cdot 785$ at 273, $\cdot 9$ at 313, $1 \cdot 072$ at 373; and the step for $1°$ C would be $\cdot 0029 \times 10^9$.

Treating the equation as a dimensional equation of the form $q = x^\alpha y^\beta z^\gamma$, where x, y, z are fundamental units of mass, length and time, we have

$$a\ \mathrm{L^2/T^2} = p\ \mathrm{ML/(L^2T^2)} \times v\ \mathrm{L^3/M}$$

and the absolute temperature as ordinarily expressed

when $a = 5 \times 10^8$ is 174 or $-99°$ C or $-146°$ F,

,, 8×10^8 ,, 278 ,, $+ 5°$ C ,, $+ 41°$ F,

,, 9×10^8 ,, 313 ,, $+ 40°$ C ,, $+ 104°$F,

,, 10^9 ,, 348 ,, $+ 75°$ C ,, $+ 167°$ F.

So an instrument for measuring the absolute temperature identical in construction with an ordinary thermometer could be made in which the graduations represent the appropriate energy of a gramme of air expressed in ergs.

This expression of the temperature of the air as energy has been referred to already in chap. VI of vol. III, where it is shown that entropy, the correlated factor of thermal energy, is a number without dimension and is the means of expressing the transference of energy from air to its environment in terms of the energy of the air by which the transference is accomplished.

The mode of treatment has not yet come into ordinary meteorological practice; but it ought not to be overlooked nor forgotten.

MOTION UNDER BALANCED FORCES

Isobaric motion and isentropic motion

One of the most fruitful and enjoyable applications of Newton's laws of motion is to be found in the case of persistent steady motion when we know from observation that the motion is uniform and can therefore conclude from Law I that there is no *vis impressa*. The equality of the motion is in itself evidence that the moving body is free from the action of any resultant force. If we are sure that some force is acting, we can be equally sure that other forces must be balancing it in order to free the body from any variation of its motion.

A motor-car, train or steamer, travelling with uniform speed along a level track is a case in point—the forces of friction, resistance, turbulence and drive must balance between themselves, leaving the car, the train or the ship, to pursue its course under balanced forces.

The example which is most familiar to us is that which forms the subject of chapters III to VIII of this volume, namely the strophic balance between pressure and the influence of the rotation of the earth when the velocity of the air has become steady within a recognised path. For that we depend on some assurance derived from observation or otherwise that the air for which the relation is sought is moving without appreciable acceleration. This was referred to in Part IV as the first law of atmospheric motion, and in vol. III in another connexion the same title has been conferred upon the principle that in the absence of a supply of heat atmospheric motion is confined to an isentropic surface.

Let us therefore re-arrange our statements and express the relation of wind to pressure under balanced horizontal forces as the law of isobaric motion:

Law IX. In the upper layers of the atmosphere at any given latitude steady horizontal motion of the air at any level is along the horizontal section of the isobaric surface at that level and the velocity is inversely proportional to the separation of the isobaric lines in the level of the section.

Whenever the adjustment of the velocity to the gradient is disturbed there is always a component of the geostrophic force tending to restore it.

The principle, that if there be no heat supplied or lost the motion of air is confined to an isentropic surface, may be expressed as the law of isentropic motion:

Law X. Air which is moving preserves a state of motion along an isentropic surface except in so far as it is interfered with by the supply or removal of heat.

With regard to both these laws it should be understood that they refer primarily to the instantaneous motion of the air and the instantaneous position of the isobar in the horizontal plane or of the isentropic surface with reference to the earth. The run of the isobar or the configuration of the isentropic surface may be continually changing. We can recognise a process of continuous adjustment of the horizontal wind to the horizontal distribution of pressure, but the adjustment of the motion of the air to the change of configuration of the isentropic surface has yet to be investigated.

Limiting velocity

A special case of steady motion under balanced forces frequently referred to in meteorological literature is that of the limiting velocity of falling raindrops. Any other falling body, an aviator hanging on to a parachute, even an aerial torpedo, is subject to similar rules. Starting from rest and allowed to fall, it finds itself under the action of gravity, with only the air-resistance, frictional or impactive, to interfere with the growth of momentum. Roughly speaking the impactive resistance which is the resultant of the air-pressure at right angles to every part of the surface varies as the square of the velocity and the frictional resistance varies as the velocity itself; hence, as the velocity increases with the time of falling the resistance also increases and, given enough range of height, a time will come when the frictional and impactive forces balance the weight and then motion becomes uniform. The formula will be given later.

Law XI. For every body allowed to fall in air or other fluid there is a limiting velocity which can be reached if there is sufficient height, but which cannot be overpassed.

THE DYNAMIC INDICATOR: WIND-FORCE OR WIND-SPEED

With these eleven laws at our service we propose next to take up wind-speed as a dynamic indicator, regarding the traditional barometric pressure as a static indicator. In vol. I we have set out the general relation between wind-speed, wind-force and Beaufort number; let us now resume our inquiry into the relation of speed to force and incidentally discharge an obligation to the law of dynamical similarity.

When a surface is exposed to a stream of air we have to consider the momentum lost by the air in consequence of the direct impact, the plain resistance, which gives a force proportional to the square of the velocity of the air, as well as that lost by viscosity, proportional to the relative velocity of the moving parts, and that lost in turbulence which is developed when the limit of stream-line motion is passed. All these contributions to change of momentum are aggregated to express the force of the air on the surface in accordance with Law II.

As regards plain resistance and viscosity the action is comparatively simple; but turbulence is very disturbing. If we take the apparently simple case of a sphere, for example, the behaviour of the front portion which meets the current gives forces which can be calculated and related to the pressure on the sphere. But besides the thrust on the front there is the loss of pressure in the rear, and the effect of that depends upon where the motion becomes turbulent; it may become turbulent anywhere where the relative motion is sufficiently brusque, and that may be anywhere beyond the zone of maximum girth. Anywhere in that region a wake of eddies may be formed which is very unstable, and its instability affects the pressure on the rear surface and alters the direction of the resultant force of resistance. It is on that account that it is found impossible to keep a straight course in towing a sphere by a single string, and indeed the same is true of towing any other solid through water.

In order to prevent the formation of eddies or to bring the turbulence under control, the shapes of the afterparts of bodies exposed to the wind are "faired" into stream-line shape so that no eddies are formed except from the tail, and the resistance is thereby regularised and reduced to a minimum. There can be little doubt that the shapes of fishes are mainly controlled by this consideration.

Questions of this kind come up in many relations; for example, from the point of view of preserving its direction in a steady wind a weather-cock should be fish-shaped rather than flag-shaped. In this connexion it should be observed that a sheet of paper or any other flat surface never falls edgewise. It swings here and there sideways and takes much longer to reach the ground than a sphere of the same weight. A fish body, properly loaded to make use of the shape to avoid forming eddies, should make even better speed than a sphere. A flat vane is likely to be very unstable. The fluctuation of a flag-shaped weather-vane is not necessarily a fair representation of the fluctuation of the wind.

These considerations have been taken into account in the more recent designs of the vane of a Dines anemometer. The whole subject of turbulence is of vital importance in aeronautics and for further particulars the reader may be referred to the text-books on that subject.

The forces on plates

A rigid plate so exposed that the front surface is always at right angles to the line of the stream of air suffers an increase of pressure upon its face which is very approximately proportional to the square of the undisturbed velocity of the stream, and at the same time a decrease of pressure over its back, in consequence of the turbulence created in the shadow, which also follows a law not much different from that of the square of the velocity.

The subject of the force upon a body in a current of air may be approached on the experimental side by citing the results of observations upon a square or circular plate kept head to wind by suitable mechanism. The best wind to use for that purpose is that of a modern wind-channel because it is free from the fluctuations which are unavoidable in the turbulent motion represented by the natural wind passing over a building.

For the thrust per unit area in absolute units on a circular plate about 1 square foot, 1000 sq. cm we get a formula $P = \cdot 56\rho V^2$ (W. H. Dines) equivalent to $\cdot 003$ lb per sq. ft when the wind is expressed in miles per hour. For an area of 2 sq. ft, or 2000 sq. cm the factor[1] is $\cdot 52$ (N.P.L.) or for 5 sq. ft to 10 sq. ft $\cdot 62$.

Regarding the barometric pressure as recording the pressure in an undisturbed current, the force on a plate exposed normal to the wind expresses the resultant force in the direction of the wind which arises from the fact that the actual pressure of the air is different at different points on the surface of the plate, greater than the barometric pressure on the front of the plate and less than the barometric pressure at the back.

The reduction of pressure at the back is created and maintained by the formation of eddies where the air passes the rim; these carry away some of the air which would otherwise find its place just behind the plate. Thus the force upon the plate is the expression of the influence of turbulence.

A corresponding effect proportional to the square of the velocity of the current is produced on a body with any other shape. The resistance of that kind for any particular body is called the "profile resistance."

[1] Advisory Committee for Aeronautics, *Reports and Memoranda*, No. 15, 1909–10, p. 35.

The same law holds with sufficient accuracy for plane obstacles of larger area. But differences occur if the surface behind the exposed plate differs from the ordinary plane because turbulent motion caused thereby depends upon the shape of the rear surface.

Naturally also the moulding of the front surface affects the result; a spherical obstacle has a different law from that of a plane obstacle.

Skin-friction

Besides the profile resistance a solid destroys part of the momentum of a current of air by the process which is known as skin-friction. That can be illustrated by the force exerted upon a thin plate placed with its edge towards the wind, which can have no ordinary profile resistance.

The subject is treated in a lecture by Prof. B. M. Jones[1] before the Royal Aeronautical Society. Quoting among other results those of Prof. Burgers of Delft, Jones explains that the nature of the "friction" upon a plate varies with the Reynolds number appropriate to the air and therefore for air of the same density and viscosity varies with the velocity of the air-stream.

Plotting the coefficient of friction k_F, i.e. $(\text{drag}/\rho V^2 E)$, where V is the relative velocity of air and plate and E is the total exposed area of both sides of the plate, against Reynolds numbers Vl/ν (fig. 1) he gets two curves

$$k_F = 0.019 \, (Vl/\nu)^{-.15} \text{ and } k_F = 0.66 \, (Vl/\nu)^{-\frac{1}{2}},$$

and he finds that the friction for a plate follows the second curve for very low Reynolds numbers and the first for high Reynolds numbers.

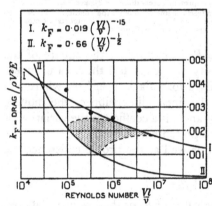

Fig. 1. Friction on a thin plate edgewise to the wind. (B. M. Jones.)

Curve I represents the results of experiments at Göttingen (shown by black dots on the diagram) for values of Vl/ν between 2×10^5 and 10^7. It refers to conditions in which the boundary layer is turbulent over the greater part of the plate.

Curve II represents conditions for small values of Vl/ν where the layers of fluid near the surface slip over each other smoothly and there is no turbulence.

In cases where the boundary layer is smooth over the front part of the surface and turbulent over the rear part the average drag coefficient will lie between the upper and lower curves. The region of transition from one curve to the other is shaded. The value at which the transition occurs depends on the steadiness of the air before reaching the plate.

This behaviour has been explained by Burgers as being due to the change in the "boundary" layer. With high velocity and consequent high Reynolds numbers the surface layer of air is turbulent over the whole plate but when velocity and consequently the Reynolds number is low then a smooth surface-layer of stream-line motion extends from the leading edge over part of the plate. For intermediate velocities there is a very uncertain curve intermediate between the two.

[1] *Journal of the Royal Aeronautical Society*, vol. XXXIII, 1929, p. 362.

The forces on spheres

For meteorology the force of an air-current upon a sphere is a question of direct practical importance because it leads to estimating the extent to which gravity is counteracted in the case of a sphere falling through air, and thus to the limiting velocity of falling drops. It furnishes a good example of the difficulties which beset even the simpler problems of atmospheric dynamics and is at the same time an excellent example of effective meteorological calculus.

In the case of the sphere we have to consider the profile resistance which depends on the square of the velocity of the sphere relative to the air, and the skin-friction which expresses the effect of turbulence and therefore depends not only upon the square of the velocity of the air and the roughness of the surface but also upon the Reynolds number for the air.

The result of experiment on the subject is clearly expressed by a diagram by L. Bairstow[1] which is reproduced in fig. 2.

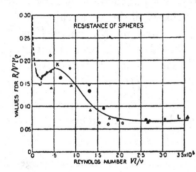

Fig. 2. Chart of observations of the resistance of spheres in air referred to JKL a curve of relation between the "drag" coefficient (i.e. resistance divided by density, velocity squared and area) and the appropriate Reynolds number. (L. Bairstow.)

⊙ ○ △ represent observations by Eiffel with spheres of diameter ·16 m, ·25 m and ·33 m.
• × represent observations by Shakespear.
The freckled line represents the conditions where l and V are so small that l^2 and V^2 are unimportant factors and the viscosity ν is the dominant influence.

We may suppose that the resistance R depending on the square of the velocity and also depending on the density of the air and the area of cross-section of the sphere can be expressed by a formula $R = k\rho V^2 l^2$, where ρ is the density, V the velocity of the air, l the linear dimension of the moving system—that is the radius of the sphere—and k is a coefficient the variation of which in varying circumstances is to be represented.

Taking the skin-friction into account as well as the profile resistance the equation might be expressed as

$$R = \rho V^2 l^2 f(Vl/\nu),$$

where Vl/ν is the Reynolds number. The value of R, or the law of resistance, changes with the variation of $f(Vl/\nu)$ although the subject of the function itself is undimensional.

In order to obtain this equation on the principle of dynamical similarity we may assume that the force can be expressed as a function of the density and velocity of the stream, the frictional force and the dimensions of the sphere. The frictional force we

---

[1] Advisory Committee for Aeronautics, 'Variation with speed of the forces due to viscosity,' *Report*, 1913–14, pp. 41–4; see also L. Prandtl, 'Der Luftwiderstand von Kugeln,' abs. *Ibid.*, p. 477.

can assume to be dependent on the viscosity of the flowing stream and we may suppose as a first approximation the force to be expressed as the product of suitable powers of the effective elements.

$$ML/T^2 = \Sigma \, (M/L^3)^a \, (L/T)^b \, (L)^c \, (L^2/T)^d,$$

and by equating coefficients of M, L and T we obtain a formula for the resistance in the form

$$\text{Resistance} = \rho V^2 l^2 f(Vl/\nu)$$

or the resistance per unit area $= \rho V^2 f(Vl/\nu)$.

The resistance depends therefore directly on the density and the square of the velocity and upon some function of the coefficient Vl/ν, the Reynolds number, which itself is undimensional.

In the first part of the curve where l is very small and the profile resistance is negligible, the coefficient decreases rapidly with the increase in the Reynolds number. For this region we have a formula originally computed by Stokes, namely $R/\rho Vl\nu = 6\pi$, which can be brought into line with the general formula as $R = 6\pi\rho V^2 l^2 \, (\nu/Vl)$.

In the other extreme of the curve the value of k has become practically constant and the force varies as the square of the velocity for variations of the Reynolds number beyond $2 \cdot 5$.

The diagram recalls to mind that of B. M. Jones for a system of any smooth shape already given in fig. 1. The first part of the curve corresponds with the curve of little turbulence and the second part with the curve of great turbulence; the intermediate part JK is the unstable and irregular transition curve.

From this curve we can obtain the limiting velocity of the fall of cloud-drops and the limitation of their size with the modification that is introduced when the size of the particles increases and the cloud-drops become rain-drops. We have already given the results in a table on p. 336 of vol. III, and on p. 333 also an estimate of the rate of fall of small dust-particles in accordance with Stokes's formula.

L. F. Richardson[1] has proposed the determination of wind-velocity in the upper air by the observation of the horizontal travel of steel spheres shot upward. The theory of the method involves the consideration of the questions of the influence of turbulence but leads to a more complicated expression than we can pursue here.

Experiments on small spheres falling through water have been described by R. G. Lunnon[2] who gives a formula $R = av^2 + b\dfrac{dv}{dt}$, where the coefficient b indicates the effect of the carried mass. Experiments[3] on spheres falling through the air of a deep coal mine or from a high tower can also be referred to by those who wish to pursue this complicated subject beyond the limits which we are asking the reader to accept.

---

[1] 'The aerodynamic resistance of spheres, shot upward to measure the wind,' *Proc. Phys. Soc. London*, vol. XXXVI, part II, 1924, pp. 67–80.

[2] *Proc. Roy. Soc.* A, vol. CXVIII, 1928, pp. 680–94; *Science Abstracts*, 1928, p. 523.

[3] R. G. Lunnon, 'Fluid resistance to moving spheres,' *Proc. Roy. Soc.* A, vol. CX, 1926, p. 302; G. A. Shakespear, *B.A. Report*, 1913, p. 402; and *Phil. Mag.*, vol. XXVIII, 1914, p. 728.

Anemometers

The force exerted by an air-current upon a plate, a sphere or some other shape is the natural avenue towards the measurement of wind-velocity. The description which we have given of the effect of air-currents upon plates and spheres shows that in a wind of variable turbulence no simple formula is likely to be found for the velocity of wind in terms of the angular deflexion of a plate or the speed of rotation of a fan-wheel.

The arrangement which has found most general acceptance for the measurement of wind-velocity is the Robinson anemometer in which as a standard pattern four hemispherical cups are mounted at the extremities of four 2-foot arms, and the wind is measured by recording the rotation of the spindle to which the arms are fixed.

A good deal has been written about the cup-anemometer. The assumption that underlies the measurement is the steady revolution of the spindle in a steady wind, and that is so far from being theoretically calculable that in these days the empirical graduation of the instrument in a wind-channel is the practical method of dealing with it.

In recent years J. Patterson[1] has developed an anemometer with three cups instead of four, and obtained results which have commended the instrument for general use in Canada and the United States.

In 1880 Sir George Stokes, a near relative of Robinson's, devised a modification of the cup-anemometer in order to obtain a record of the force of the wind in gusts. The design was prompted by the Tay Bridge disaster of 28 December 1879 when a train was lost by the collapse of the bridge during a storm. The instrument was called the bridled anemometer.

The object which this form of anemometer is designed to accomplish is the measurement of the varying force of the wind and particularly of the strongest gusts; a question on which great doubt exists and the solution of which involves considerable difficulties which is of no little practical importance in connexion with the stability of engineering works. *Report of the Meteorological Council*, 1888–9.

The force of the wind on the cups produced a couple which turned the spindle through an angle controlled by a counter-balancing couple produced by weights operating over pulleys on a spiral disc with a grooved edge so that the angle of deviation of the disc might be proportional to the velocity of the wind.

Subsequently when the instrument was calibrated by W. H. Dines five cups at different angles and different levels were substituted for the four cups at the same level.

The spindle with its cups had practically no momentum, the instrument indeed as installed at the anemometer-station at Holyhead used to excite the derision of the locality because, unlike its four-cup neighbour, it was never recognised as moving.

[1] J. Patterson, 'The cup anemometer,' *Trans. Roy. Soc. Canada*, Third Series, vol. xx, Sect. III, 1926. See also E. L. Davies and N. K. Johnson, 'The cup-anemometer,' *Meteor. Mag.*, vol. LXII, 1927. p. 184; T. Okada and E. Miura, *Geophysical Magazine*, vol. III, Tokyo, 1930, p. 87.

It is a matter for regret that the suggestion has not been further developed because the instrument showed the gusts with fidelity and the records presented many points of interest. A portion of a record is reproduced in the Aeronautical Society's Journal[1]. The single instrument that was constructed was bulky and heavy and was only brought into action when winds were strong, but there is no reason why a much lighter instrument should not have been constructed on the same principle that might have done good service.

The Tay Bridge disaster led others to consider various questions in the measurement of wind, among them W. H. Dines, who turned his attention to pneumatic recorders of wind of which his own pressure-tube anemograph is typical, and which with modification in some cases is now in general use.

The instrument depends upon the increase of pressure upon the mouth of a tube which faces the wind. It is a recording modification of the pressure-tube introduced by Pitot early in the nineteenth century. To obtain the measure of the increase of pressure upon the tube which faces the wind provision has to be made for a manometer to show the effect of the increased pressure and some arrangement is necessary to prevent the other limb of the manometer being affected irregularly by the wind. The device adopted is to bring the second limb of the manometer into communication with an outer casing of the support of the exposed tube.

In the Pitot tube the outer casing is a parallel tube and the communication with the outside is by holes drilled through the casing past which the air flows. In the Dines instrument communication from the second limb of the manometer is led to the outside casing of the vertical tube which forms the spindle upon which the head turns, and the effect of the air upon a ring of openings in the casing is a diminution of pressure in the second limb of the manometer which increases the difference shown on it. A special arrangement transforms the manometric difference into a record of the velocity of the wind.

Again the final appeal for the interpretation of the record is obtained by the aid of a wind-channel.

The addition of an apparatus to record the changes in the direction of the wind on the same time-scale as that of the velocity completes the representation of the horizontal component of the air-motion experienced at the position of the vane. The dynamical interpretation of the record is a definite challenge to the power of meteorological calculus.

When the wind has been successfully dealt with taking into account the exposure of the anemometer as well as the weather, there will be little left for the meteorologist to explain: here therefore we place a selection of records typical of various kinds of exposure and of weather as a statement of the problem of dynamical meteorology more eloquent than words. Each record carries one or more small inset charts taken from contemporary weather-maps to illustrate the barometric distribution.

[1] W. N. Shaw, 'Wind-gusts and the structure of aerial disturbances,' *Aeronautical Journal*, vol. XVIII, 1914, p. 173, fig. 1.

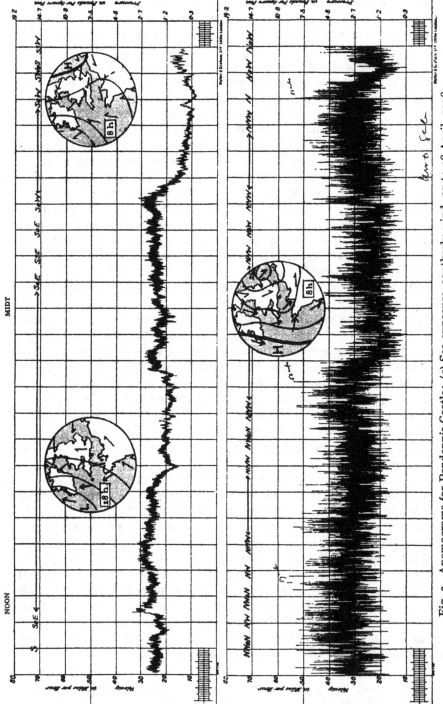

Fig. 3. Anemograms for Pendennis Castle: (*a*) Sea exposure, southerly wind, 27 to 28 April, 1908; (*b*) Land exposure, north-westerly wind, 29 February to 1 March, 1908.

List of wind-records forming the statement of the problem of dynamical meteorology. Figs. 3–7 are from Reports and Memoranda No. 9 of the Advisory Committee for Aeronautics; figs. 8 and 9 from the original records of the Meteorological Office.

Fig. 3. Pendennis Castle, Cornish coast. Sea-wind and land-wind.

Fig. 4. Southport, Lancashire coast, estuarial plain, persistent fluctuations of direction and force.

Fig. 5. Kew Observatory, Richmond, Thames valley. Southerly squall intervening between the light southerly air preceding and light easterly or northerly airs of six hours' duration following a thunderstorm.

Fig. 6. Scilly Isles, English Channel. Temporary squalls of one-half to three-quarters of an hour's duration.

Fig. 7. Gibraltar—summit of the rock. Sudden drop of velocity preceding change of direction from SW to NW.

Fig. 8. A gale at Spurn Head, east coast of England. Sudden increase of velocity with transition from W (a land-wind) to N by W (a sea-wind).

Fig. 9. A gale at South Kensington, London. A roof exposure, with transient variations ranging over 50 miles an hour.

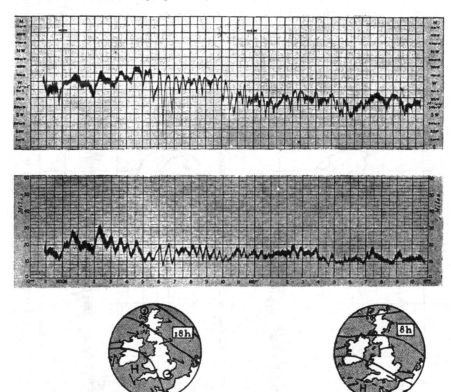

Fig. 4. Anemogram from Marshside, Southport, 6 to 7 January, 1907, showing oscillatory changes. Upper panel—direction, lower panel—velocity.

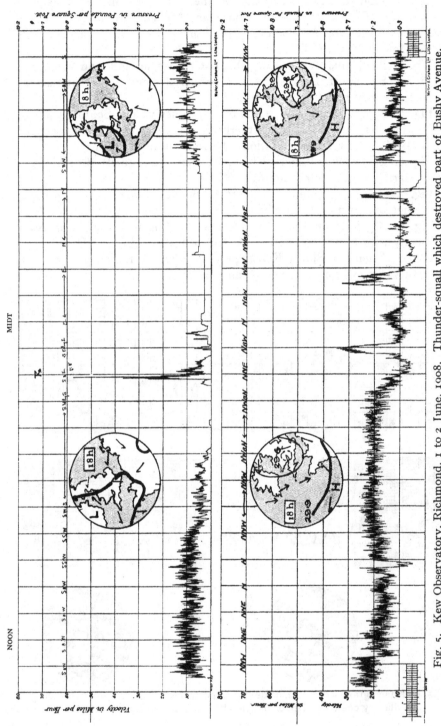

Fig. 5. Kew Observatory, Richmond, 1 to 2 June, 1908. Thunder-squall which destroyed part of Bushy Avenue.
Fig. 6. St Mary's, Scilly Isles, 3 to 4 March, 1908. Consecutive squalls.

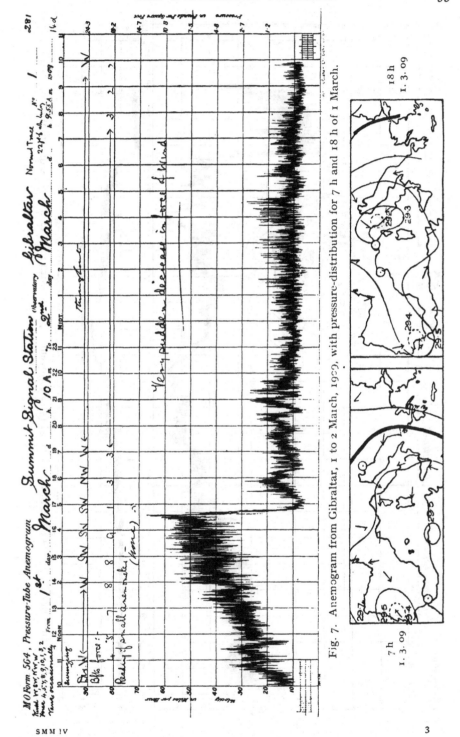

Fig. 7. Anemogram from Gibraltar, 1 to 2 March, 1909, with pressure-distribution for 7 h and 18 h of 1 March.

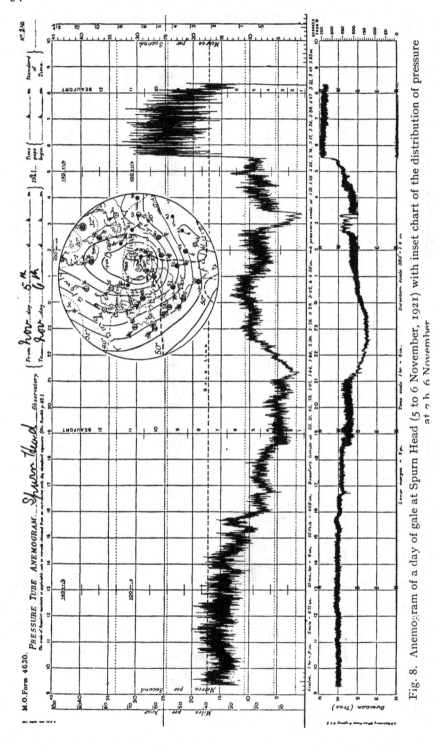

Fig. 8. Anemogram of a day of gale at Spurn Head (5 to 6 November, 1921) with inset chart of the distribution of pressure at 7 h. 6 November

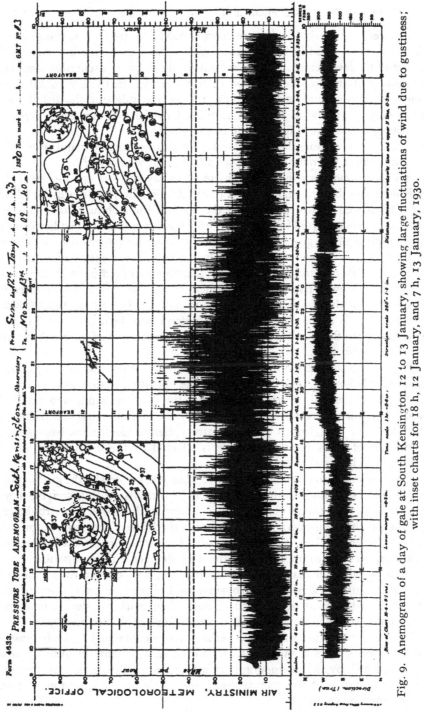

Fig. 9. Anemogram of a day of gale at South Kensington 12 to 13 January, showing large fluctuations of wind due to gustiness; with inset charts for 18 h, 12 January, and 7 h, 13 January, 1930.

STATEMENT OF THE PROBLEM OF DYNAMICAL METEOROLOGY

The records of wind which are reproduced here state for us the problem which has to be faced in this volume—it is nothing more nor less than, with the aid of the laws which we have set out, to describe in intelligible language the motion of the air, the dynamic indicator, of which the horizontal component is recorded, and its connexion with the static indicator, the distribution of pressure. In so far as the flow is related to the pressure every layer of the atmosphere above it has a share in the effect.

It is not improbable that the motion is analysable into flow and spin. The questions which the records ask is, what flow? and what spin?

To those questions the traces invite us to say that we can ascertain the equivalent flow by drawing through the middle of the ribbon a line such as might have been drawn to represent the trace of the Robinson anemometer, that the line is converted into a ribbon by irregular spin due to eddies caused by the friction of the ground. The fluctuations of next larger duration, of the order of an hour or so, are probably due to spin about axes which may be vertical or horizontal and due to irregular distribution of temperature or water-vapour in the flow. The major disturbances of the flow which can be directly correlated with the distribution of pressure may be the spin of wave-motion or rotation about an axis the direction of which has to be defined; and finally the undisturbed flow itself is to be associated with the spin of the air about the axis of the earth.

It was with this view of the problem in mind that the text of this volume was originally written. The conception of the undisturbed flow as the expression of the horizontal geostrophic wind, or the gradient wind, on a sea-level map, apart from the interference caused by the contours and the frictional effect of sea or land, brings the whole atmosphere under review because the distribution of pressure at the surface is affected by the distribution in the succession of layers above it.

The first step is to examine that interference as expressed by the gustiness of the trace and such laws of turbulence as can be formulated, thus arriving at an opinion as to what the flow would be if it were undisturbed by turbulence, and where in the atmospheric structure undisturbed flow or the nearest approach to it is to be found. Next to form a picture of the structure or stratification of the atmosphere and to trace the relation of its changes with the distribution of temperature, and thus having ascertained the original flow and the relation of flow to structure we arrive at the consideration of the spin on a large scale as associated with curved isobars and through them with the depressions and anticyclones as the advective and divective regions of the weather-map.

To-day the weather offices are perhaps more concerned with air-masses and less with isobars than they were before the war; but the fundamental question asked by the anemograph in association with the weather-map is the same to-day as it was twelve years ago. It is still a question of flow and its disturbance by spin.

CHAPTER II

THE GENERAL EQUATIONS OF MOTION OF A PARCEL OF AIR IN THE FREE ATMOSPHERE SUBJECT TO THE FORCES INCIDENTAL TO GRAVITY AND FRICTION

"Having defined our variables [pressure, mass, temperature, humidity or salinity and motion] we can thus concisely state the problem of meteorology and hydrography: *To investigate the five meteorological and the five hydrographic elements as functions of co-ordinates and time*" (V. Bjerknes and others).

EQUATIONS of motion represent the attempt of mathematicians to provide means for calculating the movement of any part of the atmosphere from a knowledge of the forces which are acting upon that part, including necessarily the reactions between the particular part selected and its environment. The choice of that method of approaching a difficult subject arises from the tacit understanding that atmospheric motion of every kind is subject to the three simple laws which Newton enunciated, and the solution of the dynamical problem of the atmosphere, however complicated it may be, should be approached by developing the laws of motion to include the expression of all the forces of every kind which act upon any isolated portion of the atmosphere.

The quantities involved and the mode of expression

Note on symbols. The equations of motion, as developed for the calculus of the motion of the air in three dimensions upon a rotating earth, deal with a large number of physical quantities, with mass and with velocity; with momentum, representing the instantaneous value of the product of mass and velocity, with acceleration and with forces, all with three components along the selected axes, with moment of momentum, angular momentum, which also may have components in three dimensions. And when account is taken of the earth's rotation, and the components of momentum, velocity and acceleration are referred to axes moving with the earth, the expression of the laws of motion is further complicated.

The independent quantities which are introduced into the calculation are thus: mass, length, time, velocity, momentum, angular velocity, angular momentum, co-ordinates of position on the earth's surface or in the atmosphere above it, acceleration and its components, the forces acting and their components referred to the co-ordinate axes. All these quantities are introduced into the equations as algebraical symbols and the representation makes a considerable demand upon the alphabets which we have set out as preliminary to vol. III. And as we are contemplating the use of the equations for the solution of meteorological problems in which all the various meteorological elements may be involved it is desirable to use some discretion about the symbols chosen. The various authors whose work we shall quote in setting out the sequence have not been hampered by the consideration that the fluid for which the equations were devised might be the atmosphere with all the implications of its variables.

We have accordingly thought it best to use one system of symbols for the different forms of the equations. The symbols which we find most appropriate are as follows:

x, y, z for rectilinear co-ordinate axes. t for the time.

$\dot{x}, \dot{y}, \dot{z}$ $\left(\dfrac{dx}{dt}, \dfrac{dy}{dt}, \dfrac{dz}{dt}\right)$ for the components of velocity in the direction of the axes.

$\ddot{x}, \ddot{y}, \ddot{z}$ $\left(\dfrac{d^2x}{dt^2}, \dfrac{d^2y}{dt^2}, \dfrac{d^2z}{dt^2}\right)$ for the components of acceleration in the direction of the axes.

m for the mass of a parcel or particle of air.

ρ for the density or mass of unit volume.

Γ for the geopotential which defines the level or niveau at which events take place.

U, V, W for the components of momentum along the three axes.

V for resultant velocity.

X, Y, Z for the components of force along the three axes.

r for the radius vector of polar co-ordinates and for the radius of a small circle on the earth.

ϕ for latitude, measured from 0° to 90° in the northern and from 0° to $-$ 90° in the southern hemisphere.

λ for longitude measured positively towards the east.

ϵ for the radius of the earth.

ω for the angular velocity of the earth's rotation.

g for the acceleration of gravity.

G has already been assigned in vol. III for the constant of gravitation. G is used here for gravitational attraction and also for the geostrophic wind.

e, n, z for co-ordinates measured towards east, towards north and vertically upwards.

u, v, w for the components of velocity along the co-ordinate axes.

E, N, R for the components of force, other than gravity, towards east, towards north and vertically upwards.

∂ for partial differentiation. δ for a small finite increase of a variable.

$\dfrac{D}{D\!t} = \left(\dfrac{\partial}{\partial t} + u\,\dfrac{\partial}{\partial x} + v\,\dfrac{\partial}{\partial y} + w\,\dfrac{\partial}{\partial z}\right)$, the total change in a quantity in an infinitesimal time.

bb for gradient of pressure. tt for tercentesimal temperature.

θ for the angular radius of the small circle which represents the curvature of the path.

Symbols of operation and constants are in roman type, variables in italic.

The fundamental Newtonian equation from which all the others are derivable by sufficient mathematical and physical ingenuity is applicable to a finite mass m concentrated in a point at a position which may be defined by a single co-ordinate x, the rectilinear distance of the point from the origin of co-ordinates, and under the action of a single force X in the direction of the co-ordinate.

If we use dx/dt or $\dot{x}$ to represent the velocity of the mass m at the time t, dx being the infinitesimal change in x in the corresponding infinitesimal change dt in the time; and d^2x/dt^2 or $\ddot{x}$ to represent the acceleration or rate of change of the velocity of m per unit of time in the direction of the co-ordinate at the same instant, the original fundamental expression of the laws of motion when the force X is regarded as tending to increase x is $m\ddot{x} = m\,\dfrac{d^2x}{dt^2} = X$. If the force were in the opposite direction and tended to reduce x the sign of X would be reversed:

$$m\ddot{x} = m\frac{\mathrm{d}^2x}{\mathrm{d}t^2} = -X.$$

In this form the equation assumes that the mass m is invariable and that is not necessarily the case with a parcel of air defined by $\delta x, \delta y, \delta z$, because it may increase or decrease its mass by the alteration of its density. We can provide for this contingency by considering the alteration of momentum $\frac{\mathrm{d}\mathsf{U}}{\mathrm{d}t}$ where U is the variable momentum replacing $m\dot{x}$.

If the force which is acting is not in the direction of the existing motion $\mathrm{d}x/\mathrm{d}t$ along the co-ordinate x, but in any direction whatever, we require three co-ordinates at right angles x, y and z to define the position of the mass m, and have to take account of three initial velocities $\mathrm{d}x/\mathrm{d}t$, $\mathrm{d}y/\mathrm{d}t$ and $\mathrm{d}z/\mathrm{d}t$, and three components of acceleration $\mathrm{d}^2x/\mathrm{d}t^2$, $\mathrm{d}^2y/\mathrm{d}t^2$, $\mathrm{d}^2z/\mathrm{d}t^2$ in the directions of the co-ordinates x, y, and z; also three components of the force X, Y and Z. And consequently we require three equations

$$m\frac{\mathrm{d}^2x}{\mathrm{d}t^2} = X, \quad m\frac{\mathrm{d}^2y}{\mathrm{d}t^2} = Y, \quad m\frac{\mathrm{d}^2z}{\mathrm{d}t^2} = Z.$$

From these simple equations, systems of equations have been elaborated which are designed as the first step in the calculation of the motion of a parcel of air.

A large number of examples of the general method of procedure for the solution of geophysical problems, with which the problems of atmospheric motion may be brigaded, will be found in a work by Harold Jeffreys[1] with the comprehensive title *The Earth*. But the most deliberate attempt to use the method directly in the solution of the general problem of atmospheric motion is to be found in a work on *Weather Prediction by Numerical Process* by L. F. Richardson[2].

We have explained that the simple equations here quoted require development in order to meet the special requirements of the atmospheric problem which is concerned, of course, with the particular case of the motion of air on a rotating earth.

The extent of the necessary development is itself impressive. This will be understood if we quote the developed form with which Richardson has endeavoured to attack the problem of determining the weather of to-morrow from its condition to-day, or the change in some other selected interval, under the forces to which any selected part of the atmosphere is subject. The chief of these forces are universal gravitation with its indirect expression pressure, and the quasi-forces due to the continuous rotation of the earth; but the effect of all the physical processes, radiation, convection, friction, viscosity and turbulence, must also come in.

[1] Cambridge University Press, 1924.
[2] Cambridge University Press, 1922.

The equations by which Richardson endeavours to obtain a quantitative description of the movements of the atmosphere are:

$$-\frac{\partial \mathbf{W}}{\partial t} = g\rho + \frac{\partial p}{\partial z} + \frac{\partial}{\partial e}\,(\mathbf{W}u) + \frac{\partial}{\partial n}\,(\mathbf{W}v) + \frac{\partial}{\partial z}\,(\mathbf{W}w) - 2\omega \mathbf{U} \cos \phi$$
$$+ (2\mathbf{W}w - \mathbf{U}u - \mathbf{V}v - \mathbf{W}v \tan \phi)/\epsilon,$$

$$-\frac{\partial \mathbf{U}}{\partial t} = \frac{\partial p}{\partial e} + \frac{\partial}{\partial e}\,(\mathbf{U}u) + \frac{\partial}{\partial n}\,(\mathbf{U}v) + \frac{\partial}{\partial z}\,(\mathbf{U}w) - 2\omega \mathbf{V} \sin \phi + 2\omega \mathbf{W} \cos \phi$$
$$+ \frac{3\mathbf{U}w}{\epsilon} - \frac{2\mathbf{U}v \tan \phi}{\epsilon},$$

$$-\frac{\partial \mathbf{V}}{\partial t} = -g_n\rho + \frac{\partial p}{\partial n} + \frac{\partial}{\partial e}\,(\mathbf{V}u) + \frac{\partial}{\partial n}\,(\mathbf{V}v) + \frac{\partial}{\partial z}\,(\mathbf{V}w) + 2\omega \mathbf{U} \sin \phi + \frac{3\mathbf{V}w}{\epsilon}$$
$$+ \tan \phi \,(\mathbf{U}u - \mathbf{V}v)/\epsilon.$$

The axes are directed to the east, to the north and vertically upwards. The equations are appropriate to a fluid under the action of the forces of pressure and gravity; no account is taken of diffusive forces. In the first equation g represents the resultant of the attraction of gravity and of the vertical component of the centrifugal force of the earth's rotation $-\rho\omega^2 \cos^2 \phi\,(\epsilon + z)$; in the third equation g_n includes the centrifugal force towards the north $-\rho\omega^2 \sin \phi \cos \phi\,(\epsilon + z)$, and the component of the acceleration of gravity due to the ellipticity of the earth.

In deriving the equations Richardson makes use of the equation of continuity of mass, which, when expanded in spherical co-ordinates, becomes, as we shall see later,

$$\frac{\partial \rho}{\partial t} = -\frac{\partial}{\partial e}\,(\rho u) - \frac{\partial}{\partial n}\,(\rho v) + \frac{\rho v}{r} \tan \phi - \frac{\partial}{\partial z}\,(\rho w) - \frac{2\rho w}{r}.$$

It will be understood from the equations set out above that Richardson includes among the possible contingencies the variation of the mass of the parcel of moving air, and consequently his formulae equate the components of the forces to the change of momentum in the corresponding directions. Not many writers on dynamical meteorology have used equations of that form, they generally assume the mass of the moving parcel to be invariable; in that case the equations are somewhat simplified.

W. Ferrel set them out in 1858–60 in a paper reprinted subsequently as a *Professional Paper of the Signal Service*, Washington, 1882. In the notation which we have adopted, Ferrel's equations for motion relative to the earth, expressed in geographical co-ordinates, take the form

$$\ddot{r} - r\dot{\phi}^2 - r \cos^2 \phi \dot{\lambda}\,(\dot{\lambda} + 2\omega) - r\omega^2 \cos^2 \phi = -\frac{1}{\rho}\frac{\partial p}{\partial r} - \frac{\partial \Omega}{\partial r},$$

$$r \cos \phi \ddot{\lambda} + 2\,(\dot{\lambda} + \omega)\,(\dot{r} \cos \phi - r\dot{\phi} \sin \phi) = -\frac{1}{\rho}\frac{\partial p}{r \cos \phi \partial \lambda} - \frac{\partial \Omega}{r \cos \phi \partial \lambda},$$

$$r\ddot{\phi} + 2\dot{r}\dot{\phi} + r \sin \phi \,.\, \cos \phi \,.\, \dot{\lambda}\,(\dot{\lambda} + 2\omega) + r\omega^2 \sin \phi \,.\, \cos \phi = -\frac{1}{\rho}\frac{\partial p}{r\partial \phi} - \frac{\partial \Omega}{r\partial \phi},$$

where Ω is the potential of the attractive force of the earth.

A second form is obtained from the first by considering that for a particle in equilibrium

$$r\omega^2 \cos^2 \phi \cdot - \frac{\partial \Omega}{\partial r} = \frac{1}{\rho}\frac{\partial p}{\partial r} = -g; \quad \frac{1}{\rho}\frac{\partial p}{\partial \lambda} = -\frac{\partial \Omega}{\partial \lambda} = 0,$$

and

$$r\omega^2 \sin \phi . \cos \phi + \frac{\partial \Omega}{r\partial \phi} = -\frac{1}{\rho}\frac{\partial p}{r\partial \phi} = 0.$$

These equivalents can be included in the equations on the understanding that gravity g includes also the effect of the vertical component of the earth's rotation, and that the horizontal component due to the rotation is accounted for by the elliptical shape of the earth's surface.

F. H. Bigelow[1] has developed the equations in slightly different form in a general review of the mathematics of weather from which we quote the following:

$$\frac{Dw}{Dt} - \frac{(u^2 + v^2)}{r} - 2\omega u \cos \phi - r\omega^2 \cos^2 \phi = -\frac{1}{\rho}\frac{\partial p}{\partial r} - G,$$

$$\frac{Du}{Dt} + \frac{uw}{r} - \frac{uv}{r}\tan \phi + 2\omega(w \cos \phi - v \sin \phi) = -\frac{1}{\rho}\frac{\partial p}{r \cos \phi \partial \lambda},$$

$$\frac{Dv}{Dt} + \frac{vw}{r} + 2\omega u \sin \phi + \frac{u^2}{r}\tan \phi + r\omega^2 \sin \phi . \cos \phi = -\frac{1}{\rho}\frac{\partial p}{r\partial \phi}.$$

The equations take account only of the forces of pressure and gravity, they refer to motion on a spherical earth rotating with angular velocity ω.

Similar equations have been developed by many other writers, including F. M. Exner in his work on *Dynamische Meteorologie*, and by H. Jeffreys.

We must leave the reader to give expression to his own opinion of the rather surprising fact that the physical meaning of these elaborate equations only differs from that of the simple equations from which they were derived by being made applicable to air on a rotating earth.

THE EVOLUTION OF THE EQUATIONS

In this work we are concerned to make clear, and as far as possible to study, the physical conceptions upon which mathematical reasoning has been based, rather than to follow in detail the several examples. It is to that purpose that the following paragraphs are addressed.

We propose to set out the stages in the transformation of the simple Newtonian equations into the forms adopted by W. Ferrel, F. H. Bigelow, L. F. Richardson, F. M. Exner and H. Jeffreys.

[1] Report of the international cloud observations, 1896–7, *Report of the Chief of the Weather Bureau*, 1898–9, Washington.

Polar co-ordinates

Beginning with the expression of the equations of linear motion of a single mass referred to rectangular co-ordinates which we have already quoted, viz. $m\ddot{x} = X$, $m\ddot{y} = Y$, $m\ddot{z} = Z$, the first stage to be noticed is the change of the co-ordinates from rectangular axes to what are known as spherical or polar co-ordinates. The transformation is not limited to terrestrial problems; but for our purposes it will be good to think of the centre of gravity of the earth as the origin, the plane of the equator as one plane of reference and the plane of the Greenwich meridian as another. On this understanding the position of a particle is defined by (1) its distance r from the earth's centre, called the radius vector, (2) the angle ϕ between the radius vector and the equatorial plane, which corresponds with the latitude, and (3) the angle ψ which the plane through the radius vector and the polar axis makes with a zero plane. It is longitude λ (taken as positive when on the eastern side of the standard meridian) if the zero plane coincides with the Greenwich meridian. We cannot yet regard this quantity as a substitute for the variable ψ; we shall have to take account of the movement of the plane of the Greenwich meridian itself with the earth, we have to regard λ as $\psi - \omega t$. This distinction has always to be kept in mind because we have to remember that the kinetic energy of a particle moving along a line of latitude on the earth's surface with angular velocity λ depends in fact upon $(\lambda + \omega)^2$ not merely upon $\dot{\lambda}^2$.

For the two dimensions x and y the transformation to polar co-ordinates of the equations $m\ddot{x} = X$, $m\ddot{y} = Y$ can be carried out algebraically if we remember that $x = r \cos \phi$, $y = r \sin \phi$. By differentiation we obtain

$$\ddot{x} = \ddot{r} \cos \phi - 2\dot{r} \sin \phi \cdot \dot{\phi} - r \cos \phi \cdot \dot{\phi}^2 - r \sin \phi \cdot \ddot{\phi},$$

$$\ddot{y} = \ddot{r} \sin \phi + 2\dot{r} \cos \phi \cdot \dot{\phi} - r \sin \phi \cdot \dot{\phi}^2 + r \cos \phi \cdot \ddot{\phi}.$$

Hence the acceleration along the radius vector OP, which may be written $\ddot{x} \cos \phi + \ddot{y} \sin \phi$, becomes $\ddot{r} - r\dot{\phi}^2$; and the acceleration perpendicular to the radius vector becomes $2\dot{r}\dot{\phi} + r\ddot{\phi}$ or $\dfrac{1}{r}\dfrac{d}{dt}(r^2\dot{\phi})$.

The transformation can be extended to three dimensions if we recognise that with co-ordinates centred at the centre of the sphere O, Oy along the earth's axis and Ox and Oz in the equatorial plane,

$$x = r \cos \phi \cdot \sin \psi, \qquad y = r \sin \phi, \qquad z = r \cos \phi \cdot \cos \psi.$$

The differentiation of these quantities to obtain the acceleration along the radius vector, and at right angles to it in the meridian plane and perpendicular to the meridian plane, is a useful exercise which may await the reader's desire for algebraical practice and his leisure. The result should be the same as that which, following D. Brunt, we can obtain more readily by considering the motion of any point P as made up of (1) motion *in* its meridian plane and (2) motion *across* its meridian plane.

(1) The components of the motion of P in the meridian plane give an acceleration $\ddot{r} - r\dot{\phi}^2$ along OP, and an acceleration $\frac{1}{r}\frac{d}{dt}(r^2\dot{\phi})$ perpendicular to OP in the direction of increasing ϕ, i.e. to north.

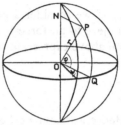

(2) The motion across the meridian plane with angular velocity $\dot{\psi}$ gives an acceleration along NP equal to $- PN\dot{\psi}^2$, and an acceleration perpendicular to NP in the direction of increasing ψ, i.e. to east, equal to $\frac{1}{PN}\frac{d}{dt}(PN^2\dot{\psi})$ or

Fig. 10. Spherical co-ordinates.

$$r\cos\phi \cdot \ddot{\psi} + 2\dot{\psi}(\dot{r}\cos\phi - r\dot{\phi}\sin\phi).$$

Hence the resultant accelerations become:

along OP, $\quad\quad\quad\quad \ddot{r} - r\dot{\phi}^2 - r\cos^2\phi \cdot \dot{\psi}^2,$

perpendicular to OP towards east, $\quad r\cos\phi \cdot \ddot{\psi} + 2\dot{\psi}(\dot{r}\cos\phi - r\dot{\phi}\sin\phi),$

perpendicular to OP towards north, $\quad \frac{1}{r}\frac{d}{dt}(r^2\dot{\phi}) + r\sin\phi\cos\phi \cdot \dot{\psi}^2.$

Modifications for a rotating earth

If now we wish to consider motion relative to the earth, which is what we have in mind in thinking about the motion of air, we may regard the angular velocity of a particle about the earth's axis as made up of $\dot{\lambda}$ its angular velocity in longitude relative to the earth, and ω the angular velocity of the earth; if the longitude of a moving body is increasing, its total movement may be regarded as made up of movement with reference to the earth's surface superimposed upon the motion of the earth's surface itself. In the above equations we write therefore $(\dot{\lambda} + \omega)$ for $\dot{\psi}$, and $\ddot{\lambda}$ for $\ddot{\psi}$ since ω is constant. We shall not enter into the question as to whether the motion of air or other matter on the earth's surface is sufficient to invalidate the assumption that ω is constant. In an interesting work on *The Energy System of Matter* (Longmans, 1912) the late James Weir, basing his conclusions on very wide experience as an engineer, developed the idea that the transformations of energy in the earth's field are derived from the rotation of the earth in a manner similar to the development of energy in the armature of a dynamo, but as a matter of observation the rotation remains sufficiently constant for meteorological calculus.

We obtain therefore for the accelerations of a particle relative to axes fixed in the earth:

along OP, $\quad \ddot{r} - r\dot{\phi}^2 - r\cos^2\phi\,(\dot{\lambda} + \omega)^2,$

perpendicular to OP towards east, $\quad r\cos\phi \cdot \ddot{\lambda} + 2(\dot{\lambda} + \omega)(\dot{r}\cos\phi - r\dot{\phi}\sin\phi),$

perpendicular to OP towards north, $r\ddot{\phi} + 2\dot{r}\dot{\phi} + r\sin\phi \cdot \cos\phi\,(\dot{\lambda} + \omega)^2.$

Vertical and horizontal co-ordinates

If we consider a particle of unit mass at rest on the earth's surface and subject to no force but gravitational attraction G, we write $\dot\lambda = \dot\phi = \dot r = 0$ in the above expressions of the accelerations and obtain

$$\ddot r - r\cos^2\phi \,.\, \omega^2 = -G,$$
$$r\cos\phi\,.\,\ddot\lambda = 0,$$
$$r\ddot\phi + r\sin\phi\cos\phi\,.\,\omega^2 = 0.$$

Hence with spherical co-ordinates, considering a sphere with radius r, equilibrium is not obtained for a particle on the sphere; if instantaneously, its velocity is zero, it remains subject to acceleration $\ddot r$ and $\ddot\phi$. In order that the particle may rest on the earth these accelerations will have to be compensated by a force which has a component $r\sin\phi\cos\phi\omega^2$ towards the north tending to increase ϕ, and a force tending to diminish that of gravitation by $r\omega^2\cos^2\phi$.

Hitherto we have used a radius vector drawn from the earth's centre which is not *vertical* except at the equator and poles. By vertical we mean the direction of the plumb-line which is controlled partly by the direction of the force of gravitation and partly by the centrifugal force of the earth's rotation. It is usual to cover these by calling the resultant force terrestrial gravity. The vertical is normal along its length to the level or horizontal surfaces of which one, namely sea-level, represents the surface of the open ocean at rest.

If then we use a vertical axis with two other axes at right angles to it, and call g the acceleration of "terrestrial gravity" along the vertical, the numerical value of g will take account of $r\omega^2\cos^2\phi$ and there will be no component to the north along the true horizontal; that will be accounted for by the inclination of the vertical to the radius vector. We can then use equations with g representing the observational value of the *vertical* acceleration and omit $r\omega^2\cos^2\phi$; and an equation without any term $r\omega^2\sin\phi\cos\phi$ in the direction of ϕ. But now no horizontal component of gravity appears in the equations and we must understand that g varies along a meridian in accordance with the equation on p. 5 ; the direction is normal to horizontal surfaces and the vertical is not strictly speaking a straight line. The south–north axis is still perpendicular to the vertical and to the line of latitude so that the acceleration λ east or west remains zero for a body initially at rest.

The angle between the true vertical and the radius vector to the centre of the earth is $68\cdot8''\sin 2\phi$. This angle is so small that no appreciable error is introduced when we treat the accelerations derived above as being in the horizontal plane and perpendicular to it, instead of in a plane tangential to the sphere at P, and along OP.

Referred to these axes the three equations of motion take the form used by Ferrel, namely:

$$\ddot r - r\dot\phi^2 - r\cos^2\phi\dot\lambda(\dot\lambda + 2\omega) = -g + R \qquad \ldots\ldots(1),$$
$$r\cos\phi\ddot\lambda + 2(\dot\lambda + \omega)(\dot r\cos\phi - r\dot\phi\sin\phi) = E \qquad \ldots\ldots(2),$$
$$r\ddot\phi + 2\dot r\dot\phi + r\sin\phi\,.\,\cos\phi\dot\lambda(\dot\lambda + 2\omega) = N \qquad \ldots\ldots(3),$$

where E, N, R represent the external forces, other than gravity, directed to the east, to the north, and perpendicular to the level surface. In the general case E, N and R must take account of forces due to pressure, viscosity, turbulence, etc.

It is the value of g embodying the effect of the earth's rotation which is employed in the equation $d\Gamma = gdz$ to compute the geopotential Γ corresponding with any selected level surface, dz being an element of distance upward along a vertical line.

Equation (2) may be written $\dfrac{1}{r \cos \phi} \dfrac{d}{dt} \{r^2 \cos^2 \phi \, (\dot\lambda + \omega)\} = E$ and if $E = 0$ the angular momentum $r^2 \cos^2 \phi \, (\dot\lambda + \omega)$ is constant; which means that the angular momentum of a particle remains constant whatever be the alteration of r or ϕ due to R and N. Thus the law of conservation of angular momentum is endorsed.

Expression in terms of motion along or perpendicular to the earth's surface

Since in general we have to consider linear velocities relative to the earth, it may be convenient to write the equations in terms of these velocities.

If u, v, w represent components relative to the earth directed to the east, to the north and vertically upwards, we may write u for $r \cos \phi \dot\lambda$, v for $r\dot\phi$ and w for $\dot{r}$. Then

$$\dot{u} = \dot{r} \cos \phi \,.\, \dot\lambda - r \sin \phi \,.\, \dot\phi \,.\, \dot\lambda + r \cos \phi \ddot\lambda, \quad \dot{v} = \dot{r}\dot\phi + r\ddot\phi, \quad \dot{w} = \ddot{r},$$

and we obtain in place of equations (1), (2), (3)

$$\dot{w} - \frac{v^2}{r} - \frac{u^2}{r} - 2u\omega \cos \phi = -g + R \qquad \qquad ......(4),$$

$$\dot{u} + \frac{uw}{r} - \frac{uv}{r} \tan \phi + 2\omega \, (w \cos \phi - v \sin \phi) = E \qquad ......(5),$$

$$\dot{v} + \frac{vw}{r} + 2u\omega \sin \phi + \frac{u^2}{r} \tan \phi = N \qquad \qquad ......(6).$$

The change of geopotential for a difference of height z is $\int gdz$. In the equations of motion we retain the use of height z as the vertical co-ordinate though it may possibly be urged that the vertical being normal to successive horizontal surfaces is not actually a straight line.

Application to a parcel of air

We have dealt so far with the accelerations of a single particle relative to a rotating earth; when we wish to deal with a fluid we must select a small parcel and express each of the three components of the acceleration at a point $\dot{u}$, $\dot{v}$, $\dot{w}$, as the sum of (1) the changes with time in the velocity of the fluid at points moving with the earth, $\dfrac{\partial u}{\partial t}$, $\dfrac{\partial v}{\partial t}$, $\dfrac{\partial w}{\partial t}$, and (2) the changes in the velocity of the point in consequence of the fact that the fluid velocity varies from point to point as referred to the earth's surface.

We write therefore $\dfrac{Du}{Dt} = \dfrac{\partial u}{\partial t} + u\dfrac{\partial u}{\partial x} + v\dfrac{\partial u}{\partial y} + w\dfrac{\partial u}{\partial z}$ in place of $\dot{u}$ and similarly for the other components. The equations then become:

$$\frac{\partial w}{\partial t} + u\frac{\partial w}{\partial x} + v\frac{\partial w}{\partial y} + w\frac{\partial w}{\partial z} - \frac{v^2}{r} - \frac{u^2}{r} - 2u\omega\cos\phi = -g + R,$$

$$\frac{\partial u}{\partial t} + u\frac{\partial u}{\partial x} + v\frac{\partial u}{\partial y} + w\frac{\partial u}{\partial z} + \frac{uw}{r} - \frac{uv}{r}\tan\phi + 2\omega(w\cos\phi - v\sin\phi) = E,$$

$$\frac{\partial v}{\partial t} + u\frac{\partial v}{\partial x} + v\frac{\partial v}{\partial y} + w\frac{\partial v}{\partial z} + \frac{vw}{r} + 2u\omega\sin\phi + \frac{u^2}{r}\tan\phi = N.$$

The equation of continuity

So far our equations have applied to a particle of unit mass. Richardson, as we have seen, deals with the variation of the mass of the parcel of moving air. In order to derive from those which we have developed the equations used by him we have to make use of the principle of conservation of mass to which we have already referred, p. 6.

In rectilinear co-ordinates we can obtain a comparatively simple algebraical expression of the principle by considering an element of volume $\delta x \cdot \delta y \cdot \delta z$.

If u, v, w are the components of the velocity in the direction of the co-ordinate axes, the excess of fluid which flows into the element of volume in time δt across the face $\delta y\,\delta z$ nearest the origin, over that which flows out across the opposite face, is given by the difference of the product of the density, the velocity and the area, over the two faces, or in algebraical symbols by $-\dfrac{\partial}{\partial x}(\rho u\,\delta y\,\delta z)\,\delta x\,\delta t$ and a similar expression is applicable for the other pairs of faces. This increase of fluid in the element of volume can be expressed as the product of the rate of increase of mass and the element of time, i.e. by $\dfrac{\partial}{\partial t}(\rho\delta x\,\delta y\,\delta z)\,\delta t$ and by equating these expressions we obtain the equation

$$-\frac{\partial}{\partial x}(\rho u\,\delta y\,\delta z)\,\delta x\,\delta t - \frac{\partial}{\partial y}(\rho v\,\delta z\,\delta x)\,\delta y\,\delta t - \frac{\partial}{\partial z}(\rho w\,\delta x\,\delta y)\,\delta z\,\delta t$$
$$= \frac{\partial}{\partial t}(\rho\,\delta x\,\delta y\,\delta z)\,\delta t,$$

or
$$\frac{\partial}{\partial x}(\rho u) + \frac{\partial}{\partial y}(\rho v) + \frac{\partial}{\partial z}(\rho w) + \frac{\partial\rho}{\partial t} = 0.$$

Separating the variation of ρ from those of the velocities we get

$$\frac{1}{\rho}\frac{D\rho}{Dt} + \frac{\partial u}{\partial x} + \frac{\partial v}{\partial y} + \frac{\partial w}{\partial z} = 0.$$

When we wish to obtain the equation in terms of spherical co-ordinates the expression is somewhat more complicated. In this case u, v, w represent components of the velocity in the direction of increasing λ, increasing ϕ and increasing r. We cannot now regard the element of volume as rectangular but must consider it as bounded by four planes, passing through the origin, and two spherical surfaces. The areas of the faces are given by expressions of the form

$$r\delta\phi \,.\, \delta r, \qquad r\cos\phi\delta\lambda \,.\, \delta r, \qquad r\delta\phi \,.\, r\cos\phi\delta\lambda,$$

We obtain therefore for the increase of fluid within the element of volume due to the difference of the flow across opposite faces the three expressions

$$-\frac{\partial}{r\cos\phi\partial\lambda}\left(\rho u\, r\,\delta\phi \,.\, \delta r\right)r\cos\phi\delta\lambda\,.\,\delta t,$$

$$-\frac{1}{r\partial\phi}\left(\rho v \,.\, r\cos\phi\delta\lambda \,.\, \delta r\right)r\delta\phi\,.\,\delta t,$$

$$-\frac{\partial}{\partial r}\left(\rho w \,.\, r^2\cos\phi\delta\phi \,.\, \delta\lambda\right)\delta r\,\delta t,$$

and we can equate the sum of these three terms to the product of the rate of increase of mass multiplied by the element of time δt, i.e. to

$$\frac{\partial}{\partial t}\left(\rho r^2\cos\phi \,.\, \delta r \,.\, \delta\phi \,.\, \delta\lambda\right)\delta t.$$

The equation of continuity in spherical co-ordinates therefore runs as follows:

$$-r\frac{\partial}{\partial\lambda}(\rho u)-r\frac{\partial}{\partial\phi}(\rho v\cos\phi)-\cos\phi\frac{\partial}{\partial r}(\rho w r^2)=r^2\cos\phi\frac{\partial\rho}{\partial t},$$

or

$$\frac{\partial\rho}{\partial t}+\frac{1}{r\cos\phi}\frac{\partial}{\partial\lambda}(\rho u)+\frac{1}{r}\frac{\partial}{\partial\phi}(\rho v)-\frac{\rho v}{r}\tan\phi+\frac{\partial}{\partial r}(\rho w)+\frac{2}{r}\rho w=0,$$

which in Richardson's notation may be expressed in the form

$$\frac{\partial\rho}{\partial t}+\frac{\partial}{\partial e}(\rho u)+\frac{\partial}{\partial n}(\rho v)-\frac{\rho v}{r}\tan\phi+\frac{\partial}{\partial z}(\rho w)+\frac{2\rho w}{r}=0.$$

The expression of the forces

We pass on to consider the algebraical or numerical expressions of the forces. We have shown already that the effect of the gravitational attraction is included in the term $-g$, and we proceed therefore to the consideration of the forces which we have denoted by R, N, E.

Pressure. Instead of a particle let us again consider a parcel of air in the free atmosphere. As the parcel for this purpose let us take a small rectangular block with its sides $\delta x, \delta y, \delta z$ along the three axes.

We have already taken account of gravity. The parcel of air will be subject to the effect of pressure which may vary from point to point. If p is the pressure at the centre of one side, and $p+\dfrac{\partial p}{\partial z}\delta z$ the pressure at the opposite side, the difference between the two, directed towards the origin, acting on the area $\delta x\,\delta y$ is $\dfrac{\partial p}{\partial z}\,.\,\delta z\,.\,\delta x\,.\,\delta y=\dfrac{m}{\rho}\,.\,\dfrac{\partial p}{\partial z}.$

Hence, if m is the mass of the parcel, the forces of pressure upon it are equivalent to $-\dfrac{1}{\rho}\dfrac{\partial p}{\partial x},\quad -\dfrac{1}{\rho}\dfrac{\partial p}{\partial y},\quad -\dfrac{1}{\rho}\dfrac{\partial p}{\partial z}.$

The frictional forces. In this calculation we have assumed that the pressure upon a plane surface is normal to the surface—that is never actually true for a fluid in motion; in any real fluid there are frictional forces depending on the viscosity or the turbulence. It is usual in hydrodynamics to get over the difficulty by limiting the application of the equations to what is technically known as a "perfect" fluid in which any tangential force that there may be is regarded as negligible.

For a number of problems the atmosphere can be so regarded, but the restriction is an unreal one; we require therefore to obtain some expression for the viscous forces and for this we shall refer to the hydrodynamical equations of motion referred to fixed axes. This is perhaps permissible with frictional forces because the calculation of their influence is necessarily restricted to the limited area "within sight" of the observer.

According to Lamb[1] the viscous forces may be expressed in each of the three equations by two expressions of the form

$$\tfrac{1}{3}\mu \frac{\partial}{\partial x}\left(\frac{\partial u}{\partial x} + \frac{\partial v}{\partial y} + \frac{\partial w}{\partial z}\right) + \mu\left(\frac{\partial^2 u}{\partial x^2} + \frac{\partial^2 u}{\partial y^2} + \frac{\partial^2 u}{\partial z^2}\right),$$

and the equations of motion for a viscous fluid are three equations of the type

$$\rho\frac{Du}{Dt} = \rho X - \frac{\partial p}{\partial x} + \tfrac{1}{3}\mu \frac{\partial}{\partial x}\left(\frac{\partial u}{\partial x} + \frac{\partial v}{\partial y} + \frac{\partial w}{\partial z}\right) + \mu\left(\frac{\partial^2 u}{\partial x^2} + \frac{\partial^2 u}{\partial y^2} + \frac{\partial^2 u}{\partial z^2}\right).$$

As we have seen, the equation of continuity is

$$\frac{\partial \rho}{\partial t} + \frac{\partial}{\partial x}(\rho u) + \frac{\partial}{\partial y}(\rho v) + \frac{\partial}{\partial z}(\rho w) = 0,$$

and for an incompressible fluid this becomes $\frac{\partial u}{\partial x} + \frac{\partial v}{\partial y} + \frac{\partial w}{\partial z} = 0$, hence for an incompressible fluid the effect of viscosity may be expressed by the terms of the form $\mu\left(\frac{\partial^2 u}{\partial x^2} + \frac{\partial^2 u}{\partial y^2} + \frac{\partial^2 u}{\partial z^2}\right)$.

Even in the atmosphere $\frac{\partial u}{\partial x} + \frac{\partial v}{\partial y} + \frac{\partial w}{\partial z}$ or $-\frac{1}{\rho}\frac{D\rho}{Dt}$ will always be a small quantity and its effect will in practice be negligible by comparison with those of the term $\mu\left(\frac{\partial^2 u}{\partial x^2} + \frac{\partial^2 u}{\partial y^2} + \frac{\partial^2 u}{\partial z^2}\right)$.

If further we may neglect the variation of velocity of the air in the successive horizontal layers and consider only the variations in the motion from one layer to those above, the expression may be further simplified and we may represent the viscosity by two terms $\mu\frac{\partial^2 u}{\partial z^2}$ and $\mu\frac{\partial^2 v}{\partial z^2}$.

---

[1] *Hydrodynamics*, 4th edition, p. 573, Cambridge, 1916.

We obtain therefore for the equations of motion

$$\frac{Dw}{Dt} - 2\omega u \cos \phi - \frac{u^2}{r} - \frac{v^2}{r} = -g - \frac{1}{\rho}\frac{\partial p}{\partial z} + F_z \qquad \qquad ......(7),$$

$$\frac{Du}{Dt} + 2\omega \left(w \cos \phi - v \sin \phi\right) + \frac{uw}{r} - \frac{uv}{r} \tan \phi = -\frac{1}{\rho}\frac{\partial p}{\partial x} + F_x ......(8),$$

$$\frac{Dv}{Dt} + 2\omega u \sin \phi + \frac{vw}{r} + \frac{u^2}{r} \tan \phi = -\frac{1}{\rho}\frac{\partial p}{\partial y} + F_y \qquad \qquad ......(9),$$

where F_z, F_x, F_y are the components of forces due to viscosity or turbulence.

The terms printed in small type are those which according to the *Skipper's Guide* (p. 51) can be neglected except in the immediate neighbourhood of the poles; there the terms containing tan ϕ may become appreciable.

An examination of the several terms

Before going further we may endeavour to form a conception of the physical implications of the different terms, which have an appearance of complication that is rather oppressive when we consider that they are only intended to express what in the beginning was $m\ddot{x} = X$ and two other equations like it; but the complication is unavoidable in the study of the atmosphere because it arises partly from the rotation and partly from the spherical, or approximately spherical, form of the earth upon which the subject of investigation rests.

Exner gives the result of the inter-operation of the various components in a form from which the following is derived.

Gravity, gravitational attraction and weight. If we consider equation (1), p. 44, which represents the vertical acceleration of a body, the weight of unit mass may be expressed as the downward acceleration $-\ddot{r}$, which when no force but gravity is acting is given by

$$\ddot{r} = -g + r\dot{\phi}^2 + r\dot{\lambda}^2 \cos^2\phi + 2\omega\dot{\lambda}r \cos^2 \phi.$$

The terms $r\dot{\phi}^2$ and $r\dot{\lambda}^2 \cos^2 \phi$ represent centrifugal forces due to the motion of the body relative to the earth, both of which tend to reduce the weight in whatever direction the motion may be. For actual motion of air on the earth's surface where the velocities are of the order of 20 m/sec or 2×10^3 c, g, s units, the effect is very small in comparison with g. g is of the order of 1000 c, g, s units, whereas $r\dot{\phi}^2$ and $r\dot{\lambda}^2 \cos^2\phi$ are both of the order of $(2 \times 10^3)^2/(6 \times 10^8)$ or 10^{-2}.

The term $2\omega\dot{\lambda}r \cos^2 \phi$ arises from the fact that the effect of the west–east motion on the vertical acceleration of the body depends on the square of the sum of the earth's rotation and the angular velocity relative to the earth; an addition to the west to east velocity has therefore a greater effect than a diminution of similar amount. We find therefore that when λ is positive, i.e. when the body is moving from west to east, its weight is less than when it is moving from east to west. Since the effect depends upon cos ϕ, it is greatest at the equator. For a wind of 20 m/sec, $2\omega\dot{\lambda}r$ is $2 \times 7\cdot3 \times 10^{-5} \times 2 \times 10^3$, or $\cdot03$ per cent of the value of gravity—an east wind of 20 m/sec is therefore o6 per cent heavier than a wind of similar composition and velocity moving to the east.

The effect of the earth's rotation on horizontal motion. If now we neglect the vertical motion so that w or $\dot{r}$ is zero, and consider motion under no forces but gravity, equations (2) and (3) become:

$$r \cos \phi \ddot{\lambda} - 2\dot{\lambda}r\dot{\phi}\sin\phi - 2\omega r\dot{\phi}\sin\phi = 0;$$

$$r\ddot{\phi} + r\cos\phi \,.\, \sin\phi\dot{\lambda}^2 + 2\omega r\sin\phi \,.\, \cos\phi \,.\, \dot{\lambda} = 0,$$

or, in terms of u and v,

$$\dot{u} - \frac{uv}{r}\tan\phi - 2\omega v\sin\phi = 0,$$

$$\dot{v} + \frac{u^2}{r}\tan\phi + 2\omega u\sin\phi = 0.$$

The terms $u^2 \tan\phi/r$ and $uv \tan\phi/r$ are small in comparison with $2\omega u \sin\phi$ and $2\omega v \sin\phi$, the ratio of either pair is $u/(2\omega r\cos\phi)$ or $\cdot 02 \sec\phi$ for a wind of 20 m/sec. As a first approximation we may neglect $u^2 \tan\phi/r$ in comparison with $2\omega u \sin\phi$, except in the polar regions where $\sec\phi$ becomes large. For a wind of 20 m/sec the two terms are of equal magnitude in latitude 89°.

The effect of the earth's rotation on the horizontal motion may be very approximately expressed therefore by an acceleration $2\omega V \sin\phi$ at right angles to the motion. This is in fact the basis of the equation of the geostrophic wind of this volume, $bb = 2\omega V\rho \sin\phi$.

We can picture it qualitatively. A body moving towards the north will be passing to regions of smaller east–west velocity and will appear to be accelerated to the east.

For a body moving to the east, since the surface of the earth is horizontal and the body is moving faster than the earth, the direction of the resultant force is not quite vertical, there is an acceleration towards the equator and the body will tend to move towards the equator.

The order of magnitude of the several terms. The Skipper's Guide

As an indication of the relative importance of the several terms we may endeavour to give some idea of their order of magnitude in normal conditions on the earth's surface. For this purpose we may regard the horizontal velocities u and v as of the order of 20 m/sec or 2×10^3 c, g, s units, the vertical velocity w is not often greater than 10^2 c, g, s units but it may on occasion reach 10^3 c, g, s units. We have already seen that in the equation for vertical motion the terms $r\dot{\phi}^2$ and $r\dot{\lambda}^2\cos^2\phi$, i.e. v^2/r and u^2/r, are of the order 10^{-2}, and $2\omega\dot{\lambda}r\cos^2\phi$ or $2\omega u\cos\phi$ is of the order $3 \times 10^{-1}\cos\phi$, and that both therefore are small in comparison with gravity.

If we turn now to the terms expressing the horizontal motion:
uw/r and vw/r are of the order 10^{-3} but depend on the magnitude of w,
$uv \tan\phi/r$ and $u^2 \tan\phi/r$ are of the order $10^{-2}\tan\phi$,
$2\omega u \sin\phi$ and $2\omega v \sin\phi$ are of the order $3 \times 10^{-1}\sin\phi$,
$2\omega w \cos\phi$ is of the order $10^{-2}\cos\phi$ but may reach $10^{-1}\cos\phi$ if w is large.

With regard to the magnitude of the accelerations Dw/Dt, Du/Dt and Dv/Dt and of the terms expressing the forces of pressure and viscosity, a good deal of discussion has arisen. We do not wish to enter into the details here; readers who are interested in the subject may be referred to a paper by Hesselberg and Friedmann[1], "Die Grössenordnung der meteorologischen Elemente und ihrer räumlichen und zeitlichen Ableitungen," and to a later discussion by Brunt and Douglas[2] on "The modification of the strophic balance for changing pressure distribution, and its effect on rainfall," to which we shall refer later. It explains that some of the magnitudes included in the *Skipper's Guide* require reconsideration as they depreciate unduly the flow of air along isobars.

EQUATIONS OF MOTION OF A RIGID SOLID

The equations which we have set out are intended to provide the means of calculating the future of a single material particle or of a parcel of air. Perhaps a morsel of air would give a better idea of its peculiar isolation. There is nothing explicit in the equations to associate one morsel of air with others of its environment to form an entity to the behaviour of which the mind of the reader can be directed.

The suggested isolation is of course the expression of logical hypothesis rather than the exposition of nature. For the observant meteorologist a morsel of air is always part of a natural scheme or system. It is part of a pool of stagnant air or of an air-current which has a definite direction, or of a wave, an eddy or a vortex, and he thinks of the motion of the morsel in relation to that of the system of which it forms part.

This hypothetical isolation of the morsel from its environment marks a conspicuous difference between the hydrodynamical equations of our meteorological calculus and the equations of motion of a rigid body.

In these equations the principle of rigidity is expressed verbally by the axiom that the morsels of a rigid body always maintain their distance one from the other, and dynamically by the axiom known as d'Alembert's principle that the equal and opposite forces between the component morsels counteract one another and produce no effect upon the motion of the body as a whole.

From this we learn to distinguish between the motion of the body as a whole as indicated by the motion of its centre of gravity, which is assumed to be the same as if the mass were concentrated there and all the operating forces acted there, and the motion of rotation or spin with reference to the centre of gravity which may be resolved into rotation about three axes at right angles through that centre.

The last introduces the important ideas of the moment of force about an axis and its equivalent, the rate of change of angular momentum. Moment of force, the product of the force multiplied by its distance from the axis, has the

---

[1] *Veröff. d. Geophys. Inst. d. Univ. Leipzig*, Zweite Serie, Heft 6, 1914.
[2] *Memoirs of the Roy. Meteor. Soc.*, vol. III, No. 22. London, 1928.

dimensions of work or energy. Angular momentum with reference to an axis is the product of the momentum of a particle in the direction perpendicular to an axis, by the distance of the line of motion from the axis.

This gives us a new equation of motion for all the particles of a rigid body rotating about a fixed axis, viz.

$$\Sigma ml^2 \ddot{\theta} = Fr,$$

where Σml^2 is the sum of the product of the mass of each particle multiplied by the square of the distance l from the axis of rotation and Fr is the moment of the force acting about the axis of rotation—that is to say the product of the component of the force in the plane perpendicular to the axis and the distance of the component from the axis, or, if the force is distributed over the mass, f being the force upon an elementary mass m at distance r from the axis, Σfr is the whole effective moment—this corresponds with the simple Newtonian equation $m\ddot{x} = X$; Σml^2, which depends on the mass of the solid and its distribution with regard to the axis, is called the moment of inertia about the axis.

For a solid rotating about a fixed axis therefore we have the law of motion of rotation, namely that the rate of change of angular momentum is equal to the moment of the impressed couple.

If the solid is free to move about any axis which passes through its centre of gravity the motion of rotation in three dimensions is represented by Euler's equations

$$A\dot{\omega}_1 - (B - C)\,\omega_2\omega_3 = L,$$
$$B\dot{\omega}_2 - (C - A)\,\omega_3\omega_1 = M,$$
$$C\dot{\omega}_3 - (A - B)\,\omega_1\omega_2 = N,$$

where A, B, C are the moments of inertia of the solid about the three principal axes, $\omega_1, \omega_2, \omega_3$ are the angular velocities of rotation about those axes, and L, M, N the impressed couples.

On these are based a number of computations with reference to the spinning of a top about its point, of the earth about its centre, of the gyro-compass about its centre, or of the heavenly bodies about their moving centres. This analysis however carries us beyond our limit; in what we have to say later on we do not propose to go beyond the rotation about a single fixed axis.

These considerations are important for us because air sometimes behaves as though it were rigid, that is to say the particles always maintain their distance apart, and we get the important principle of the conservation of moment of momentum or angular momentum. As the Latin poet Lucretius expressed it, the adjacent parts of air, though held together very lightly, still have a relation one to the other.

The vitality of spin

A ring of air which is rotating about its own axis may in a sense be regarded as having properties of a solid and therefore preserving its moment of momentum if there is no couple tending to increase or diminish its spin. So if by some suitable distribution of forces the particles of the ring can be moved so that its radius is diminished from R to r, the moment of momentum $mR^2\omega$ will be represented in the new position by $mr^2\Omega$, and the angular velocity of the ring will be increased in inverse proportion to the square of the change of radius; or if we consider the linear velocity, that will be increased in the inverse ratio of the change of radius. We can imagine such a case on the earth's surface if there are forces which move a ring of air to expand or contract at the equator or to increase its latitude north or south. The motion is possible if there are forces which can draw the ring along the surface so that its mass is nearer to the axis in the higher latitudes. And in that case we get an increase in the velocity of rotation which will appear to an observer on the rotating earth as a motion from west to east. The force in that case may be regarded as a force towards the axis of rotation and the work done by that force will be expressed as the increase in the kinetic energy of the ring which is $\frac{1}{2}mu^2$; u is here the velocity in rotation round the polar axis, which we have supposed equal to that of the earth in the original latitude.

We shall see later on that Helmholtz imagines the atmosphere near to the earth's surface to be divided into a series of rings, such as we have imagined, rotating each in its own latitude, the rings being divided by isentropic surfaces, and he uses the proposition which we have cited to determine the relative velocity corresponding with different displacements. But obviously the only condition that we ask is that a particular mass of air should have its motion corresponding with a part of a ring; it need not extend round the whole circle of latitude. And so, wherever on the earth's surface we can find air which could form part of a rotating ring, we can use the principles of the motion of a solid to aid us in discussing the motion of the air.

Such rings as we have indicated are not the only form of atmospheric motion to which the analogy of the motion of a solid can be applied. A parcel of air, so small that the effect of the rotation of the earth is not apparent and may be left out of the calculation, may be made to spin by a suitable distribution of forces such as are apparent when the motion of the air is turbulent. Any mass of air spinning about an axis with the velocity proportional to the distance from the axis will in suitable circumstances behave like a solid and can be treated as such. External forces which do not affect its rotation may move its centre of gravity as if it were a solid, and thus we get an idea of rotation about an axis endowing the air with a certain vitality which we propose to refer to on occasions as the vitality of spin.

Spin, as we have indicated, is a characteristic of turbulence, and when turbulence is operative we may obtain portions of the atmosphere in a state of spin quite temporarily with the assurance that while it is spinning like a solid it can be treated to some extent as if it were a solid.

We may remember that as a deduction from the maps of the normal distribution of pressure at different levels in vol. II, we concluded (pp. 279, 372) that a belt of the upper air from latitude 30° to 60° N was found to rotate like a solid. So perhaps we may consider the normal condition of the upper air to be a spin, different at different levels, completing a revolution in 38 days at the 4 km level in summer, 33 days at 6 km and 23 days at 8 km.

THE APPROACH BY THE DOCTRINE OF ENERGY

The process for the expression of motion of the atmosphere which we have sketched in the preceding pages has been based upon the assumption that we understand what is meant by force and can assign values to the components of the forces acting in the directions of the co-ordinate axes. There is another point of view which is sometimes employed in the discussion of dynamical problems. From that point of view the sequence of events is regarded as the process of transformation of energy, the motion expresses kinetic energy and the configuration of the system is an indication of potential energy. Thence we get a generalised conception of force as the change in kinetic energy corresponding with change in the configuration. Some system must be employed to specify the configuration and express the energy but we are not limited to co-ordinate axes.

The development of the method is due to Lagrange. Its application to dynamical problems can be found in the text-books of dynamics. The method was employed by J. J. Thomson[1] for the study of transformations of energy of different kinds in physics and chemistry and is well worth the attention of meteorologists.

The name of Lagrange is introduced into meteorological literature on another count by V. Bjerknes[2] as the inventor of a special form of the equations of motion of a particle of fluid in which the co-ordinates of position of the particle at any time are expressed as functions of their value at a past epoch and the time.

Der Übergang von einer Meteorologie der Druck-, Temperatur- und Bewegungs-*felder* zu einer der *bewegten Luftmassen* bedeutet aber hydrodynamisch der Übergang von den Eulerschen zu den Lagrangeschen Gleichungen. Zurückgehalten durch den traditionellen Schrecken vor den letzteren, suchte ich aber möglichst lange mit den Eulerschen auszukommen. Das Einpassen der Probleme in ungeeignete Form zeigte sich jedoch immer unbefriedigender, und führte mich zuletzt—leider erst ganz neulich—zu einem eingehenden Vergleich der beiden Gleichungssysteme. Das Resultat wurde mir eine grosse Überraschung: es kann nur als ein historisches Vorurteil bezeichnet werden, wenn man *allgemein* das eine Gleichungssystem als einfacher als das andere hinstellt.

The analysis involves very elaborate manipulation of symbols.

A cursory examination may inspire the reader with the hope that the establishment of some recognisable entity intermediate between the co-ordinates and the atmospheric circulation, that might appear as a generalised co-ordinate, would make the situation more easily comprehensible.

---

[1] *Applications of Dynamics to Physics and Chemistry*. London, Macmillan and Co., 1888.

[2] 'Über die hydrodynamischen Gleichungen in Lagrangescher und Eulerscher Form und ihre Linearisierung für das Studium kleiner Störungen,' *Geofysiske Publikasjoner*, vol. v, No. 11, Oslo, 1929.

THE APPLICATION OF THE EQUATIONS

We have now completed our recital of the equations which have been evolved for determining the motion of the atmosphere from our knowledge of the forces in operation. We began by citing the forms evolved by Richardson because they represent the most elaborate representation of the problem. We have no example of the practical use of the Lagrangian method.

In the ordinary practice of Newtonian dynamics the next step should be to integrate the equations and so express the values of all the variable elements at any one time, the present, in terms of the values at any selected previous epoch of the past.

The equations are however too complicated for general solution by direct integration. The nearest approach is that of Richardson, who would substitute for the impracticable general solution an approximate solution by the method of finite differences which he had found serviceable in a different dynamical problem, namely that of the stresses in masonry dams. He obtains a solution by introducing finite changes due to the effect of the various forces in a finite time, assigning approximate values to the various forces for the interval.

We do not propose at this stage to follow Richardson's approximate solution. We shall pass on to cite some of the applications of the general equations of motion to the solution of atmospheric problems.

THE CLASSICS OF METEOROLOGICAL CALCULUS

Let us consider more or less in chronological order the contribution of the equations of motion to the solution of the problem of weather.

Those contributions which refer especially to the origin or behaviour of the group of phenomena associated with the Daily Weather Chart and are concerned with the origin or development of cyclonic depressions should find their place at the end of the volume. In this chapter we propose to deal only with contributions which embody dynamical principles of general application.

Our expressed purpose in setting out the laws of atmospheric motion in chap. I and the derivation from those laws of the various forms of the general equations of motion in chap. II was to cite the efforts which have been made to apply the various equations to the general weather-problem and, without going into the details of the mathematical operations employed, to set out the results arrived at by the authors, and the auxiliary assumptions by which those results have been achieved, with a view to satisfying ourselves and the reader as to whether a natural sequence of events has really been expressed by the manipulation of the mathematical symbols and their numerical equivalents.

To do this for the whole number of contributions would be too much of a labour for the author as for the reader. To make them intelligible to the reader of ordinary mathematical attainments, amplification and expansion of the original would be required instead of abbreviation and compression. We have seen it suggested that a test of efficiency of a contribution to science in these days is that the result can be given within four lines of print. That test often fails in the case of a mathematical contribution.

We have endeavoured to select from the large number of meteorological memoirs which use differential equations of motion those only which have made some impression upon the meteorological world and which form, or have formed, part of the stock-in-trade of the exponents of the practical dynamics of weather. From this point of view it is not necessary that we should re-examine the literature of the subject, our criterion will obviously be satisfied if an author of acknowledged competence has already exercised his judgment and made his selection.

In our summary of the history of the subject in vol. I we have cited two compilations of memoirs bearing on dynamical meteorology, one by Cleveland Abbe[1] for the Smithsonian Institution, and the other by Marcel Brillouin[2] which is sufficiently comprehensive for the years preceding its publication to serve the purpose that we have in view. We do so with great confidence because M. Brillouin was led by one or other of the papers which are included in his summary to the further development of the subject in a substantial volume of the *Annales du Bureau Central Météorologique* devoted to *Vents contigus et nuages*.

A large number of the papers translated by Abbe are concerned with the thermodynamics of the atmosphere, especially von Bezold's contributions and Neuhoff's equations of saturation adiabatics, a subject which we have treated separately in vol. III.

Let us therefore turn our attention to the papers which Brillouin cites.

Halley to Helmholtz

In the volume dated 1900 Brillouin passes in review the memoirs that deal with the general circulation of the atmosphere from Halley (1686) to Helmholtz (1889). The end of the nineteenth century is a suitable epoch in the history of dynamical meteorology because the vital division of the atmosphere into troposphere and stratosphere dates from about that time. Before that it may be said to have been usual for the purpose of theory to regard the atmosphere either as isothermal or as isentropic. The systematic investigation of the upper air has put an end to that kind of simplification.

Brillouin's selection includes, besides the initial contributions of Halley and the final contributions of Helmholtz, those of George Hadley (1735), 'Concerning the cause of the general trade-winds'; M. F. Maury (1855), 'The physical geography of the sea and its meteorology'; W. Ferrel (1856–61), 'The winds and the currents of the ocean,' 'The motions of fluids and solids, etc.'; James Thomson (1857, '92), 'On the grand currents of atmospheric circulation'; W. Siemens (1886), 'Über die Erhaltung der Kraft im Luftmeere

---

[1] 'The Mechanics of the Earth's Atmosphere,' *Smithsonian Miscellaneous Collections*, No. 843, Washington, 1893. Other collections are dated 1877 and 1910.

[2] *Mémoires originaux sur la circulation générale de l'atmosphère*. Paris, 1900.

der Erde'; M. Möller (1887), 'Der Kreislauf der atmosphärischen Luft zwischen hohen und niederen Breiten, die Druckvertheilung und mittlere Windrichtung'; A. Oberbeck (1888), 'Über die Bewegungserscheir. ngen der Atmosphäre.'

In a brilliant introduction of ten pages he notices also the contributions of de Tastes (1865–78), Teisserenc de Bort (1883), Duclaux (1891), and Duponchel (1894).

About these we may hint that Halley finds a physical basis for the atmospheric circulation in a map of the winds of the intertropical regions and suggests the influence of solar radiation upon the surface-temperature as their cause—the winds seek the sunny regions of the earth and pursue the track of the sun from east to west.

He was therefore a pioneer in the determination of the forces acting upon the atmosphere by observation of its motion. The same basis of progress may be traced in Dove's *Law of Storms* and, with the aid of Buys Ballot's relation of motion to the distribution of pressure, in the work of de Tastes and Duclaux.

With Hadley may be said to begin the other aspect of the subject, namely the calculation of the motion from the recognised forces. He accounts for the westward motion of the trade-wind by the dynamical effect of the earth's rotation upon a parcel of air which is free to move over the earth's surface, and for the motion of the counter-trades by what amounts to an equation of continuity.

The development of the idea of calculating the effect of forces upon the motion finds expression in the general equations of motion of a particle on a rotating earth which are given in Ferrel's second memoir after some general reasoning, physical and dynamical, in his first memoir.

The peculiarity of Hadley's treatment is that, having found a sufficient cause for the deviation of the motion of air over the earth's surface from southward to south-westward, he allows his mind to rest in that conclusion. More rigorous dynamical treatment traces the path of a particle, free to move, through a variety of curves of interesting shapes which are set out for the curious reader by F. J. W. Whipple[1]. In so far as they omit to take account of the environment with which the moving air is associated they are not, strictly speaking, meteorological.

As a matter of fact, the maps which are reproduced in vol. II may be cited as evidence that the circulation is better described with reference to the distribution of pressure than as a direct inference from terrestrial rotation.

From the hypothetical operation of known forces, Maury, Ferrel, James Thomson and Oberbeck obtain schemes of general circulations arranged in zones of latitude ignoring the effect of the distribution of pressure and of that of land and water. Maury brought in hypothetical electric and magnetic forces when his dynamical courage failed. As a deduction from his equations in one of his schemes, Ferrel regarded the polar regions as devoid of atmosphere assuming that there was no frictional effect.

[1] *Phil. Mag.*, vol. XXXIII, 1917, p. 457.

We must therefore regard the contributions as interesting speculations about a planet which does not represent the earth as we know it; nor can the application to the real earth be developed by any slight modification of the hypothesis.

L'influence des terres et des mers dans notre hémisphère nord, le seul bien connu, est tellement prépondérante qu'il n'y a aucune raison d'admettre la moindre ressemblance, entre la distribution de la pression et de la température déduites des obser-- vations en faisant les moyennes par parallèles, et les distributions que les mêmes situations astronomiques produiraient sur un globe réellement uniforme.

(Brillouin, *loc. cit.*, p. xix.)

Siemens introduced a more general form of reasoning when he imagined the total momentum of the atmosphere and its total kinetic energy to be derived from the complete mixing of the atmosphere at rest. But the application of this principle to the actual condition of the earth is beset with difficulty on account of the thermal energy.

Helmholtz: Stratification of the atmosphere

Helmholtz, on the other hand, opens out a new province by using his reasoning to develop certain processes which are operative in any part of the atmosphere in accordance with the laws of hydrodynamics, and reaches conclusions which one may expect to find illustrated wherever the atmosphere admits the prescribed assumptions. The memoirs[1] cited by Brillouin are the real basis of the productive part of modern mathematical treatment of the problems of meteorology. Here therefore we reproduce the main lines of his argument.

First, in relation to the general circulation of the atmosphere, we conclude that if the atmosphere, at rest on a smooth earth, were completely mixed the result would be convective equilibrium or, as we prefer to say, equality of entropy throughout. There would be a zone of calms at the surface near the tropic where the pressure would reach a maximum, a gradient of pressure on the north side for westerly winds and on the south side for easterly winds.

Next he shows that if a belt of air with negligible viscosity were moved from a calm zone at the equator either northward or southward, retaining its moment of momentum, the earth's rotation 10° away from the initial position would give wind with a velocity of 14 m/sec, 20° away 58 m/sec and 30° away 134 m/sec. Since these velocities are beyond any that are recorded for regular winds of the globe, he considers whether the reduction to practical values could occur in consequence either of the viscosity of air or the thermal conductivity; by consideration of the hydrodynamical equations of a viscous fluid (p. 48) he finds that for a fluid with Maxwell's value of the viscosity of air at 0° C it would take 42,747 years to reduce the original velocity at the top of the homogeneous atmosphere to one-half; and using a corresponding formula for the equalising of temperature it would require 36,164 years to

---

[1] 'Ueber atmosphärische Bewegungen,' *Sitzungsber. Akad. Wiss. Berlin*, 1888, S. 647–63; 1889, S. 761–80.

reduce the original difference of temperature between the upper and lower layers to one-half. Consequently in considering the motion of the air in the atmosphere we can leave out of account the effect of simple viscosity and conductivity and must seek the cause of the reduction of velocity, which we recognise as taking place, in some other activity such as that of eddies, by which the area of layers in contact is multiplied enormously and an opportunity afforded for the natural viscosity of the air to become effective.

We have already referred (vol. II, p. 295) to Brunt's estimate of the rate of decay of atmospheric motion, from which it would appear that if left to itself for six days the kinetic energy would be reduced to one-hundredth of its normal value.

We are therefore invited to consider the action between two layers of air which are themselves in relative motion, each of them being isentropic in itself but with a different value of the entropy.

To elicit the properties at the boundary surface between two such layers with finite differences of entropy and velocity we take the hydrodynamical

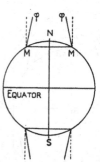

Fig. 11 a. The "normal" inclination to the direction of the polar axis of a surface of discontinuity between two isentropic layers for gradual diminution of temperature with latitude (Brillouin).

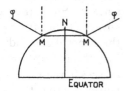

Fig. 11 b. The inclination of a surface of discontinuity in an "abnormal" case when there is a local increase of temperature northward.

equations in relation to axes drawn from the centre of the earth, that of x through the poles and y and z at right angles thereto, ψ is the angular rotation of a particle when at a distance s from the polar axis, Ψ its angular momentum. ω is the angular rotation of the earth, P the potential of the impressed forces, so that $\partial P/\partial x$ is the component of the impressed force in the direction of x. We get two equations,

$$\frac{\partial P}{\partial x} + \frac{1}{\rho}\frac{\partial p}{\partial x} = 0 \quad \text{and} \quad \frac{\partial P}{\partial s} + \frac{1}{\rho}\frac{\partial p}{\partial s} = \frac{\Psi^2}{s^3}.$$

This gives us surfaces of equal pressure which bridge a zone of calms (where $\psi = \omega$) and if there are two isentropic rings in contact one with the other we get an equation for the curve of separation in a section through the pole, which is linear with respect to the inverse of the distance from the centre of the earth and quadratic with respect to s.

At the surface of separation the pressure is continuous and Helmholtz deduces that in the normal case (fig. 11 a), where the entropy is continuously diminishing along the earth's surface from the equator to the pole, the surface of separation cuts the earth's surface at a smaller angle than that between the pole star and the earth's surface; but in the abnormal case (fig. 11 b), in which

locally the entropy increases northwards, the angle between the dividing surface and the earth's surface is greater than the angle between the pole star line
and the earth's surface. That is to say, in the normal case the isentropic surface
is within the cylinder bounded by lines parallel to the polar axis and in the
abnormal case it is outside that cylinder.

Consequently in the normal case, on the range from the pole towards the
equator, a series of isentropic rings of latitude with successively greater
entropy will cut the earth in a series of circles; and the surfaces, starting upwards from the circles, will slope towards one another from the two sides of
the polar axis. The limiting ring which touches at the equator will coincide in
the equatorial region with the cylindrical surface itself, and thus the series of
surfaces will be as represented in Brillouin's diagram (fig. 12), which we have
amplified to show the surfaces passing over the poles; in the abnormal case a
surface which starts from a point in the abnormal condition will turn round
and cut the earth again in a normal condition.

Helmholtz considers only the inclination of the surfaces at the earth, i.e.
the fringes of the surfaces which cut the earth's
surface. Since the memoir was written the discovery of the stratosphere and of the structure of
the atmosphere leading up to it has made it clear
that the entropy at every part of the globe increases in the vertical, and after passing the tropopause into the stratosphere increases very rapidly.
Thus it must ultimately reach a value as great as
that of the entropy of any part of the earth's surface, and consequently the Helmholtz isentropic
surfaces, which start from a circle of the earth,
must, if produced upwards, form caps over the
pole; the cylindrical form, which passes tangen-
tially from the equator, must also be developed in
the upper regions so as to cover the poles.

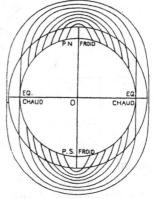

Fig. 12. A series of isentropic
lines representing the surfaces of contact of atmo-
spheric layers of successively
larger entropy from pole to
equator (after Brillouin).

We are therefore reminded of the division of
the atmosphere into underworld and overworld
(vol. III, chap. VIII). The surfaces which Helmholtz
suggests in order to calculate the effect of one layer
on another are idealised forms of the actual surfaces which are described in
the chapter referred to. First the one tangential at the equator as separating
the overworld from the underworld, and the others, called normal, which
cut the surface between the equator and the pole we have regarded as
separating the underworld of the atmosphere into successive zones.

Helmholtz's surfaces cut the earth in rings of latitude, with the isentropic
condition as an hypothesis; we have taken the actual pressures and temperatures and computed therefrom the lines of equal entropy on the earth's surface.
Helmholtz uses his assumptions in order to limit the motion to that along
parallels of latitude, but his reasoning about the effect of one layer upon

another of different entropy is not thereby invalidated, and the memoirs give us in fact an introduction to the study of the underworld as defined in vol. III.

Helmholtz also considers the disturbance of the equilibrium between the isentropic layers by heating and friction.

Heating is only dynamically effective if the heat is supplied to the lower side of the stratum or removed from the upper side. In either of these conditions the propagation through the layer is relatively rapid until a new isentropic condition of equilibrium obtains; in the reverse case, when heat is supplied at the top and removed at the bottom, the propagation depends on conduction and radiation and is very slow.

In the case of friction the disturbance operates differently according as to whether the motion at the surface is to the east or to the west. If we consider the normal inclination of the layers when the upper end of the layer is nearer to the axis than its lower end then, if there is a west wind at the surface, friction will tend to decrease the moment of momentum of the wind and the air at the surface will slide outwards towards the upper end of the layer in order to reach a position of equilibrium. In course of time therefore the moment of momentum of the whole layer will be diminished.

If however the surface-wind is from the east the effect of friction is to increase the moment of momentum of the wind; the air tends to find its new place of equilibrium by moving along the surface towards the equator and pressing into the stratum in front of it. In this case the change is confined to the surface-layer, and is relatively more effective there than in the previous case where the effect is distributed through the whole thickness.

Helmholtz then refers to the winds caused by the cooling of the earth at the poles. The cold layers flow outwards at the surface and form an east wind; the effect of friction is to increase the moment of momentum; so the advance continues and the air remains at the surface. He attributes the irregularity of the polar east wind partly to the fact that the pole of maximum cold does not coincide with the geographical pole, and partly to the effect of mountain-ranges on the shallow layer of cold air.

The general conclusion of his study is that

the principal obstacle to the circulation of our atmosphere, which prevents the development of far more violent winds than are actually experienced, is to be found not so much in the friction on the earth's surface as in the mixing of differently moving strata of air by means of whirls that originate in the unrolling of surfaces of discontinuity. In the interior of such whirls the strata of air, originally separate, are wound in continually more numerous, and therefore also thinner layers spirally about each other, and therefore by means of enormously extended surfaces of contact there thus becomes possible a more rapid interchange of temperature and equalisation of their movement by friction.   (Helmholtz, tr. C. Abbe.)

Helmholtz: Atmospheric billows

In the second memoir we are first introduced to the theory of waves in a surface of discontinuity which depends upon hydrodynamical theory without any consideration of the spherical form of the earth or its rotation. In the latter part we are brought again to consider the conditions at the bounding surface between two layers of air of different entropy and moving with different velocity. Helmholtz discusses the mixing of the air derived from the juxtaposition of the two layers at the bounding surface, the ascent of the mixture and consequent development of discontinuity at the new surface of juxtaposition.

It is this view of the mutual action between two layers of different entropy which forms the starting-point of Brillouin's work on *Vents contigus et nuages*. The upward motion which gives rise to clouds takes place in the boundary between two layers of different entropy and the rising air is formed by those parts of the two adjacent layers that have become mixed. In this way pockets of mixed and cloud-forming air are developed between the two layers in the upper air and are used by Brillouin in the explanation of many forms of cloud.

We give some account of the reasoning of Helmholtz' paper.

The picture which is presented is that of gravity-waves in the surface of separation between two layers of air of different densities each invariable in its own layer. The waves may be produced by wind as represented by relative motion between the two layers expressed as a geometrical difference of velocity between two limits, upper and lower, sufficiently far from the surface of separation to be regarded as undisturbed. The waves may be regarded as stationary in the surface of separation.

I have at first treated the simplest case of the problem, namely, the movement of rectilinear waves which propagate themselves with unchanged forms and with uniform velocity on the plane boundary surface between indefinitely extended layers of two fluids of different densities and having different progressive movements. I shall call this kind of billows stationary billows, since they represent a stationary motion of two fluids when they are referred to a system of co-ordinates which itself advances with the waves.

In any case of real relative motion of air-currents the waves will be carried with the common boundary at a velocity equal to the difference between the undisturbed velocities if they are in the same line, or to the difference of the components in the direction opposite to the smaller velocity if they are inclined at an angle.

The method of presentation may be indicated by the following quotations from Abbe's translation.

As soon as a lighter fluid lies above a denser one with well-defined boundary, then evidently the conditions exist at this boundary for the origin and regular propagation of waves, such as we are familiar with on the surface of water.

.　　.　　.　　.　　.　　.　　.　　.　　.

The calculations performed by me show further that for the observed velocities of the wind there may be formed in the atmosphere not only small waves, but also those whose wave-lengths are many kilometres which, when they approach the earth's surface to within an altitude of one or several kilometres, set the lower strata of air into violent motion and must bring about the so-called gusty weather. The peculiarity of such weather (as I look at it) consists in this, that gusts of wind often accompanied by rain are repeated at the same place, many times a day, at nearly equal intervals and nearly uniform order of succession.

The reader may remember the suggestion of fig. 11 of vol. III as a possible example of breaking waves in air.

A section already referred to is interposed to explain that the discontinuity necessary to produce a surface for waves is developed between the isentropic layers discussed in the previous section and represented in fig. 12. The waves are in fact waves belonging to the surface of separation of two isentropic layers. Mixing due to heating may take place at the surface of contact and will cause the withdrawal of the mixed air from the ground-limit upwards along the surface of separation. This will result in the accentuation of the discontinuity at the new boundary between the strata.

Hence results the important consequence that all newly formed mixtures of strata that were in equilibrium with each other must rise upwards between the two layers originally present, a process that of course goes on more energetically when precipitations are formed in the ascending masses.

While the mixed strata are ascending, those parts of the strata on the north and south that have hitherto rested quietly approach each other until they even come in contact, by which motion the difference of their velocities must necessarily increase since the strata lying on the equatorial side acquire greater moment of rotation with smaller radius, while those on the polar side acquire feebler rotation with a larger radius. If this occurs uniformly along an entire parallel of latitude, we should again obtain a new surface of separation for strata of different rates of rotation whose equatorial side would show stronger west winds than the polar side, which latter might occasionally show east winds. On account of the numerous local disturbances of the great atmospheric currents there will, as a rule, be formed no continuous line of separation, but this will be broken into separate pieces which must appear as cyclones.

But as soon as the total mixed masses have found their equilibrium the surfaces of separation will again begin to form below, and new wave-formations will initiate a repetition of the same processes.

From these considerations it follows that the locality for the formation of billows between the strata of air is to be sought especially in the lower parts of the atmosphere, while in the upper parts an almost continuous variation through the different values of rotation and temperature is to be expected. The boundary surfaces of different strata of air, along which the waves travel, have one edge at the earth's surface and there the strata become superficial. Experience also teaches, as does the theory, that water-waves that run against a shallow shore break upon it, and even waves which originally run parallel to the shore propagate themselves more slowly in shallow water. Therefore waves that are originally rectilinear and run parallel to the banks will in consequence of the delay become curved, whereby the convexity of their arcs is turned toward the shore; in consequence of this they run upon the shore and break to pieces there.

$\cdot\qquad\cdot\qquad\cdot\qquad\cdot\qquad\cdot\qquad\cdot\qquad\cdot\qquad\cdot$

I therefore believe there is no reason to doubt that waves of air which in the ideal atmospheric circulation symmetrical to the axis could only progress in a west-east direction, must, when once they are initiated in the real atmosphere, turn down toward the earth's surface and break up by running along this in a north-westerly direction (in the northern hemisphere).

Another process that can cause the foaming of the waves at their summits is the general increase in velocity of the wind. My analysis also demonstrates this: it shows that waves of given wave-length can only co-exist with winds of definite strength. An increase in the differential velocities within the atmosphere indeed often happens, but one can not yet give the conditions generally effective for such a process.

If we confine ourselves to the search for such rectilinear waves as advance with uniform velocity without change of form, we may, as before remarked, represent such a movement as a stationary one, by attributing to both the media a uniform rectilinear velocity equal and opposite to that of the wave. It is well known no change is thereby introduced into the relative motions of the different parts of the masses.

Helmholtz applies the principle of dynamical similarity to enable him to infer the motion at the surface between two layers with a discontinuity of temperature of say $10°$ C from the motion at the surface of discontinuity of air and water.

The ordinary hydrodynamical equation is employed, assuming stream-line motion, incompressible fluid, a velocity potential and no friction, viz.:

$$p = C - g\rho z - \tfrac{1}{2}\rho \, (u^2 + v^2).$$

The application of the principle leads to the conclusion that in order that the motion in the two cases may be similar $\dfrac{\sigma}{1-\sigma} \cdot \dfrac{b_1^2}{n}$ and $\dfrac{1}{1-\sigma}\dfrac{b_2^2}{n}$ must remain unchanged, where σ denotes the ratio of the densities on either side of the discontinuity, b_1 and b_2 the velocities parallel to the surface of discontinuity in the two media, and n the linear dimension. From this it follows:

(1) If the ratios of the densities are not changed, then, in geometrically similar waves, the linear dimensions increase as the squares of the velocities of the two media; the velocities therefore will increase in equal ratios.

(2) For stationary waves when σ the ratio of the densities is varied, the quantities $\sigma \dfrac{b_1^2}{b_2^2}$ must remain constant, that is to say, the ratio of the kinetic energies of the two streams remains unchanged.

(3) If in geometrically similar motion the ratio of the densities of the two media is different in the two cases, then if the waves are to have the same wave-length the velocities in the two media must be increased in the same ratio as $\sqrt{\dfrac{1}{\sigma} - 1}$ and $\sqrt{1 - \sigma}$ increase.

Further, for a given wind-velocity the ratio of the length of waves at the boundary between two air-masses with a temperature difference of $10°$ C to the length of geometrically similar water-waves is 2630·3 to 1 (for a given form of wave whose store of energy is equal to that of the rectilinear flow along a plane boundary surface). This gives a wave-length of more than 500 metres for a wind of 10 metres per second.

Since the moderate winds that occur on the surface of the earth often cause water-waves of a metre in length, therefore the same winds acting upon strata of air of $10°$ difference in temperature, maintain waves of from 2 to 5 kilometres in length. Larger ocean-waves from 5 to 10 m long would correspond to atmospheric waves of from 15 to 30 kilometres, such as would cover the whole sky of the observer and would have the ground at a depth below them less than that of one wave-length, therefore comparable with the waves in shallow water, such as set the water in motion to its very bottom.

The principle of mechanical similarity, on which the propositions of this paragraph are founded, holds good for all waves that progress with an unchanged form and constant velocity of progress. Therefore these propositions can be applied to waves in shallow water, of uniform depth, provided that the depth of the lower stratum in the [model] varies in the same ratio as the remaining linear dimensions of the waves.

The velocity of propagation of such waves in shallow water depends on the depth of the water. For water-waves of slight height and without wind it can be computed by well-known formulae. When we indicate the depth of the water by h and put $n = 2\pi/\lambda$ [if b is the velocity of propagation], then

$$b^2 = \frac{g}{n} \frac{e^{nh} - e^{-nh}}{e^{nh} + e^{-nh}} = \frac{g\lambda}{2\pi} \tanh \frac{2\pi h}{\lambda},$$

which for $h = \infty$ becomes

$$b^2 = \frac{g}{n} = \frac{g\lambda}{2\pi},$$

[for $h = \lambda$ the same equation holds with an accuracy of one part in a thousand] and for small values of h it becomes $b^2 = gh$.

When however the depth of the water is not small relatively to the wave-length, then the retardation is unimportant, thus for

$h/\lambda = \frac{1}{2}$, the speed of propagation diminishes as $1 : 0\cdot95768$,

$h/\lambda = \frac{1}{4}$, the speed of propagation diminishes as $1 : 0\cdot80978$,

$h/\lambda = \frac{1}{10}$, the speed of propagation diminishes as $1 : 0\cdot39427$.

When it is calm at the earth's surface the wind beneath the trough of the aerial billow is opposed to the direction of propagation, but under the summit of the billow it has the same direction as that. Since the amplitudes at the earth's surface are diminished in the proportion $e^{-nh} : 1$ with respect to the amplitudes at the upper surface, therefore these latter variations can only make themselves felt below when the depth is notably smaller than the wave-length. Variations of barometric pressure are only to be expected when decided changes in the wind are noticed during the transit of the wave.

.   .   .   .

· *The energy of the waves.* When we investigate the energy of the waves of water raised by the influence of the wind, and compare it with that which would be appropriate to the two fluids uniformly flowing with the same velocity when the boundary surface is a plane, we find that a large number of the possible forms of stationary wave-motion demand a smaller storage of energy than the corresponding current with a plane boundary. Hence the current with a plane boundary surface plays the part of a condition of unstable equilibrium to the above described wave-motion. Besides these, there are other forms of stationary wave-motion where the store of energy for both the masses that are in undulating motion is the same, as in the case of currents of equal strength with plane bounding surfaces: and, finally, there are those in which the energy of the wave is the greater.

.   .   .   .

It is sufficient to have proven that for one form of wave, billows due to wind are possible, which billows have a less store of energy than the same wind would have over a plane boundary surface. Hence it follows that the condition of rectilinear flow with plane boundary surface appears at first as a condition of indifferent or neutral equilibrium, when we consider only the lower powers of small quantities. But if we consider the terms of higher degree, then this condition is one of unstable equilibrium, in view of certain disturbances that correspond to stationary waves between definite limits as to wave-length; but on the other hand is a condition of stable equilibrium when we consider shorter waves.

.   .   .   .

Breaking, foaming atmospheric billows cause mixture of strata in the mass of air. Since the elevations of the air-waves in the atmosphere can amount to many hundred metres, therefore precipitation can often occur in them which then itself causes more rapid and higher ascent. Waves of smaller and smallest wave-length are theoretically possible. But it is to be considered that perfectly sharp limits between atmospheric strata having different motions certainly seldom occur, and therefore in by far the greater number of cases only those waves will develop whose wave-length is very long compared with the thickness of the layer of transition.

Helmholtz's ideas of discontinuity and wave-motion find their most effective exposition in V. Bjerknes' *Atmospheric vortex*, and the practical developments that have arisen from it. These fall for consideration with the theory of cyclones. In respect of the general circulation of the atmosphere, *horribile dictu*, Bjerknes seems to arrive at a zonal arrangement of the atmosphere over the globe. It is difficult for his readers spontaneously to adjust his theory to allow for the distribution of land and water.

Frictional effects

The next section of our story deals with frictional effects. They find their consummation in the theory of turbulence foreshadowed by Helmholtz which took its present form about the year 1920.

The section which we are proposing now to present is effectively set out by C. G. Rossby[1], whose arrangement for the meteorological reader we gratefully acknowledge.

For the sake of consistency we have found ourselves obliged to alter the symbolism. μ, originally Maxwell's molecular viscosity with $\nu = \mu/\rho$, becomes a general symbol for a diffusion coefficient; k_{mm} is the coefficient of eddy viscosity in relation to momentum, k_{tt} in relation to temperature, k_θ to potential temperature and k_E to eddy energy.

We have seen in the last section how Helmholtz explained that the viscosity of air as measured in the conventional manner was quite inadequate to account for the practical limitation of wind-velocity in the atmosphere, and the conductivity of air similarly measured equally inadequate to account for the distribution of temperature. And in the previous chapter we have seen that there is a peculiar property of the motion of air past a solid body, the criterion of which was formulated by Osborne Reynolds, and was intended to express the experimental fact that if the motion past a solid is slow the air will flow in stream-lines, but when the relative velocity exceeds a critical value the stream will develop eddies that have a kind of spin and consequent vitality or energy of their own. These eddies confer upon the stream a power of stirring or mixing, depending on the linear dimension l of the solid and on the viscosity of the fluid ν as well as on the velocity of relative motion V. The relation is expressed by the so-called Reynolds number, Vl/ν.

It is not unnatural to assume that the criterion of turbulence is somehow applicable to the motion of one air-current past another, and energy of a special kind may be developed in place of the energy of wave-motion which Helmholtz obtains by the assumption of discontinuity. The attainment of that position we hope to set out in this section; but meanwhile it may obviously be understood that, if turbulence with its eddies produces mixing with a time-scale one hundred thousandth or one millionth of that of conventional viscosity, the application to the motion of the atmosphere at different levels of an equation of the same form as that of viscosity, with coefficient regarded as

---

[1] 'The theory of atmospheric turbulence—An historical résumé and an outlook,' *Monthly Weather Review, Washington*, vol. LV, 1927, pp. 6-10.

constant but unknown, should give results in agreement with observation, provided the assumption of constancy is not too crude. The coefficient so determined is legitimately called virtual viscosity.

That idea gives the key to the arrangement of this section.

But first we must refer to the work of Guldberg and Mohn[1] (1876–80), who sought to establish a relation between the distribution of surface-pressure (at sea-level) and observed winds by introducing into the equations a term expressing surface-friction as a force opposite to and proportional to the surface-wind. The investigation was subsequently developed by Sandström[2], who found that the frictional effect on the surface layer was the equivalent of a force acting at 38° to the right (in the northern hemisphere) of the reversed wind-direction.

The surface is however only one aspect of the problem, as the surface-effect is contingent upon the conditions in the air above.

The first writer to explain the variation with height of wind-velocity and direction as expressing a special form of "viscosity" near the surface was Åkerblom[3], who applied the hydrodynamical equations for viscous motion to the observations of wind-velocity at different levels of the Eiffel Tower extending to 300 metres.

He used equations which are equivalent to

$$\frac{\mu}{\rho}\frac{\partial^2 u}{\partial z^2} = -2\omega v \sin\phi + \frac{1}{\rho}\frac{\partial p}{\partial x} = -2\omega v \sin\phi + \frac{bb}{\rho}\sin mz,$$

$$\frac{\mu}{\rho}\frac{\partial^2 v}{\partial z^2} = 2\omega u \sin\phi + \frac{1}{\rho}\frac{\partial p}{\partial y} = 2\omega u \sin\phi - \frac{bb}{\rho}\cos mz,$$

to compute the virtual viscosity μ. The coefficient for the summer was found to be 700,000 times that of the conventional viscosity of standard air, and for the winter 500,000 times.

It ought here to be remembered that Åkerblom's investigation was prompted by a mathematical investigation by V. W. Ekman[4], of Göteborg, of the vertical distribution of velocity in a drift-current produced by wind, which is a classic of hydrography. One of its results is that the surface-drift is not actually in the direction of the wind that produces it.

That statement, which is often quoted, requires some consideration. It could hardly be maintained that the wind blowing over a field of corn bends the corn at any considerable angle to the line of its own motion; were that so all our weather-cocks would need reinterpretation. Nor is the march of a "cat's paw" over the water other than the motion of the air; but statistically, taking the effect of eddies into account, the resultant motion of the surface-

[1] *Études sur les mouvements de l'atmosphère*. Christiania, 1876–80 and 1883.

[2] J. W. Sandström, 'Eine meteorologische Forschungsreise in dem schwedischen Hochgebirge,' *K. Sv. Vet. Ak. Handl.*, Bd. 50, No. 9, Stockholm, 1913.

[3] F. Åkerblom, 'Recherches sur les courants les plus bas de l'atmosphère au-dessus de Paris,' *Nova Acta Reg. Soc.* Uppsala, 1908.

[4] 'On the influence of the earth's rotation on ocean-currents,' *Arkiv för Mat. Astr. och Fysik*, Bd. 2, No. 11, Stockholm, 1905.

water is not the same as the resultant motion of the surface-air, both are affected by the rotation of the earth and by eddies.

The next investigation of virtual viscosity was the application of Åkerblom's method by Th. Hesselberg and H. U. Sverdrup[1] to Stevenson's and Hellmann's observations of wind close to the surface and to the abundant data for wind-direction and velocity of the air over Lindenberg which gave a variation of the coefficient with height—a very small value (0·9) close to the surface increasing to one many times greater (50) at about 300 metres; from there upwards little variation. The last conclusion is disallowed by Rossby on the ground that in the calculation the frictional force per unit layer has been taken as $\mu \dfrac{\partial^2 u}{\partial z^2}$ instead of $\dfrac{\partial}{\partial z}\left(\mu \dfrac{\partial u}{\partial z}\right)$, thus assuming μ to be constant and neglecting the term $\dfrac{\partial \mu}{\partial z} \cdot \dfrac{\partial u}{\partial z}$, which is of the same order of magnitude as $\mu \dfrac{\partial^2 u}{\partial z^2}$.

That this approximation highly affects the results of the computation has been shown in another place[2]. There is, moreover, a principal difference between a frictional term of the type $\mu \dfrac{\partial^2 u}{\partial z^2}$ and one of the type $\dfrac{\partial}{\partial z}\left(\mu \dfrac{\partial u}{\partial z}\right)$. The former will, since μ is essentially positive, always have the same sign as $\dfrac{\partial^2 u}{\partial z^2}$, whether μ is variable or not, and will therefore always tend to annihilate existing differences of velocity. The latter expression may, however, on account of the term $\dfrac{\partial \mu}{\partial z} \cdot \dfrac{\partial u}{\partial z}$ have the opposite sign to $\dfrac{\partial^2 u}{\partial z^2}$. In this case differences of velocity already existing may under favourable conditions increase and finally develop into real discontinuities[3].

(C. G. Rossby, *Monthly Weather Review*, 1927, p. 7.)

But it has been noted by H. Jeffreys[4] that there is a strong argument in favour of the form $\mu \dfrac{\partial^2 u}{\partial z^2}$ even when μ varies with z, the allowance for the effect of varying pressure upon the momentum of the moving eddy being just such as to cancel the term $\dfrac{\partial \mu}{\partial z} \cdot \dfrac{\partial u}{\partial z}$. Taylor's treatment of the problem is two-dimensional, but he shows that the appropriate form is $\mu \dfrac{\partial^2 u}{\partial z^2}$. Unfortunately his treatment cannot be generalised for three dimensions and we are left in considerable doubt as to how close is the analogy between eddy-viscosity and ordinary viscosity.

We come next to a paper by G. I. Taylor[5], who in 1913 after the loss of s.s. *Titanic* had been engaged in investigating the physical conditions of the atmosphere in the fog-regions of the North West Atlantic. He recognised that

[1] 'Die Reibung in der Atmosphäre,' *Veröff. d. Geophys. Inst. d. Univ. Leipzig*, 1915; 'Die Windänderung mit der Höhe vom Erdboden bis etwa 3000 m Höhe,' *Beitr. Phys. frei. Atmosphäre*, Bd. VII, 1917, pp. 156–66.

[2] C. G. Rossby, *Statens Met.-Hydr. Anst. Medd.*, Bd. III, No. 1, Stockholm, 1925.

[3] C. G. Rossby, 'On the origin of travelling discontinuities in the atmosphere,' *Geog. Annaler*, vol. VI, Stockholm, 1924, p. 180.

[4] *Proc. Camb. Phil. Soc.*, vol. XXV, part I, 1929, p. 25.

[5] 'Eddy-motion in the atmosphere,' *Phil. Trans. Roy. Soc.*, vol. CCXV, London, 1915.

the eddy-motion due to turbulence would result in a redistribution of mass or of water-vapour-content in accordance with the diffusion law, and similarly a redistribution of temperature, or, with the natural restriction to adiabatic condition of the motion, of potential temperature. Such a process would lift cold surface-air and produce still colder upper air or a new mixture. An account of Taylor's work in this connexion is given in chap. IV, where it is required in connexion with the relation between surface-wind and gradient. The upward transport of water-vapour is represented by $- k_w \frac{\partial m}{\partial z}$, where k_w is a coefficient of eddy-conductivity and the transport of potential temperature θ is $- k_\theta \frac{\partial \theta}{\partial z}$, where k_θ is a coefficient determined by different data but in all probability the same as the coefficient for water-vapour.

For the vertical transport of horizontal momentum Taylor has $- k_{mm} \frac{\partial u}{\partial z}$ and $- k_{mm} \frac{\partial v}{\partial z}$ where k_{mm} is the coefficient of eddy-viscosity.

By Rossby and other workers great importance is attached to the conclusion that the result of turbulence which tends to equate potential temperature within the region of its operation is that heat is carried downward by the natural operation of the atmosphere and not upward. That is not by any means surprising. We have referred eddy clouds to it in vol. III, p. 349; it has always been understood that in the absence of condensation of vapour complete mixing of any layer to produce isentropic conditions cools the upper part and warms the lower. But it must be a very small part of the transference of heat between the surface and the upper layers. The alternative upward flow may be spasmodic but it is very effective, and the stressing of the upper layers, where entropy is higher, as a source of heat, would suggest to the reader, unless care is taken, that the upper air had access to supplies of heat that are denied to the surface. The position appears to be that dynamical convection carries heat downwards and thermal convection carries heat upwards.

In reality, pursuing the financial analogy of vol. III, we may fairly say that the increase of potential temperature with height, which implies the storage of a certain amount of energy, represents the commission which the atmosphere charges for cashing Mother Earth's cheques on Sol and Co.'s bank. The fee depends upon the humidity, which may be described as liquid assets, at the time of drawing.

Nevertheless, the transfer downwards is used by W. Schmidt as the basis of *Austausch*, practically the equivalent of eddy-conductivity, and in his investigation he uses the virtual viscosity to discuss diurnal variation of temperature, the influence of large cities on climate and many other applications of diffusion.

What is done for vertical motion by Taylor and Schmidt can be applied to the vertical stratification and consequent horizontal *Austausch* of entropy which we have represented in fig. 95 of vol. III; and another coefficient can be obtained to illustrate statistically the effect of cyclones and other disturbances of the horizontal motion of air to carry warm air to the poles in exchange for cold air to the equator. This side of the general question of eddy-diffusion has been taken up by Defant[1], Exner[2] and Ångström[3].

From these we pass to the work which Richardson imposed upon himself by his attempt to determine the "known" forces to calculate the changes of motion according to the general equations of motion in his *Weather Prediction by Numerical Process*, and in a previous paper on "The supply of energy to and from atmospheric eddies[4]."

The attempt to determine the present distribution of the meteorological elements from that of the previous day is not very encouraging. Richardson finds the explanation of its ill-success in the irregular distribution of the stations available for data. But the main part of the work is devoted to a searching analysis of the material available for the expression of the forces required to implement the equations. In the course of that endeavour he has tabulated the values of eddy-conductivity and eddy-viscosity computed by the most varied methods. The results will be quoted in chap. IV.

They indicate that the eddy-conductivity is very small close to the ground, then increases to a maximum within the first kilometre, and finally at the higher levels again decreases.

His data also show that the conditions are not the same along the stream and across it. The "eddy stress" $\left(-k_{mm}\dfrac{\partial v}{\partial z}\right)$ for the cross-wind direction is seven times as large as $\left(-k_{mm}\dfrac{\partial u}{\partial z}\right)$ that for the down-wind direction.

In his paper on the supply of energy to and from atmospheric eddies he develops the idea of automatic turbulence due to relative motion in the atmosphere itself and arrives at a conclusion summarised by Rossby as follows:

To every value of the vertical lapse-rate of temperature there corresponds a critical value for the increase of wind with height. If the increase actually observed is less than the critical value, then the turbulence (more definitely the eddy-energy) has a tendency to decrease. If the vertical increase of wind-velocity exceeds the critical value then the eddy-energy has a tendency to increase.

For an average clear day the critical value (in the troposphere) is equal to 1 m/sec for 100 m.

Rossby sums up the achievements as follows:

1. An expression has been derived for the production of eddy-energy [presumably energy of spin] from the kinetic energy of the mean motion.

2. An expression has been obtained for the loss of eddy-energy through work against a stable stratification and for the gain of energy in case of unstable stratification.

[1] 'Die zirkulation in der Atmosphäre in den gemässigten Breiten der Erde,' *Geog. Ann.*, vol. III, Stockholm, 1921, p. 209.

[2] *Dynamische Meteorologie*, Zweite Aufl. 1925, Wien, p. 239.

[3] 'Evaporation and precipitation at various latitudes and the horizontal eddy-convectivity of the atmosphere,' *Arkiv för Mat. Astr. och Fysik*, Bd. 19 A, No. 20, 1925.

[4] *Proc. Roy. Soc.* A, vol. XCVII, London, 1920, pp. 354–73.

3. A criterion has been found for the conditions under which an air-current will remain laminar or become turbulent.

4. An equation of continuity has been obtained expressing the increase of eddy-energy in a closed system as the difference between production and consumption of eddies within the system.

The shortcomings of the theory as an aid to the solution of the general problem are:

1. The state of turbulence is characterised by three different quantities—eddy-viscosity k_{mm}, eddy-conductivity k_θ, eddy-energy E.

2. Between these three characteristics there is as yet only one relation, an energy-equation. Thus two of the above characteristics of the turbulence remain undetermined.

3. The energy-equation refers to a closed system and can therefore not be used for computation of the state of turbulence at individual points.

Rossby (in two papers) makes amends for these defects by assuming that eddy-energy per unit mass, or specific eddy-energy (E), has its own coefficient of diffusion k_E so that specific eddy-energy is diffused upward per cm² sec at the rate of $-k_E \frac{\partial E}{\partial z}$.

Turbulence has thus four coefficients—eddy-viscosity k_{mm}, eddy-conductivity k_θ, diffusion of eddy-energy k_E and specific eddy-energy E—which are reduced to one by assuming that each is proportional to E, and thus a general equation is arrived at the solution of which gives the specific eddy-energy as a function of the vertical distribution of temperature and wind.

In the absence of mean motion the equation for E is combined with the equation for eddy-convection of heat, and a system is obtained containing E and the potential temperature which gives a statistical solution of the problem of thermal convection.

The theory awaits the determination of the statistical value of specific eddy-energy at different levels.

About all these equations and coefficients we would remember that the whole conception of diffusion which they employ is statistical in the same way as the law of molecular diffusion is statistical. But eddies which, so to speak, are live objects are much bigger than molecules and have individual existence which is shown for example by gusts and squalls in the record of turbulent wind on an anemometer.

It may be that there are lessons contained in the study of the individual records which their amalgamation into a statistical mathematical equation will not convey. Somewhere we must draw the line between the eddy of which the atmosphere carries an innumerable quantity to the discomfort of airships and airplanes, and the eddy that asserts its individuality and destroys a bridge or carries a roof away a mile or two.

It is not quite likely that nature will let us dispose of these phenomena by a common coefficient of diffusion; for "normal" conditions, based on means, it may do so; but for daily use some further discrimination is necessary.

It may be noted that the contributions mentioned in our summary are restricted as to the country of origin. In the past century we have noted contributions from the United States, Great Britain, France, Germany, Austria, and the Scandinavian countries.

In the latter part of the period Japan has taken a full share in the task of expressing the behaviour of the atmosphere according to the laws of mathematical physics. Many of the contributions are directed to the phenomena of the weather-map and find their place in chap. XI.

We note the following as having the general character which is appropriate to this chapter. S. Fujiwhara[1] deals with the growth and decay of vortical systems in water; S. Sakakibara[2] with the transverse eddy-resistance to the motion of air in the lower layers of the atmosphere; Y. Isimaru[3] with the eddy-motion of air near the earth's surface; these are developments of the same subject.

EXNER'S *DYNAMISCHE METEOROLOGIE*

It will be noticed that in the preceding sections we have evaded the direct application of the general equations of motion in their complete form. We have noticed in the first place work that can be done with them if we restrict the issues to conditions of no acceleration, no friction, or no terrestrial rotation.

In the last section we have dealt with turbulence by the application of the hydrodynamical equations of motion with frictional forces considered independently of the vertical motion, the same equation as that used by Helmholtz in his exposition of possible atmospheric waves.

The only attempt to use the equations in their entirety is that of L. F. Richardson already mentioned.

The general equations of motion could hardly have been intended, by those who have developed them, merely as museum specimens, and we turn therefore to the work of an acknowledged master of mathematical method in its application to meteorology, the *Dynamische Meteorologie* of the late Prof. F. M. Exner, with confidence that the achievements of mathematical physics will be adequately set out. As an indication of mathematical method we take a chapter with the title "Allgemeine Dynamik der Luftströmungen," in which he discusses the application of the general equations of motion. The relation of pressure to gradient and the development of turbulence are part of the story. Incidentally an excellent account of the work on turbulence, which we have already summarised under the heading of the effects of friction, finds a place in the chapter.

The portion which we select for the purpose of illustration is § 37, "Zwei integrale der Bewegungsgleichungen für horizontale Luftströmungen ohne Reibung." The following is a free translation; words in square brackets have been added by the translator.

---

[1] *Japanese Journ. Ast. Geophys.*, vol. I, No. 5, 1923, p. 125.
[2] *Geophysical Magazine*, Tokyo, vol. I, No. 4, 1928, p. 130; vol. II, No. 3, 1929, p. 139.
[3] *Ibid.*, vol. II, No. 2, 1929, p. 91; vol. II, No. 3, 1929, p. 133.

The equations previously developed (S. 96) give very convenient and simple formulae for treating the motion of air in a horizontal plane. On a limited portion of the earth's surface (ϕ constant), neglecting air-friction, they are:

$$\frac{du}{dt} + 2\omega \sin\phi v = -\frac{1}{\rho}\frac{\partial p}{\partial x},$$

$$\frac{dv}{dt} - 2\omega \sin\phi u = -\frac{1}{\rho}\frac{\partial p}{\partial y}.$$

I. *Velocity due to [the potential energy of] pressure-gradient.* (A calculation of Margules.)

We have already assumed (1) the motion to be horizontal [improbable if the motion is guided by the isentropic surfaces] and (2) free from viscosity. Next *in order to integrate the equations* we assume (3) that the motion of the air does not affect the pressure-distribution so that the gradient is undisturbed in spite of the motion.

If c is the resultant velocity $\sqrt{u^2 + v^2}$, we get by adding the equations

$$\frac{1}{2}\frac{dc^2}{dt} = -\frac{1}{\rho}\left(\frac{\partial p}{\partial x}\cdot\frac{dx}{dt} + \frac{\partial p}{\partial y}\cdot\frac{dy}{dt}\right).$$

Since $\quad \dfrac{dp}{dt} = \dfrac{\partial p}{\partial t} + \dfrac{\partial p}{\partial x}\cdot\dfrac{dx}{dt} + \dfrac{\partial p}{\partial y}\cdot\dfrac{dy}{dt}$, and $\dfrac{\partial p}{\partial t}$ is zero,

we get $\quad\quad\quad\quad \dfrac{1}{2}\,dc^2 = -\dfrac{dp}{\rho}.$

ρ changes with the motion, but, with Margules, one can make one or other of two simple assumptions, namely, (4) that during the change of pressure the motion is either adiabatic or that it is isothermal. In either of the two cases integration is possible, and to a first approximation

$$c^2 = c_0{}^2 + 2\,\frac{p_0 - p}{\rho_0},.$$

Taking c_0 as zero $\quad p_0 - p =$

$p_0 - p =$	1	2	5	10	20	30 mm Hg.
$c =$	14·4	20·3	32·1	45·4	64·2	78·6 m/sec.

Whence we can conclude that the annihilation of 20 m/sec of velocity would be sufficient to carry air over two millimetres of pressure-difference.

[It is difficult to formulate atmospheric conditions in which the assumptions of this proposition would be valid. As a matter of meteorological experience pressure-energy and air-motion are not alternative but collateral, they rise and fall together, the sacrifice of one does not increase the other. The hesitation which (3) naturally arouses is confirmed by the fact that in the course of the mathematical manipulation the influence of the earth's rotation disappears.

Conservation of angular momentum in steady convergence towards a centre is perhaps the nearest approach to the postulated case and in that connexion presumably if the spiral motion of air at the base of a circular vortex carries air from 760 mm to 740 mm its velocity will be increased by 64 m/sec unless friction of some sort opposes the development. That particular case is essentially superficial.]

II. *Air flow with incidental oscillations in the horizontal plane.* A quite general integration of the equations of motion is not possible, we will consequently take a special case which can be treated more closely—the flow of fluid combined with oscillations. These can be either horizontal or vertical. Here also however integration is only practicable with simplification. For the flow of the fluid besides the equations of motion the equation of continuity is appropriate. For simplification let the motion be supposed to be assumed to be stationary so that everywhere

$$\frac{\partial u}{\partial t} = \frac{\partial v}{\partial t} = 0.$$

The general solution of the problem is difficult. It consists in the integration of the equations

$$\frac{du}{dt} + 2\omega \sin \phi v = - \frac{1}{\rho}\frac{\partial p}{\partial x},$$

$$\frac{dv}{dt} - 2\omega \sin \phi u = - \frac{1}{\rho}\frac{\partial p}{\partial y},$$

$$\frac{\partial \rho}{\partial t} + \frac{\partial}{\partial x}(\rho u) + \frac{\partial}{\partial y}(\rho v) = 0 \quad \text{and} \quad p = R\rho T.$$

We will consequently assume simplifications and first of all treat the air as an incompressible fluid and neglect the rotation of the earth. Then in the stationary condition

$$u \frac{\partial u}{\partial x} + v \frac{\partial u}{\partial y} = - \frac{1}{\rho}\frac{\partial p}{\partial x},$$

$$u \frac{\partial v}{\partial x} + v \frac{\partial v}{\partial y} = - \frac{1}{\rho}\frac{\partial p}{\partial y},$$

$$\frac{\partial u}{\partial x} + \frac{\partial v}{\partial y} = 0.$$

If there is no turbulence there exists a velocity potential ϕ such that $\frac{\partial \phi}{\partial x} = u$ and $\frac{\partial \phi}{\partial y} = v$ and we obtain instead of the three equations given above the following:

$$p = P - \frac{\rho}{2}\left[\left(\frac{\partial \phi}{\partial x}\right)^2 + \left(\frac{\partial \phi}{\partial y}\right)^2\right], \quad \text{where } P \text{ is a constant,}$$

$$\frac{\partial^2 \phi}{\partial x^2} + \frac{\partial^2 \phi}{\partial y^2} = 0.$$

We choose for ϕ a solution which has as the chief component a flow along the positive x-axis, but besides an oscillation transverse to this direction, viz.

$$\phi = Ax - B \cos ax\,[e^{a(b-y)} + e^{-a(b-y)}],$$

the velocity-components are

$$u = A + Ba \sin ax\,[e^{a(b-y)} + e^{-a(b-y)}],$$
$$v = Ba \cdot \cos ax\,[e^{a(b-y)} - e^{-a(b-y)}].$$

For $y = b$, $v = 0$ and there the flow is rectilinear along the x direction; and for y greater or less than b, v differs from zero, and indeed the more so the further we get from $y = b$. We have to deal therefore with an oscillating flow. It has, with no turbulence, the property that the oscillation must increase in a direction at right angles to the flow, otherwise there would be no continuity. The stream-line is given by a stream-function ψ, where

$$\frac{\partial \psi}{\partial x} = - \frac{\partial \phi}{\partial y}, \qquad \frac{\partial \psi}{\partial y} = \frac{\partial \phi}{\partial x}.$$

One easily finds that

$$\psi = Ay - B \sin ax\,[e^{a(b-y)} - e^{-a(b-y)}] = K.$$

For a given stream-line K is constant. A and B are determined from the velocity u_0, v_0 at the origin of co-ordinates. If one introduces these magnitudes, the equation to the stream-line becomes

$$y = \frac{K}{u_0} + \frac{v_0}{au_0} \sin ax \left[\frac{e^{a(b-y)} - e^{-a(b-y)}}{e^{ab} - e^{-ab}}\right].$$

This integral, which has been won by many simplifications, shows us a stationary flow of fluid which is only rectilinear in the case of $v_0 = 0$. If this is not so, wave-motion is present in the fluid of which the wave-length and amplitude depend on the two magnitudes a and b.

These waves are no simple sine waves but such in which the crest is continually displaced further from the mean position than the trough. On one side the elongation increases continually. [In order to obtain the result in nature we have further to assume two containing surfaces, one on either side, between which the stream-line flow can take place.]

Exner then gives a picture of the wave-form. He finds an example of this kind of oscillation in a map from Sandström's memoir in which the stream-lines are drawn over Northern Europe on January 7, 1902, at 9 p.m.

Other chapters of Exner's volume are devoted to the subjects treated in succeeding chapters of this volume. Chap. VII deals with the kinetic energy of moving air, based largely upon Margules' *Energy of Storms*; and chap. VIII deals with stationary air-flow in the atmosphere. Chap. IX is concerned with the general circulation of the atmosphere.

The memoirs which Brillouin has collected are again referred to, but, apart from the necessity for a zone of high pressure somewhere about latitude 30° in either hemisphere, little practical result appears. Exner's chapter is largely devoted to the combination of the results of observation in respect of the distribution of temperature, pressure and wind-velocity, potential temperature (entropy), moment of rotation, which we have dealt with in chap. VI of vol. III, and the correlation of pressure in different parts of the northern hemisphere which we have noted in vol. II.

The calculus of isallobars

We cannot claim to have given an exhaustive summary of the literature of meteorology based on the general equations such as may be found in Sprung's *Lehrbuch der dynamische Meteorologie* or the great *Lehrbuch* of Hann and Süring. The specimens which we have chosen, however, may give the reader an idea of the mode of treatment to which the general equations lead.

As regards practical applications we may remark that the calculus of isobars based on the equations of the gradient-wind is in effect the real subject of this volume, and its implications are detailed later. The treatment is mainly graphical, but among the recent contributions to dynamical meteorology we note an analytical paper by Brunt and Douglas[1] about the dynamical implications of what are called by their originator isallobars. These are lines drawn to connect places at which the rates of change of pressure are equal, as estimated by observed changes in a definite interval previous to the epoch of the chart. The idea was suggested and the subject treated graphically by Ekholm in 1907. It has been extended over the world by the adoption of the barometric tendency (change of pressure in 3 hours) as an element of terrestrial weather.

In their memoir Brunt and Douglas show that when the pressure distribution is changing, then to a first approximation the wind in the upper air may be estimated by adding to the geostrophic wind a component blowing in a direction normal to the isallobars from high to low values of the isallobars, and

Memoirs of the Royal Meteorological Society, vol. III, No. 22, London, 1928.

having a velocity proportional to the isallobaric gradient. There is therefore a convergence of winds into the isallobaric low and a divergence of winds from the isallobaric high, which explain the frequent association of cloud and rain with low values, and of bright weather with high values on the isallobaric chart. The authors emphasise the fact that a system of geostrophic winds in which we have

$$2u\omega \sin \phi = -\frac{1}{\rho}\frac{\partial p}{\partial y}, \quad 2v\omega \sin \phi = \frac{1}{\rho}\frac{\partial p}{\partial x}$$

satisfies the equation of continuity $\frac{\partial}{\partial x}(\rho u) + \frac{\partial}{\partial y}(\rho v) = 0$ without any vertical component w. A system of winds which is purely geostrophic therefore has no ascending or descending motion within its ambit. They also show that in general the disturbance produced by surface-friction in the equilibrium of a surface of separation of cold and warm air can only give sufficient ascent to produce light drizzle.

We must also add a note on a recent contribution to the dynamics of the atmosphere by M. Gião[1], who investigates by the aid of equations of motion the rate of propagation of the various fronts which form the basis of the system of forecasting of the Norwegian school of meteorologists.

M. Gião regards a map of the distribution of pressure at sea-level and its changes as fundamental representations of the condition of the atmosphere against which no objection can be taken, and further he regards the discontinuities with which he deals as ideal mathematical discontinuities in computing numerical values from the equations.

We cannot exactly follow him in either of these assumptions. We have pointed out in vol. III that the sea-level map is based on reduction of pressure to sea-level which is always subject to error, and it must be remembered that a small error in the measurement of pressure may give rise to a large error in the dynamical computation. The only observations of temperature and wind which are associated with sea-level pressure are not sea-level values and therefore not strictly appropriate to the map; and in the present chapter we have already remarked that, so long as the effect of divergence of the discontinuity from the mathematical ideal cannot be demonstrated quantitatively, it is not safe to assume that it can be neglected.

Nevertheless M. Gião's conclusions have the great advantage that they are applicable to individual cases and are not dependent upon mean values. Moreover, in the cases that are cited the calculated motions of the fronts are apparently verified by observation although no account is taken of turbulence. We can only conclude that the motion of the front is a matter of somewhat crude and general character which does not require the highest precision in the calculation.

[1] 'La mécanique différentielle des fronts et du champ isallobarique,' *Mémorial de l'Office National Météorologique de France*, No. 20, Paris, 1929. The reader may be referred to a second paper by Gião, 'Sur les perturbations mécaniques des fluides,' *Mémorial*, etc., No. 21, Paris, 1930.

THE CONTRIBUTION OF MATHEMATICAL ANALYSIS

We have seen that an immediate solution of the general equations is not generally possible and that, for the achievement of the purpose in practically every case, some auxiliary assumption has to be made; the assumption may take the form of neglecting terms as being relatively of no importance or some specification of the initial conditions which *a priori* may seem reasonable. The assumptions once made and a solution once obtained, it is not unusual for the mathematical result to be quoted as applicable generally without any mention of the limitations of the solution. It is, however, obvious that any assumption which is introduced in this way should always be the subject of careful scrutiny, because strictly speaking the solution applies to an ideal atmosphere, and meteorological reasoning is not really pertinent unless it deals with actual conditions.

Assumptions generally characteristic of the solution of the equations of motion which are *a priori* improbable and are difficult to verify are apt to give the impression of hypothesis rather than reality. Hence the contributions of mathematics to the solution of problems of weather are apt to convey an impression of academic exercises rather than of definite contributions to practical meteorology.

We shall have occasion to remark that the assumption that the motion of air is horizontal may lead to serious misconception. The same kind of assumption sometimes takes the form of regarding the surface of separation of the underworld and the overworld as horizontal, which may perhaps be taken into account in the history of the airship R 101.

For example, the calculation of the motion of the atmosphere on the hypothesis that any part of it follows the course of a particle which is free to move under balanced gravity and the rotation of the earth, neglects the effect of pressure in guiding the motion, and is seriously defective.

The solution may of course be an approximation to reality, and is indeed useless unless it is so, but in the phenomena of weather important occurrences turn on small deviations from typical conditions and a solution which assumes typical conditions may be wide of the mark.

There is in fact some justification for a suspicion that conclusions based upon an adjusted view of nature, even if the adjustment is not very violent, have not quite the validity that attaches to the doctrine of universal gravitation, the original model of mathematical manipulation; and that the real virtue of that great instrument was not so much the process of mathematical reasoning by which it was evolved as the extraordinarily perfect representation of the facts which preceded it in the work of Kepler. It is doubtful even now with the multitude of observations which have been accumulated in meteorological libraries whether the structure and behaviour of the atmosphere are sufficiently well-known or well-expressed for the effective operation of mathematical equations. We are entitled to say that we have yet to learn a good deal from the physical integration which nature is able to carry out so much more easily than the mathematician.

Wenn man die Gleichungen der Hydrodynamik und Thermodynamik auf die Vorgänge in der Atmosphäre anwendet, wird man immer einige vereinfachende Annahmen machen müssen. Durch diese Voraussetzungen werden die Resultate mehr oder weniger beeinflust, und wenn sie nicht gut gewählt sind, kommt man sogar leicht zu ganz fehlerhaften Schlüssen. Andererseits gibt es aber in den hydrodynamischen und thermodynamischen Gleichungen sowie in den Gleichungen, die aus diesen abgeleitet werden können, Glieder, welche im Falle der atmosphärischen Vorgänge so klein sind, dass man sie vernachlässigen kann, ohne merkbare Fehler zu machen.

(Th. Hesselberg and A. Friedmann)

The subsidiary assumptions made by meteorologists in order to facilitate the solution of the problems which they have set themselves afford material for study of great interest and sometimes for serious reflexion.

Some who have endeavoured to find a general solution of the problem of the atmospheric circulation have essayed to advance by leaving out of consideration the distribution of land and water, to say nothing of the secondary effect of mountain-slopes. The result is a plan of the circulation which will not bear comparison with the original maps from which, in part, the conception took its rise. It is difficult to believe that real progress can be made in that way; it is not at all suggestive of the way in which Newton approached the investigation which found its expression in the law of universal gravitation. It gives us an ideal picture of the circulation which must be repudiated equally by the airman, the seaman, and the husbandman, when they are seeking guidance in the affairs of everyday life.

As we have seen, another condition which is postulated in meteorological theory is the existence of a discontinuity of temperature and wind. It is allowed that discontinuity in pressure is unthinkable, it would mean explosive motion, but continuity in pressure can be reconciled with discontinuity in temperature, as Helmholtz has shown, if the winds are adjusted to balance the changes in the distribution of pressure on the two sides of the discontinuity. The assumption, however, while avoiding a discontinuity of pressure, leaves a discontinuity of pressure-gradient and we have yet to learn whether a discontinuity of that kind, a sudden transition from one gradient to its opposite, does actually occur in nature. It is probably true to say that in all the known cases observation is against it. Gradient does not change suddenly either along the horizontal or along the vertical. As we may gather in chap. IV, it may take a full kilometre in the vertical to make the change from a finite gradient to its opposite, and in the horizontal, of course, a kilometre is far too small a unit for the expression of observed changes.

It is usual to get over the difficulty by the consideration that the representation of the actual facts in any particular case requires a diagram of very large scale, and when the scale is reduced to manageable dimensions the changes are so nearly represented by discontinuity that the conclusions drawn by mathematical reasoning from the assumption of discontinuity are in effect applicable to the real case.

In that claim the student who wishes to visualise the physical concepts of the real atmosphere has some difficulty in finding satisfaction. He would like to suggest to the mathematical expert that if really there is no effective diffe-rence, in result, between a mathematical discontinuity and the actual continuity of nature, he would prefer to see the solution of the problem in its natural form rather than the mathematical substitute.

Margules hat die ganze Rechnung unter der vereinfachenden Voraussetzung durch-geführt dass die Atmosphäre die Erde wie eine dünne Schale umgibt, in der nur horizontale Bewegungen vorkommen. Nur unter dieser Voraussetzung war es möglich die Integration der Differentialgleichung des Druckes durchzuführen.
(F. M Exner, *Dynamische Meteorologie*, Aufl. ii, p. 403)

Wir nehmen in einer Ebene wirbelfreie Bewegungen an, bei der ein Geschwindig-keitspotential ϕ existiert. (*Ibid. p.* 78)

A meteorological observer might fairly object with regard to the first quotation that the deviations of the motion of the air from the horizontal are in fact the very elements of weather, and an equation that assumes that they do not exist must be found in practice to equate things that are not equal. And with regard to the second, that what one finds recorded on anemometers makes the assumption so improbable as to warn off the equation instead of satisfying it.

One other point may be mentioned. On the plan of hydrodynamical analysis the motion of air requires for its description the partial differentia-tion of the pressure and velocity with regard to the co-ordinate axes and time. It would be helpful to meteorologists who are, by custom, graphically in-clined if examples of the magnitude and importance of these partial coeffi-cients were adduced from actual observations of air-changes.

The complete differential D/Dt, taking account of the motion, is

$$\partial/\partial t + u\partial/\partial x + v\partial/\partial y + w\partial/\partial z.$$

It is understood, for example, that the record of a barogram, thermogram or anemogram represents the changes $\partial p/\partial t$, $\partial tt/\partial t$, or $\partial V/\partial t$, at the position of the instrument. A map might indicate the partial coefficients with regard to x and y at least.

Watching the sky towards sunset when there is a westerly wind observers may have noticed that a succession of cloudlets advances across from the west, but each is dissolved before it reaches a certain north-south line. According to theory that might correspond with a vertical component due to loss of heat which ought to appear in a complete equation.

The form of assumption which seems most tolerable is one that allows the development of rigorous results in individual cases which occasionally present themselves and which can be tested by actual observation.

For example, there is a generalisation that simplifies the dynamical solution and is very freely used in the subsequent chapters of this volume for which we ask the tolerance of the reader, namely the assumption that in ordinary practice the horizontal motion of the air is without acceleration, that $\ddot{x}$, $\ddot{y}$ are zero. Out of that comes the relation of the geostrophic wind. The assumption amounts in fact to considering that in most of its relations we can consider the air as moving under balanced forces.

That is an assumption, not always true, which can be studied in particular cases. It implies first of all that all the minor changes of velocity due to eddies of many kinds can be neglected, an assumption which certainly raises difficulties in the comparison of the wind as observed in a pilot-balloon and the generalised geostrophic wind, but which can be tested to some extent by making out the trajectories of air as in *The life-history of surface air-currents*, and in the study of the motion of clouds. Moreover, it is distinctly utilitarian in its application; it introduces the geostrophic wind as a new meteorological element, which is found to be a convenient term of reference in many dynamical problems.

In the course of this work we shall draw attention to the fact that whenever the balance is disturbed either by change of pressure or change of velocity there is always a component acceleration due to the earth's rotation which tends to restore the balance.

So the general motion of the atmosphere is a continuous process of spontaneous adjustment of the pressure to the motion; and as the process operates at every level of the moving air, and commences automatically with any deviation whatever from the strophic balance, the adjustment of pressure to motion will be as automatic as that of the attitude of a bicyclist on the road or a skater on the ice.

For the geostrophic wind we assume that the friction as well as the acceleration can be neglected. H. Jeffreys[1] pushes the same kind of process still further and classifies other winds in like manner. He distinguishes between three different kinds of winds—geostrophic, Eulerian and antitriptic—according as the chief controlling influence to balance the pressure-gradient is terrestrial rotation (acceleration and friction being neglected), acceleration (terrestrial rotation and friction neglected), and friction sufficient to prevent acceleration (acceleration and rotation being neglected).

The last two, for which terrestrial rotation is assumed to count little, can only apply to the equatorial region where $\sin \phi$ is very small, or to phenomena which are confined to a very limited area. As Eulerian winds he cites the winds of tropical revolving storms or tornadoes, which elsewhere we have called cyclostrophic winds, in which a considerable part of the pressure-gradient represented by $V^2 \cot \theta/\epsilon$ is balanced by rotation in a small circle. Jeffreys provides a more comprehensive name because he would regard other examples as possible when they are found. As antitriptic winds ($\dot{\alpha}\nu\tau\iota\tau\rho\acute{\iota}\psi\iota\varsigma$) he has in mind what we have referred to elsewhere as anabatic or katabatic winds or more generally as slope-effect—the travel of air up and down a slope with little or no variation of its velocity. We shall have something to say about these in a chapter on graphical methods, chap. VIII.

Thus in meteorological literature we find many efforts to solve one or other of the aspects of the general weather problem which can be isolated by judicious assumptions and which provide a useful exercise for mathematical ingenuity. Some of these we have already considered. Others will find a place

---

[1] 'On the dynamics of wind,' *Q. J. Roy. Meteor. Soc.*, vol. XLVIII, 1922, pp. 29–46.

in subsequent chapters. In presenting the results of these researches we have endeavoured to make the reader acquainted with any subsidiary assumptions that have been used in aid of the development of the mathematical reasoning.

This is with no wish whatever to discourage the application of mathematics to the solution of meteorological problems, but merely a reiteration of the obvious truth that the whole atmosphere is one and indivisible. No more can be got out of the solution of mathematical equations than is put into them by the data. The general equations of motion give us an algebraical expression of the acceleration of a small parcel of air in terms of the forces which are known to be operative upon it, and if they could be integrated would give us the position of the parcel at any subsequent time. But clearly the motion of a single parcel would not give us the general circulation. Much would depend on the motion of the environment which constitutes the boundary conditions for the single parcel. Without them the solution cannot be effective and the motion of the environment is as nearly unknown as the motion of the parcel itself.

Something must be assumed in order that the boundary conditions may be taken into account. It may be noticed that in the striking result of the mathematical analysis of the effect of the surface upon a flow in the air above it, it is assumed, without explicit statement, that the motion of any parcel in the stream is the same as that of all its neighbours at the same level. So that the result of the analysis is not the motion of a single parcel but the general motion of a layer of indefinite horizontal extent at the successive levels.

And this leads us to say that progress in the analysis of atmospheric motion depends upon the segregation of a mass of air which has quasi-vitality of its own, the behaviour of which can be referred to an assumed environment such as the undisturbed atmosphere, or a stream of air which itself has quasi-vitality. The disturbance which has to be traced may consist of waves or any other of the forms of spin which holds a mass of air together.

At present we are not able to offer an accurate description of these quasi-vital masses for mathematical treatment, but everyone who thinks about the weather feels the necessity for such personification. Even the layman is cognisant of the separate existence of a gust, a squall and a cloud, the American has his tornado, the West Indian his hurricane, the Spaniard his baguio, the Chinese his typhoon, the Anglo-Indian his cyclone, the Englishman his anticyclone, the meteorologist the isobaric features of the map that he calls depression and secondary, the Swede his isallobars, the French their noyaux, V. Bjerknes his waves, other Norwegians their fronts. There is personification in them all. The difficulty is that as yet no one can describe what exactly is personified. We are beginning to understand what is meant by a gust and to guess what is meant by a squall, but the most important part of the entity of a cloud is the part one cannot see and knows little about. We are still awaiting an exact description of the structure of the revolving storm, of the relation of the front to the surface, of the true inwardness of the isallobar, of the extent and structure of the noyau. The next steps of progress really lie with the intelligent observer.

THE COMPUTATION OF THE GRADIENT-WIND

Fig. 13. Geostrophic scale for wind in m/sec corresponding with isobars at intervals of 2 mb on a map 1 : 10⁷ or 4 mb on a map 1 : 2 × 10⁷.

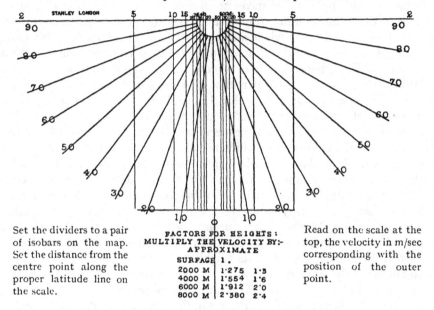

Set the dividers to a pair of isobars on the map. Set the distance from the centre point along the proper latitude line on the scale.

FACTORS FOR HEIGHTS:
MULTIPLY THE VELOCITY BY:—
APPROXIMATE

SURFACE	1 .	
2000 M	1·275	1·3
4000 M	1·554	1·6
6000 M	1·912	2·0
8000 M	2·380	2·4

Read on the scale at the top, the velocity in m/sec corresponding with the position of the outer point.

The interest of this volume in the general equations of motion arises in connexion with the relation of the kinetic index, wind, to the static index, distribution of pressure.

We have seen (p. 50) that in the absence of large vertical velocity the effect of the earth's rotation upon the horizontal motion of the air can be represented by a force $2\omega V\rho \sin\phi$ perpendicular to the direction of motion and acting from left to right. If the isobar along which the air is moving is part of a great circle, and we take its line as the x-axis, the balance of forces for a pressure-gradient bb requires that $bb = -\partial p/\partial y = 2\omega V\rho \sin\phi$.

But if the air is actually moving in a small circle of which the angular radius is θ, we must take account of the centrifugal force of that rotation also. We shall show that, of itself, it would balance a gradient $\rho V^2 \cot\theta/\epsilon$, where ϵ is the earth's radius. The direction of the centrifugal force of rotation is always away from the centre of the small circle. That is, it helps to oppose the pressure-gradient if the centre is one of low pressure; but helps the pressure to oppose the effect of the rotation of the earth if the centre is one of high pressure. The two cases are included in the general equation between the gradient bb and the wind V,

$$bb = -\partial p/\partial y = 2\omega V\rho \sin\phi \pm V^2\rho \cot\theta/\epsilon;$$

the + sign is appropriate for a cyclonic circle, the − sign for an anticyclonic circle.

We may call the first term depending on the earth's rotation the geostrophic term, and the second depending on the rotation in a small circle the cyclostrophic term.

We can accept this equation as deduced from equations 8 and 9 of p. 49. It can be derived directly on elementary dynamical principles as follows. The deduction was suggested to the author in 1904 by the late Sir John Eliot, sometime Meteorological Reporter to the Government of India.

For the geostrophic term. The rotation ω of the earth about the polar axis can be resolved into $\omega \sin \phi$ about the vertical at the place where latitude is ϕ and $\omega \cos \phi$ about a line through the earth's centre parallel to the tangent line.

The latter produces no effect in deviating an air-current any more than the polar rotation does on a current at the equator.

The former corresponds with the rotation of the earth's surface, counter clockwise in the northern hemisphere and clockwise in the southern hemisphere, under the moving air with an angular velocity $\omega \sin \phi$. We therefore regard the surface over which the air is moving as a flat disc rotating with an angular velocity $\omega \sin \phi$.

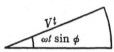

By the end of an interval t the air will have travelled Vt, where V is the "wind-velocity," and the earth underneath its new position will be at a distance $Vt \times \omega t \sin \phi$ (measured along a small circle) from its position at the beginning of the time t.

Taking it to be at right angles to the path, in the limit when t is small, the distance the air will appear to have become displaced to the right by the earth's motion to the left is $V\omega t^2 \sin \phi$.

This displacement on the "$\frac{1}{2}gt^2$" law (since initially there was no transverse velocity) is what would be produced by a transverse acceleration $2\omega V \sin \phi$. Hence the effect of the earth's rotation is equivalent to an acceleration $2\omega V \sin \phi$, at right angles to the path directed to the right in the northern hemisphere, and to the left in the southern hemisphere.

In order to keep the air on the great circle, a force corresponding with an equal but oppositely directed acceleration is necessary. The force is supplied by the pressure-distribution.

Whence $\qquad\qquad\qquad\qquad bb/\rho = 2\omega V \sin \phi.$

For the cyclostrophic term. We have the general dynamical expression V^2/r for motion in a circle radius r. The force is understood to be in the plane of the circle. In the case of motion in a small circle on the earth's surface of rectilinear diameter $2r$, or angular diameter 2θ, taking the earth as spherical, the geometry is similar to that of a small circle at latitude ϕ with the rectilinear radius r and angular radius the co-latitude θ. The rectilinear diameter is $2\epsilon \cos \phi$ or $2\epsilon \sin \theta$, and the centrifugal force $V^2/\epsilon \sin \theta$. Its direction is along the rectilinear diameter inclined at an angle θ to the horizon. Its horizontal component is therefore $V^2 \cos \theta/\epsilon \sin \theta = V^2 \cot \theta/\epsilon$.

It must be understood that the small circle which is here supposed to be described is that of the *path*. It is in fact the radius of curvature, the radius of the circle "osculating" the path at the point. It touches the isobar but is only indicated by the isobar or its radius of curvature if the path coincides with the isobar.

Combining the geostrophic and cyclostrophic components we get—

For a path concave to the low pressure side: $bb/\rho = 2\omega V \sin \phi + V^2 \cot \theta/\epsilon$;

and for one concave to the high pressure side: $bb/\rho = 2\omega V \sin \phi - V^2 \cot \theta/\epsilon$.

If we measure the pressure-gradient bb on a map, the equation may be used to determine V which we call the "gradient-wind," or if we measure the wind V

and the radius of curvature of the path, we may regard the equation as determining the pressure-gradient *bb*. Let us proceed to the solution of the equations regarding them as quadratics in V with double roots according to the laws of algebra.

Cyclonic curvature. The equation may be written:

$$V = \pm \sqrt{\left(\frac{\omega\epsilon\sin\phi}{\cot\theta}\right)^2 + \frac{\epsilon bb}{\rho\cot\theta}} - \frac{\omega\epsilon\sin\phi}{\cot\theta}.$$

Taking the positive sign there is a finite positive value of V for every value of *bb* and ϕ; for $\theta = 0$, V becomes also zero. Taking the negative sign, V is always in the negative direction and is numerically greater than before for the same value of *bb*. That is to say that if the rate of rotation is fast enough a cyclone could rotate clockwise but that would mean greater kinetic energy than is necessary to balance the gradient.

At the equator, where $\phi = 0$, $V = \pm \sqrt{\epsilon bb/(\rho\cot\theta)}$ and a whirl can exist with the same velocity in either direction.

Anticyclonic curvature. The equation now becomes:

$$V = \pm \sqrt{\left(\frac{\omega\epsilon\sin\phi}{\cot\theta}\right)^2 - \frac{\epsilon bb}{\rho\cot\theta}} + \frac{\omega\epsilon\sin\phi}{\cot\theta}.$$

V is always positive. If the positive sign is taken, the velocity is greater than if the negative is taken. Hence the negative sign gives the stable form of motion.

The solution is only real if $\omega^2\rho\epsilon\sin^2\phi > bb\cot\theta$. For a given value of θ there must be a limiting value of the gradient *bb*, and consequently of V.

For zero gradient V has two values, viz. 0 and $2\omega\epsilon\sin\phi/\cot\theta$.

If $\cot\theta$ becomes very small, that is to say if motion approximates to that in a great circle, V becomes proportional to the gradient.

If ϕ is zero, there is no solution: there cannot be a local anticyclone at the equator.

These are the equations for wind and pressure-distribution. As we shall see in the following chapters the relation cannot be recognised effectively if the pressure-gradient be taken from an ordinary map of the distribution of pressure and the value of V be taken from an anemometer. Nor can we find any general formula which will connect those two observations so long as they refer to the immediate neighbourhood of the earth's surface where frictional forces are practically the controlling factors.

The position that we take up is that the condition of no acceleration and the relation of the gradient to the geostrophic or to the gradient-wind, is probably very nearly satisfied within the limits of observation at a distance of 500 m from the earth's surface[1].

In regions of the earth near the equator, where the smallness of the factor $\sin\phi$ makes the geostrophic term negligible, we get only the cyclostrophic term and the balanced wind of the equatorial regions, if there should be one, is a cyclostrophic wind round a centre of low pressure.

We note that the geostrophic wind $(bb/2\omega\rho\sin\phi)$ is greater than the gradient wind for a cyclonic path and less for an anticyclonic.

---

[1] E. Gold, *Barometric Gradient and Wind-Force*, M.O. Publication, No. 190, 1908.

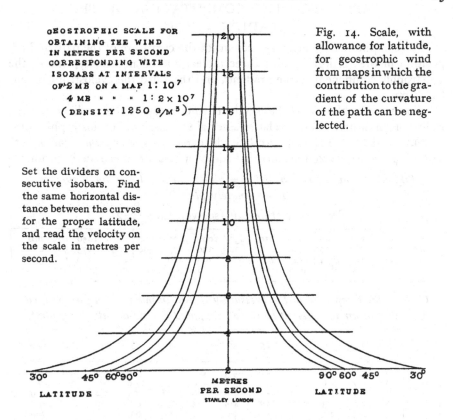

GEOSTROPHIC SCALE FOR
OBTAINING THE WIND
IN METRES PER SECOND
CORRESPONDING WITH
ISOBARS AT INTERVALS
OF 2 MB ON A MAP 1 : 10⁷
    4 MB "   " 1 : 2 × 10⁷
(DENSITY 1250 G/M³)

Fig. 14. Scale, with allowance for latitude, for geostrophic wind from maps in which the contribution to the gradient of the curvature of the path can be neglected.

Set the dividers on consecutive isobars. Find the same horizontal distance between the curves for the proper latitude, and read the velocity on the scale in metres per second.

LATITUDE   30°   45° 60°90°

METRES
PER SECOND
STANLEY LONDON

90° 60° 45°   30°   LATITUDE

Measure of the geostrophic wind

The wind which corresponds with a given gradient depends partly upon the earth's rotation and partly on the curvature of the path. The determination of the latter requires a succession of maps; it is not practicable to obtain it from the map which is sufficient to indicate the gradient. It is not unusual to assume that the path is indicated by the isobar; but that is only the case when the barometric system is stationary.

The examples of trajectories which we show on p. 244 suggest that, except in the northern parts of cyclones travelling rapidly, the curvature of the path is less than that of the isobar. Hence we may assume in many cases that the curvature of the path is so slight that the contribution of the curvature to the gradient can be neglected and an approximation to the wind can be obtained by measuring the distance of separation of consecutive isobars. The result will depend on the latitude, so that for general use something more than a linear scale is required. We give two scales (figs. 13, 14) which with a pair of dividers will give the geostrophic wind for isobars in any latitude.

TABLES FOR THE COMPUTATION OF THE GRADIENT-WIND

For computations which take account of the curvature of the path and the position in the vertical we give three tables, one of air-density, one of the geostrophic component to the gradient, and the third the cyclostrophic contribution.

To use the tables to get the gradient-wind, measure the gradient and the radius of curvature. From the latter ascertain the series of cyclostrophic components, and for the appropriate latitude choose a velocity *the same in both tables* which will give components that add or subtract to make up the gradient.

(*By the formula* $bb = bb_\omega + bb_r = 2\omega V\rho \sin \phi \pm V^2\rho/r$ *for a surface approximately plane.*)

The density of dry air (see vol. III, p. 242).

p	900	900	900	950	950	950	1000	1000	1000	1050	1050	1050 mb
tt	250	275	300	250	275	300	250	275	300	250	275	300 tt
ρ	1254	1140	1044	1323	1202	1102	1393	1266	1160	1463	1329	1218 g/m³
$\rho/1200$	1·04	·95	·87	1·10	1·00	·92	1·16	1·06	·97	1·22	1·11	1·01

Geostrophic component of the pressure-gradient in mb/100 km for specified velocities of air in different latitudes computed for air-density 1200 g/m³.

ϕ \ V	5	10	15	20	25	30	35	40	45	50	55	60 m/s
90°	·9	1·8	2·6	3·5	4·4	5·3	6·1	7·0	7·9	8·8	9·6	10·5
80°	·9	1·7	2·6	3·4	4·3	5·2	6·0	6·9	7·8	8·6	9·5	10·3
70°	·8	1·6	2·5	3·3	4·1	4·9	5·8	6·6	7·4	8·2	9·0	9·9
60°	·8	1·5	2·3	3·0	3·8	4·5	5·3	6·1	6·8	7·6	8·3	9·1
50°	·7	1·3	2·0	2·7	3·4	4·0	4·7	5·4	6·0	6·7	7·4	8·0
40°	·6	1·1	1·7	2·2	2·8	3·4	3·9	4·5	5·1	5·6	6·2	6·8
30°	·4	·9	1·3	1·7	2·2	2·6	3·1	3·5	3·9	4·4	4·8	5·2
20°	·3	·6	·9	1·2	1·5	1·8	2·1	2·4	2·7	3·0	3·3	3·6
10°	·2	·3	·5	·6	·8	·9	1·1	1·2	1·4	1·5	1·7	1·8
0°	·0	·0	·0	·0	·0	·0	·0	·0	·0	·0	·0	·0

Cyclostrophic component of the pressure-gradient in mb/100 km for specified velocities of air, and specified values of the radius, computed for air-density 1200 g/m³.

r \ V	5	10	15	20	25	30	35	40	45	50	55	60 m/s
km												
10	3·0	12·0	27·0	48·0	75·0	108·0	147·0	192·0	243·0	300·0	363·0	432·0
20	1·5	6·0	13·5	24·0	37·5	54·0	73·5	96·0	121·5	150·0	181·5	216·0
40	·7	3·0	6·7	12·0	18·7	27·0	36·7	48·0	60·7	75·0	90·7	108·0
60	·5	2·0	4·5	8·0	12·5	18·0	24·5	32·0	40·5	50·0	60·5	72·0
80	·4	1·5	3·4	6·0	9·4	13·5	18·4	24·0	30·4	37·5	45·4	54·0
100	·3	1·2	2·7	4·8	7·5	10·8	14·7	19·2	24·3	30·0	36·3	43·2
200	·2	·6	1·3	2·4	3·7	5·4	7·3	9·6	12·1	15·0	18·1	21·6
500	·1	·2	·5	1·0	1·5	2·2	2·9	3·8	4·9	6·0	7·3	8·6
1000	·0	·1	·3	·5	·7	1·1	1·5	1·9	2·4	3·0	3·6	4·3
2000	·0	·1	·1	·2	·4	·5	·7	·9	1·2	1·5	1·8	2·2

CHAPTER III

THE FOOT OF THE ATMOSPHERIC STRUCTURE

"The figure which represents the normal thermal structure of the atmosphere is made up of a foot with diurnal variation, a limb bending towards the horizontal at the tropopause, and a trunk in the stratosphere sometimes stooping forward towards higher temperatures" (Vol. III, chap. VII).

THE DETERMINATION OF PRESSURE-GRADIENT AND WIND

WE have seen in the description of the general circulation of the atmosphere (vol. II, chap. VI) that the wind as observed in the usual way near the surface is generally related to the distribution of pressure in accordance with Buys Ballot's law. In the northern hemisphere pressure is lower on the left of the flow of the wind and higher on the right. In the southern hemisphere the reverse is the case; and in either hemisphere the closer the isobars are together the greater, as a rule, is the velocity of the wind.

The relation which these words express is rather vague. At well-exposed stations on our coasts the inclination of the observed wind to the run of the isobars is generally about 30° towards the side of lower pressure but it varies between wide limits; sometimes the wind follows even the local deviations of the line of the isobars with surprising fidelity as in the map for 8 a.m. of 9 April, 1908, but occasionally the wind may cross the isobar at 45° or more. At inland stations, particularly those in hilly or mountainous districts, the law has many exceptions. The relation of the velocity to the gradient of pressure also shows variations within wide limits so that if the surface-wind be taken as the basis of computation no satisfactory expression can be given for a relation between wind and gradient as representing the underlying principle upon which Buys Ballot's law depends.

But we have already seen that under certain conditions there ought to be a relation between the flow of air and the distribution of pressure in the atmosphere of a rotating globe such as, without serious error, we may consider our earth to be. The relation which has been established is that expressed by the equation

$$bb = 2\omega V \rho \sin \phi \pm V^2 \rho \cot \theta / \epsilon,$$

where bb is the pressure-gradient, or tangent of the slope of the isobaric surface,

V the velocity of the wind,

ρ the density of the air,

ϕ the latitude of the place of observation,

θ the angular radius of the small circle which marks the path of the air at the moment,

ϵ the radius of the earth, ω its angular velocity of rotation.

(87)

The double sign $\pm$ means that the formula is different according as the path of the air deviates from a great circle to the left, or to the right. If it turns to the left the small circle which identifies the curvature of the path for the moment has low pressure at its centre. It therefore belongs to a cyclonic system, and the $+$ sign is to be used: but if the air deviates to the right the small circle in which it is moving has higher pressure at its centre; the circulation is anticyclonic and the $-$ sign is to be used.

The equation must be understood to refer to horizontal motion only and to be subject to the condition that the resultant of the horizontal forces acting upon the air is perpendicular to the direction of motion so that there is no immediate acceleration or retardation of the speed and the resultant force must be of the proper magnitude to balance the kinematic effect of the spin of the earth and the spin of the air in the small circle which marks its path.

These conditions clearly cannot be satisfied in the case of the air that flows along the surface because there is always the friction of the surface which absorbs momentum from the flowing stream and consequently appears as a retarding force in the line of motion. It is by the sacrifice of its momentum that wind produces the effects upon land and water with which we are familiar. The energy of sea-waves comes from the motion of the wind, as does also the energy that is displayed in the destructive effects of a gale, a hurricane or a tornado. Consequently, in so far as Buys Ballot's law is a manifestation of the relation of the wind to the distribution of pressure indicated by the equation for gradient wind, it is under the most unfavourable conditions that the manifestation takes place at the surface. The fact that a relation can be recognised so frequently in conditions so unfavourable certainly implies that the relation of wind to pressure in some form or other is an important principle in the structure of the atmosphere.

William Ferrel[1] had already used the effect of the earth's spin as the basis of a scheme for the general circulation of the atmosphere. Guldberg and Mohn[2] attributed the difference between the computed results and those observed for the surface-wind to friction and a coefficient of friction was evaluated; but when the soundings of the upper air made with kites by W. H. Dines from 1903 onwards were referred to our maps the fact of the nearer approximation to the computed value both as regards direction and velocity shown by the winds of the upper air at once challenged attention.

The question was examined by E. Gold in a *Report on Barometric Gradient and Wind-force*[3]. The practical conclusion of the examination was that if the wind at the 500 metres level was compared with the surface-gradient there was on the average close agreement, the deviation being sometimes one way and sometimes another. And Gold added in the same report a theoretical proposition easily verified by common observation and of great importance.

[1] *A popular treatise on the winds.* New York, 1889.

[2] 'Studies of the movements of the Atmosphere' (original paper revised by authors and translated by Cleveland Abbé), *Smithsonian Miscellaneous Collections*, vol. 51, No. 4, XI, p. 122.

[3] M. O. Publication, No. 190, 1908.

He showed that if the negative sign be taken in the gradient equation and the equation be solved as a quadratic, for the determination of the wind-velocity V corresponding with a given gradient bb, the roots become imaginary if the curvature is above a certain limit. From which it follows that if the balance of wind and gradient is the principle upon which the structure of the atmosphere is based, the winds near the centre of an anticyclone, where the curvature of the isobar must be relatively great, cannot exceed a certain small limiting value, whereas no such limitation is applicable in the case of a cyclonic depression. This gave a dynamical explanation of the well-known fact that the central region of an anticyclone is always an area of very light winds, whereas there is practically no limit to the velocity of the winds in the small circles near the core of a tropical revolving storm or a tornado.

Up to that time it had been customary to consider all such dynamical problems from the point of view of the motion of the air resulting from the effect of a finite difference of pressure, and Gold gave a calculation of the time which would be required for the velocity to adjust itself to the pressure-gradient, but if we consider that the process of adjustment is constantly going on and that in most unfavourable circumstances Buys Ballot's law gives evidence of the adjustment, we are left to surmise what uncompensated differences of pressure are likely to be found in the free atmosphere where there are no disturbing forces such as are found at the surface and where an infinitesimal change in the distribution of pressure or in velocity at once sets up the process of readjustment. We are precluded from supposing that the retarding force of "friction' at the surface extends to great heights in the atmosphere by the direct action of viscosity because Helmholtz has shown that any such effect is negligible[1].

The law of isobaric motion

The investigation of the *Life-history of Surface Air-currents*[2] showed that in actual practice air must be regarded as travelling over long tracks of sea and land with very little change of velocity from hour to hour or even, on occasions, from day to day and suggested that such incidents as being caught in the ascending current of a rain-storm or shower, the most likely disturbance of an even progress, must form a very small part of the life-history of air-currents; that the greater part must be made up of such steady motion as we see in the case of clouds in common weather which travel for long distances with very little change of speed.

If we consider other cases in which long continuous journeys are made, such as that of a ship pursuing a similarly even course from one port to another three thousand miles away, we are not, as a rule, much concerned with the acceleration of the start or the retardation of the landing but with the circumstances of the motion of the long voyage so nearly uniform that we may regard the ship as moving under balanced forces. So with the winds in meteorology, it is not that portion of their history in which they are starting or stopping, accelerated from quiescence or brought to rest by some sudden dynamical

[1] See chap. II, p. 58.   [2] M. O. Publication, No. 174, 1906.

influence operating for a short time, that we have to think about so much as the longer period when they keep on their uniform way under balanced forces like a train or a motor-car that has got its speed and is making a run on a long stretch. There will, of course, be variations in the gradient of the road that will alter the rate of motion, sometimes slowing it down, sometimes speeding it up, but the chief feature of the motion is the balance between the motive forces of the engine and the resisting forces of the air and the wheels.

We may reserve the cases of disturbed motion and of vertical motion for special treatment in accordance with the laws of entropy set out in vol. III. Considering the air as having arrived at an established state of motion, without being obliged to trace the original causes of its motion or speculate upon its arrest, we may limit our consideration to the progress of its motion; in ordinary circumstances this will show very little variation of speed in the course of hours in comparison with the possible effects of the force of pressure which has been operative during those hours, in a direction transverse to the air's motion; pressure-distribution seems to adjust itself to the motion of the air rather than to speed it or stop it. So it will be more profitable to consider the "strophic" balance between the flow of air and the distribution of pressure as an axiom or principle of atmospheric motion, which, at the moment, cannot be either directly verified or directly contradicted, and examine the phenomena from that point of view. This principle was enunciated in a paper before the Royal Society of Edinburgh[1] in 1913 as follows: *In the upper layers of the atmosphere the steady horizontal motion of the air at any level is along the horizontal section of the isobaric surface at that level and the velocity is inversely proportional to the separation of the isobaric lines in the level of the section.*

In this statement the effect of the curvature of the path of the moving air was disregarded and the cases under consideration were therefore limited to those in which the air is moving approximately along a great circle, and this limitation requires notice. There are two terms on the right-hand side of the equation for gradient-wind, so that for a given value of the velocity V two causes contribute to the making of the balance of the gradient bb. One part which we may write bb_{ω} is represented by the term $2\omega V\rho \sin \phi$ and is therefore due to the earth's rotation. We may call that part the "geostrophic" component. The other part which we may denote by bb_r is due to the spin in a small circle of angular radius θ; this we may call the "cyclostrophic" component. It is evident that as the latitude ϕ becomes less, that is, as we move nearer to the equator, the geostrophic component becomes relatively of less importance because the factor $\sin \phi$ diminishes and at the equator, where $\phi = 0$, the effect of the rotation of the earth has vanished, the only possible balance of wind and pressure is the cyclostrophic balance such as we may appreciate in rapidly rotating whirls of fluid. On the other hand, if the motion is in greater and greater "small circles" the relative importance of the cyclostrophic component becomes less and less, until when the motion is along a great circle $\cot \theta$ becomes

[1] *Proc. R.S.E.*, vol. XXXIV, 1913, p. 78. For recent development see chap. XI.

zero and the rotation of the earth alone is operative to maintain the balance between velocity and pressure. The wind computed according to the complete formula we call the "gradient-wind" and if the curvature of the path, rightly or wrongly, be disregarded and the velocity computed as balancing the pressure with the aid of the earth's spin alone we call the computed value the "geostrophic wind."

The strophic balance

To assume that this balance of wind and pressure in the upper air is an operative principle of atmospheric structure may be thought a hazardous mode of procedure and it requires the most scrupulous examination; but the proper course seems to be to accept it at least until the proved exceptions are numerous enough to show that, under the prescribed conditions of motion approximately in a great circle, finite differences of pressure do exist in the air without the compensating velocity in the air-currents. It need not be supposed that the balance is always strictly perfect but only that in ordinary circumstances the accelerating forces operating in the air are so small in relation to the pressures that we measure, that they are beyond our powers of observation. In the remainder of this chapter we propose to examine the question of the direct comparison of the actual wind with the wind computed from the gradient.

The difficulties of measurement

First it must be remembered that the determination either of the gradient or of the actual wind at any level is attended with great practical difficulty. The gradient is determined by plotting observations of pressure on a map and drawing isobars across the area. The pressures are observed by different observers and accuracy in the collection of simultaneous values depends in the first place upon punctuality, too soon being as bad as too late, and upon the careful attention to the organisation of the network of stations. The observed pressures have to be reduced to sea-level by a process which may introduce quite appreciable errors if the observing station is some hundreds of feet up. The stations of the Meteorological Office are about fifty miles apart over the British Isles and the seas have to be bridged by connexion with stations on either side or by wireless reports from distant ships. With stations so wide apart isobaric lines can be drawn with reasonable certainty so far as their general run is concerned but there may be details of variation which the observations do not show. A local gradient, for example, for a width of say five miles along a coast may correspond truly with a local wind; the local wind would be obtained by direct observation, the local gradient could only be obtained by the most elaborate process because the pressure-difference over five miles necessary to balance a geostrophic wind of gale force is not more than half a millibar. Slight variations shown on a barogram by fluctuations in the record which we call embroidery cannot be represented on the map without elaborate investigation quite beyond the scope of the daily service that provides the maps from which gradients are taken.

Gradients for higher strata than sea-level are practically unattainable at present except by calculation, and for adequate calculation accurate information of the temperature at different levels above the places of observation is required but is not yet available.

But if there is difficulty about obtaining the gradient, the difficulty of obtaining a measure of the wind to compare with it is still greater. If the map is drawn for sea-level, winds at sea-level are wanted. The nearest approach to those are winds at the surface and these are subject to interference of so many kinds that it is difficult even to say what a measure of wind exactly means.

The reader may be surprised to learn that measuring the wind is really a most difficult operation, but he may realise the truth of the statement when he understands that thirty years ago, on account of the inherent difficulty of the subject, the Meteorological Office had to be content to publish values of wind-velocity which were known to be in error by 25 per cent, and even now the quotation, without reference, of a reading of an anemometer from a considerable number of meteorological publications is no guarantee that an error of that order of magnitude is not involved. By the turn of the century much had been done and the accuracy of wind measurements has been still further improved in recent years by the use of wind-channels set up for aeronautical investigations. In working conditions an ordinary measure of wind as represented on a weather-map of the British Isles and the neighbouring parts of the continent and the islands of the Atlantic is obtained for the most part by estimation according to the Beaufort scale, and the same is true universally of the measures of wind at sea which are generally available for meteorological study. In order to express these estimates in terms of velocity the equivalents of the numbers of the Beaufort scale given below on p. 94 or in chap. II of vol. I are employed. The scale was obtained[1] by taking the mean value of many individual observations which show very considerable diversity among themselves and, strictly speaking, it can only be regarded as applicable when in like manner the means of a large number of estimates are under consideration. Any individual estimate of wind-force is liable to the various causes, partly peculiar to the locality or the special conditions of weather and partly personal to the observer, which account for the large range of estimates included under the same number of the scale in the formation of the table of equivalents. At a few observatories measures of the wind for use in the preparation of weather-maps are obtained by anemometers and the figures are transformed into the numbers of the Beaufort scale for transmission by telegraph by the use of the same table of equivalents. In such cases there are no uncertainties in the measurement such as are inseparable from personal estimates but the peculiarities of the exposure of the anemometers are at present even more disturbing than the vagueness of personal estimates. The older observatories were provided with recording cup-anemometers of the Robinson type and these are very heavy instruments which have to be supported on substantial structures and are generally installed on large buildings which, with

---

[1] *The Beaufort Scale of Wind-Force*, M.O. Publication, No. 180, 1906.

their surroundings, affect the flow of air past the instrument with eddies of various kinds (fig. 9). The subject is at present not at all fully investigated and it has only recently been pointed out[1] in a summary of the occurrence of gales in different localities that, for this reason, the observatories of the Meteorological Office are characterised by a singular and anomalous freedom from gales which is really attributable to the exposure of the anemometers. No doubt the position of the question can be improved by a discussion of the available data when a standard of reference can be agreed upon. At present the only standard that seems in any way adequate is the geostrophic wind of straight isobars and that is acceptable only if we may assume that the geostrophic wind for straight isobars is a real equivalent of the actual flow of air undisturbed by surface effects and local eddies.

With the newer tube-anemometers of the Dines type the position is considerably improved. The vane is generally exposed on a slender mast forty feet above the ground and can be set up in an open situation with only a small hut at its base, so that the whole structure offers very little cause for interference with the wind. In that case it is only the ground and the irregularities of the relief of the region which produce disturbing eddies; but even these effects are at present unknown in detail, nor is it possible to suppose that any simple relation can exist which will make the records of anemometers at different places strictly comparable and afford a satisfactory standard of reference. Allowance must be made for differences in height; and the variation of wind with height is not only very large but it is in itself an extraordinarily complicated question. Much light has recently been thrown upon it by investigations of eddy-motion in the atmosphere by G. I. Taylor and others. The theory[2] confirms what has all along been dimly foreshadowed by observation, that the variation of wind with height depends upon the locality and the nature of the surroundings: it is different for sea and for land, for a town and for the open country, and so on: it also depends upon the undisturbed velocity of the wind in the upper air, upon the surface-temperature of the ground, and hence upon the time of day and the season of the year. For the purpose of his investigations Taylor relies upon the geostrophic wind as representing the undisturbed wind in the upper air and no other standard of reference seems possible. Here we may also note that in certain circumstances the wind at or near the surface may show virtual independence of the general distribution of pressure in consequence of the direct effect of the gravitation of cold air down the slopes of hills at night, a form of wind which in vol. II we have called *katabatic*, and equally in the daytime air-currents near the surface may represent the levitation due to the warming of slopes in the sun which gives rise to winds which we have called *anabatic*, and which are illustrated by the familiar phenomena of land- and sea-breezes. (See vol. III, p. 319.)

Hence it will be seen that a measure of wind by estimate or by anemometer

[1] *The Weather of the British Coasts*, M. O. Publication, No. 230, p. 34, 1918.
[2] 'Phenomena connected with turbulence in the lower atmosphere,' *Proc. Roy. Soc.* A, vol. XCIV, 1918, p. 137.

near the surface is subject to so many local influences and other possibilities of discrepancy that it is not practicable to build upon it any reasonable picture of the structure of the atmosphere. It offers us a number of problems which we may seek to explain by tracing the effect of local causes upon the upper wind that is part of the structure of the atmosphere. Some examples are given in chap. I and the subject is considered further in chap. IV.

In recent years direct observations by pilot-balloons or other similar means have given us measurements which enable us to approach the question of the relation of wind-velocity to pressure-distribution from a new starting point. We might take the wind at 500 metres or some other height as a standard of reference and compare it with the gradient-wind; but the relation which has been assumed, and which we seek to justify, refers to wind and gradient at the same level and we have no precise measure of the gradient at the level of 500 metres; and if we had, we have still to decide whether the influence of the surface ceases at that level. Hence we come back to the method of assuming the correspondence between actual wind and geostrophic wind in the free air for straight isobars and tracing all the consequences that can be inferred therefrom. In pursuance of this plan we shall put together in the next section the facts of the relation of the surface-winds to the distribution of pressure at sea-level so far as we know them. We may pass on then in the following chapters to consider the facts about the variation of wind with height and subsequently see how far these facts can be explained upon the hypothesis which we have set out as an axiom or law of isobaric motion in level surfaces.

The equivalents of the Beaufort scale

The question of the velocities to be assigned to the different numbers of the scale used as a code for the purpose of daily international reports of weather from sea and land has recently been under consideration by the International Organisation on a report by Dr G. C. Simpson[1].

At Vienna in 1926 the following was approved:

(*a*) It is important that the force (and direction) of wind in reports for synoptic purposes should give a good representation of the general current of air over the surface of the earth in the region where the reporting station is situated.

(*b*) It is important that the basis taken for reports should be such as to give values from different stations which are inter-comparable.

(*c*) The Beaufort scale is in accordance with the conditions of paragraphs *a* and *b* and this scale should be the scale of wind for international weather telegraphy.

(*d*) The velocity equivalents of the Beaufort scale should be the values in the Table VI of Dr Simpson's report (see below) with the addition of a note that these values for the Beaufort scale correspond on land with the speeds at a height of approximately 6 m above a level surface free from all obstructions. Such an exposure of an anemometer would be called the standard exposure for synoptic purposes.

(*e*) In cases where the exposure of the anemometer differs from the standard in a manner such that the effect is known either by direct experiment or by straightforward deduction from well-established results, the records of the instrument should be con-

---

[1] 'The velocity equivalents of the Beaufort scale,' *Professional Notes*, No. 44, M. O. 273 d, 1926.

verted to the Beaufort scale by an appropriate table. For example, if an anemometer were exposed at a height of 20 m above level ground free from obstructions, the conversion would be made by a table in which the speeds of paragraph (d) were increased by 23 per cent. If the correction is less than 5 per cent the standard table should be used.

(f) Anemometers and air-meters in general use usually begin to record only when the wind is 1 m/sec or more. In cases of Beaufort force 0 and 1 the observer at a station equipped with an anemometer should, as a rule, not use the anemometer record but should estimate the force and direction of the wind for synoptic reports.

The table quoted in the report is given below; with it we have incorporated that given in the British Daily Weather Report as applicable to "about 30 feet above the ground."

Beaufort number	Limits of velocity			Beaufort number	Limits of velocity		
	International		British D.W.R.		International		British D.W.R.
	m/sec	mi/hr	mi/hr		m/sec	mi/hr	mi/hr
0	0– 0·5	0– 1	1	7	12·5–15·2	28–33	32–38
1	0·6– 1·7	2– 3	1– 3	8	15·3–18·2	34–40	39–46
2	1·8– 3·3	4– 7	4– 7	9	18·3–21·5	41–48	47–54
3	3·4– 5·2	8–11	8–12	10	21·6–25·1	49–56	55–63
4	5·3– 7·4	12–16	13–18	11	25·2–29·0	57–65	64–75
5	7·5– 9·8	17–21	19–24	12	> 29·0	> 65	75
6	9·9–12·4	22–27	25–31				

THE SURFACE-WIND AND THE GEOSTROPHIC WIND AT SEA-LEVEL. GEOSTROPHIC WIND-ROSES

Until the question is examined in its various aspects, some of which have been referred to in the preceding section, the relation between the surface-wind and the gradient-wind for sea-level seems a simple question. It used to be treated simply as a question of determining the relation between wind-force and the gradient or the separation of the isobars, which, for the geostrophic wind, is inversely proportional to the wind-velocity. It was on those lines that values were obtained in 1882 for the average relation of the wind at Kew Observatory to the barometric gradient[1]. Subsequently the relation of the surface-wind to the geostrophic wind as watched from day to day in the study of weather-maps was formed into a sort of working rule that the current of air of the upper regions lost one-third of its velocity over the sea, and two-thirds over the land; so that, for the same distribution of pressure, the wind recorded in open inland country would have one-half of the velocity appropriate to the same distribution at sea, a conclusion which was exemplified by an occasion when parallel isobars from the west-south-west covered the whole country and gave force 8 at the exposed stations on the western coasts and force 5 at the stations inland and on the eastern coasts. Naturally the coast stations if they are on a well-exposed flat shore belong to the régime of the sea for off-sea winds, and to the inland for off-shore winds.

As a rough working rule this is still a useful form of note to carry in the memory, but in the critical examination of the question of the relation of the observed wind to the geostrophic wind at sea-level no such simple generalisation can be allowed.

[1] G. M. Whipple, Q. J. Roy. Meteor. Soc., vol. VIII, 1882, p. 198.

The mere consideration of the diurnal variation of wind-velocity as shown by the anemometers at the observatories is sufficient to make it clear that no single number can express the relation of the observed wind to the geostrophic wind for any locality. A marked diurnal variation of wind-velocity is shown in all the hourly normals of wind-velocity even in winter and still more in summer. There is, on the other hand, no observational evidence for the existence of a diurnal variation of barometric gradient; and we therefore assume that for the average of a large number of observations the surface-winds of the day-hours and night-hours may be referred to the same geostrophic wind[1]. The hourly values of wind-velocity are available in tabular form for the four observatories, Aberdeen, Cahirciveen (Valencia Observatory), Falmouth, and Richmond (Kew Observatory), since 1868 and for Eskdalemuir since 1911. The average factor of relationship between the observed wind and the geostrophic wind at different times of the day in different seasons of the year will vary to the same extent as the average values of the observed wind-velocities represented in the accompanying figure of isopleths for Aberdeen, Falmouth, Eskdalemuir and Ben Nevis; corresponding diagrams for Cahirciveen and Richmond are reproduced in vol. I, p. 267 and for the Eiffel Tower in vol. II, p. 284.

It should be noted that these figures are averages: there are days when the diurnal variation of the wind is obliterated by conditions of weather and there must also be days in which it is much more marked than the average, and in consequence the factor of relation of surface-wind to geostrophic wind will certainly have a wider range than that shown in the diagrams. When the averages are expressed as fractions of the geostrophic wind, there will obviously be large variations depending upon the time of day and the season of the year for which comparisons are made. For the British observatories the lowest values will be obtained for the night or the early hours of the morning and the highest with almost unanimous concurrence at 14 h, two hours after Greenwich noon. For the Eiffel Tower, the opposite is the case.

The diurnal variation of the velocity of the wind was explained by Espy[2] and Köppen[3] as due to the effect of the temperature of the surface upon the

[1] The differences of corresponding hourly values between Richmond (K.O.) and Aberdeen or between Richmond (K.O.) and Cahirciveen show that there is no appreciable diurnal variation in the general gradient for the westerly component or the southerly component of the winds, but it should not be forgotten that there may be a local gradient on crossing a coast line due to the difference in the régime of temperature distribution in the vertical over the land as compared with that over the sea for which no measurements are available. An example of a strong local wind attributable to the coastal gradient is cited in *Barometric gradient and wind-force* (see vol. III, p. 321). E. L. Hawke has also shown that there is a small difference of pressure at 7 a.m. between the average of mean values of inland stations and coast stations in England in the same belt of latitude which may be attributed to the dynamical effect of the land upon the flow of air from the westward. The possibility of local coastal gradients deserves further investigation. The tendency of winds to set along the coast line is a phenomenon which has often been remarked upon in the course of the marine work at the Meteorological Office.

[2] *Philosophy of Storms*, p. xiv; *B. A. Report*, 1840, p. 345.

[3] *Meteor. Zeitschr.*, vol. XIV, 1879, p. 333.

mixing of the surface layers of the air with the upper layers whereby the lower layers acquire momentum at the expense of the upper layers, and the qualitative explanation thus provided has been put into a dynamical form which allows of a quantitative estimate of the effect by G. I. Taylor in the paper referred to in note 2, p. 93. The results will be considered later. Here we need only call attention again to the fact that the ratio of the surface-wind to the geostrophic wind is dependent upon the diurnal and seasonal change of tem-

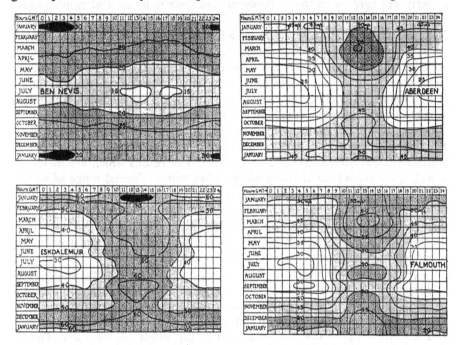

Fig. 15. Normal isopleths of seasonal and diurnal variation of wind-velocity on Ben Nevis, 1884–1903; at Aberdeen, 1881–1910; at Eskdalemuir, 1915–1924; and Falmouth, 1881–1910. The lines for Aberdeen and Falmouth are from *The Weather Map*, M. O. 225 i.

The figures for Ben Nevis are on a special scale of personal estimates adopted by the Observatory. As tested by a Robinson anemometer in the summer months the velocity equivalents of the forces 1 to 3 are: Force 1, 6 mi/hr, 2·7 m/sec; 2, 12 mi/hr, 5·4 m/sec; 3, 21 mi/hr, 9·4 m/sec. For the other diagrams the lines are drawn according to the number of metres per second marked on the diagram.

perature of the surface of the ground. The actual cause of the diurnal variation of the wind is the temperature of the ground which undergoes a periodic change with this general physical result, that colder ground means less surface-wind and therefore a smaller fraction of the geostrophic wind, while warmer ground means more surface-wind and therefore a greater fraction of the geostrophic wind. The rule must be universal and is not simply applicable to the case of diurnal variation. Representing the surface-wind by W and the geostrophic wind by G we may say that wherever there is a cold surface over which warmer air is passing the cooling of the air by the ground will reduce the factor W/G and conversely the warming of the air by the ground will increase the factor.

We may pursue this line of thought a little further. As long as there is a diurnal variation of wind-velocity without a corresponding variation of gradient we cannot expect agreement between W and G, that is, W/G will not become unity; and therefore, if we wish to find a region where the wind-velocity will agree with the geostrophic wind, we must seek a region where there is no diurnal variation of wind without a corresponding change of gradient. It is possible that such a region may be found somewhere in the upper air.

We have, at present, no satisfactory statistics of the diurnal variation of wind-velocity at high levels in the atmosphere except for observatories at high levels on mountains, the sites of which are themselves subject to diurnal variation of temperature. The number of observations of pilot-balloons is perhaps now numerous enough to supply in some degree the required information but it is not yet reduced to manageable form. The best observations for the purpose are those from the Eiffel Tower summarised by A. Angot, Director of the Bureau Central Météorologique de France[1]. The results which are included in vol. II are taken from that summary. They show a diurnal range of wind-velocity with a maximum in the night and a minimum in the day. The nightly maximum is common to all high-level observatories and is already noticeable for light winds though not for strong winds on the anemometer at Potsdam at a height of only 41 metres. It is also indicated at a height of 16 metres in the observations of Hellmann over flat meadow land near Nauen[2].

The minimum of velocity in the upper air in the day corresponding with a maximum of the surface velocity may be attributed to the effect of the eddy-motion which, as explained in previous chapters, is the real process indicated in the explanation of the diurnal variation of wind-velocity at the surface[3], and we may suppose that the minimum of the winds on the Eiffel Tower is due to the dilution of the current by the action of eddy-motion which mixes it with air of less momentum from below. In the night-hours there is at any rate less dilution, and at the level of the top of the tower, 305 m above the ground, probably no dilution at all, so that we may regard the wind at that level (about 1000 ft) as directly comparable with the geostrophic wind at the same level if we knew it. And, if that be so, then it is clear that in the day-hours the wind at 305 m over Paris is less than the geostrophic wind and is indeed on the average only about 80 per cent of the geostrophic wind at 14 h in

[1] 'Études sur le Climat de la France, Régime des Vents,' *Mémoires du B.C.M.*, 1907.

[2] Hellmann, 'Ueber die Bewegung der Luft in den untersten Schichten der Atmosphäre,' *M. Z.*, January, 1915.

[3] A discussion of the diurnal variation of the wind at low altitudes (a minimum in the day with backing and a maximum in the night with veering) is given by G. Reboul in *Comptes Rendus*, vol. CLXVI, 1918, p. 295, and by Rouch in vol. CLXIX, 1919. The subject is discussed also by J. Durward in M. O. *Professional Notes*, No. 15, using data for Noyelle-Vion, near Arras. Tables of the diurnal variation of wind in the free air over Lindenberg for summer and winter are given in vol. II, p. 282.

January, and only 60 per cent at 9 h in July. It cannot on the average exceed these percentages at the hours named because we have regarded the night values as giving the full equivalent of the gradient. Thus a rise of 1000 feet does not take us above the reach of surface disturbance in the daytime in a region such as Paris. Meanwhile we have a note in the Meteorological Office[1] that Mr S. P. Wing in 1915 found by observations on towers of open work at Ballybunion, Co. Kerry, that the average velocity of the wind at 500 ft was 90 per cent of the geostrophic wind at sea-level, whereas at 15 ft the wind was 62 per cent of the same. We have no record of the hours of the day or the precise dates when the observations were taken, but considering the relative positions the results are not far out of line with the results for Paris.

So far we have drawn no distinction between the various orientations of the isobars and the wind-directions, but these clearly give another element of variation in the relation of the observed wind to the geostrophic wind at the surface which must enter into the consideration of individual observations. Whether or not there be any difference in the ratio from purely meteorological causes such as the increase of gravity for air moving westward as compared with that moving eastward, there is certainly a difference due to the peculiarities of exposure in relation to surrounding buildings, trees, etc., and to the general relief of the district.

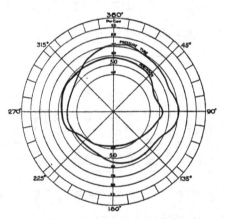

Fig. 16. The surface-wind of Cahirciveen as recorded on a Robinson anemometer and a pressure-tube and expressed as percentages of the geostrophic wind for different directions.

It will perhaps be convenient at once to illustrate the uncertainty which depends upon the exposure of the anemometer if an instrument is used to measure wind.

For that purpose we give a diagram (fig. 16) which represents the recorded wind as percentage of the geostrophic wind for two anemometers at Cahirciveen, Co. Kerry, Ireland, for which we are indebted to Mr L. H. G. Dines.

Of the two anemometers one is a pressure-tube exposed on the usual vertical column in the open ground of the observatory. The other is one of the original instruments of the Meteorological Office of the Robinson pattern mounted on the roof of the observatory building.

The range of the observations upon which the curves are constructed is in metres per second, for the pressure-tube 5·1 to 10·0 and for the Robinson 5·5 to 7·9.

The orientation of the wind used as the basis of reference is shown by the angle in degrees from North which is marked as 360 on the diagram.

[1] M. O. Circular, No. 5; *Meteor. Mag.*, vol. LVI, 1921, p. 263.

TABLE I. THE RELATION OF SURFACE-WIND TO GEOSTROPHIC WIND FOR A NUMBER OF STATIONS IN THE BRITISH ISLES

The Values of W/G expressed as the Percentage ratio of the Surface-Wind to the Geostrophic Wind for 16 points of the compass. The highest percentage ratio for each station is indicated by the symbol (:) the lowest by (-).

Station	No. of years' obsn	Forces comp'd	Height of vane or cups above ground (ft)	Nature of site	Height of ground (ft)	N 0°	NNE 22½°	NE 45°	ENE 67½°	E 90°	ESE 112½°	SE 135°	SSE 157½°	S 180°	SSW 202½°	SW 225°	WSW 247½°	W 270°	WNW 292½°	NW 315°	NNW 337½°	Mean of all
Falmouth	1	?	41	Between cliff & harbr.	167	36:	36:	33	30	31	32	32	31	29	28-	29	33	34	35	34	36:	32
Pendennis Castle	1	all	65	Conical headland	256	58	66	74	83	85:	83	74	69	63	58	54	52-	53	56	54	54	65
Pyrton Hill	2	all	98	Slope: hills to plain	500	39	38	48	48	40	37-	37-	42	48	40	42	43	47	49:	44	40	43
Southport	5	all	62	Flat sand shore	18	69:	60	53	49	48	47	47	45	41	38-	38-	42	51	63	65	69	52
Dungeness	12	12 m/s	B	Spit of shingle	21	49	57	70:	68	61	56	48	48	49	53	56	48	41	39	38-	46	52
Jersey	12	12 m/s	B	Railway station	25	54	55	56	54	56	48	46	47	47	47	48	56	57	50	56	56	52
Holyhead	12	12 m/s	B	Flat island	15	81:	75:	43	43	35	17	14-	34	44	49	40	40	49	60	68	80:	48
London, Brixton	4	12 m/s	B	Town garden	77	28	29	36	38	31	23	17	23	24	25	24	23	23	20	23	—	23
" St James's Pk.	6	12 m/s	B	Town park	27	25	29	29	29	24	13	11	12	13	23	24	22	28	18	24	19	21
Shetland, Lerwick	4	4	B	Low cliff	54	44	70:	52	61	50	57	48	47	43	40-	40-	44	45	53	45	52	49
" Dunrossness	8	4	B	Flat ridge	112	48:	57	63:	58:	51	52	48-	49	60	57	52	51	50	48-	49	50	53
Stornoway	8	4	B	Sloping shore of bay	51	50	49	48	48	48	50	58:	52	53	50	50	46-	51	57	53	54	51
Castlebay	8	4	B	Rocky island	37	72	73:	60	39	55	66	51	44-	46	48	56	56	57	67	70	68	58
Malin Head	8	4	B	Headland knoll	208	56	69	78:	54	64	58	56	52	54	48	47	45-	45-	48	48	53	55
Blacksod Point	8	4	B	{ Low shore / Low ridge }	37 / 330	58	57	55	47	55	42	39	38	37-	40-	41	54	54	60:	59	57	55
Cahirciveen	8	4	45	Mouth of glen	30	58	45	59	47	53	42	39	38	37-	40-	41	54	54	60	59	57	50
Holyhead	8	4	B	Flat island	15	85:	68	51:	57	55	37	34-	49	51	61	51	48	60	69	79	87:	59
St Ann's Head	8	4	B	Level headland	140	64	61	59	63	55	65	59	56	54	55	57	61	64	63	66:	59	60
St Mary's, Scilly	8	4	32	Hilly island	118	74:	63	73:	59	67	70	59	47	51	45-	49	49	54	63	65	61	58
Portland Bill	8	4	B	Low point of headland	19	65	68	69:	80:	68	56	53	53	48-	51	56	59	64	67	71	73	63
Dungeness	8	4	B	Spit of shingle	21	62	64	61:	68	61	55	61	52	56	52	57	51	49-	49-	67	62	58
Aberdeen	8	4	74	College roof	46	35	45	43	57	51	45	43	35	36	32	32	31-	39-	46	54	43	43
Spurn Head	8	4	40	Spit of sand	26	80:	85:	64	72	65	72	64	54	49-	49-	49-	53-	59	76	76	73	65
Yarmouth	8	4	40	Spit of sand	13	72:	62	66	72:	62	51	54	63	38	37	36	34-	39	43	46	43	51
Paisley	8	4	B	Inland station	—	37	33	35	40	47	33	28-	35	33	35	36	36	41	36	42	47	38
Carnforth	8	4	B	"	—	30	37	35	40	43:	39	28	25-	30	35	31	31	31	36	43:	—	33
Belvoir Castle	8	4	B	"	—	50	59:	59:	44	50	58	54	50	44	36-	37	37	31	40	43:	36-	43
Geldeston	8	4	B	"	—	30	30	32	—	42:	31	61:	31	29	26	29	29	29	29	30	21-	30
Woburn	8	4	B	"	—	49	49	54	57	40	45	45	42	31-	43	43	44	49	50	49	46	46

The relation of the surface-wind W to the geostrophic wind G for different orientations has been examined for a number of localities:

For Falmouth and the neighbouring Pendennis Castle[1], for Pyrton Hill and Southport[2], for Dungeness, Jersey, Holyhead and London (Brixton and Westminster)[3], and recently figures have been obtained, for eight years of observations (1908–1915) at 7 a.m., for fourteen of the telegraphic reporting stations of the Meteorological Office. The comparisons have been made in the first two sets for a miscellaneous assortment of wind-forces but in the others for selected wind-forces. Two stations, Holyhead and Dungeness, appear in Mr Fairgrieve's selection as well as in the larger group of fourteen included in the inquiry at the Meteorological Office and it will be noted that the average percentage of the geostrophic wind is different for both stations, for Holyhead 48 per cent as against 59 per cent, and for Dungeness, 52 per cent as against 58 per cent. The observations are for different groups of years and the differences may perhaps be attributed to changes of observers at the stations. With few exceptions the same kind of difference for different orientations is displayed in both sets of results. The results for these various investigations are given in Table I. At some of the stations an anemometer is in operation which has been used either directly or indirectly in the estimations of the velocity of the wind. In these cases the height of the vane of the anemometer is given in a special column of the table. In the other cases, indicated by the letter B in the same column, the velocities of the wind have been derived from estimates in the Beaufort scale converted into velocities by the table given in vol. I.

The hour of observation, 7 a.m., is selected because for that hour the most complete map is available for the determination of the gradient, a dominant consideration; but a reference to the table of diurnal variation of wind shows that it is not a good hour if the observations of all months are to be combined. Seven o'clock is before sunrise in the extreme winter months and some hours after sunrise in the extreme summer months. The average for that hour therefore gives a composite result which will ultimately require further analysis. It cannot give either the maximum or minimum value, but allowing for one hour's lag in the temperature it may give a serviceable approximation to the mean value for the whole year.

The figures which are given in Table I make no reference to the deviation of the direction of the wind from the run of the isobars. On Taylor's theory the deviation is related numerically to the ratio W/G. The deviation is given in the original results for Pyrton Hill and Southport (see figs. 17 and 18) and also for the fourteen stations of the Meteorological Office.

The general meaning of the figures which are given in Table I can be most easily conveyed by plotting those for each station on a diagram.

[1] Shaw, *Forecasting Weather* (Constable and Co.), 1911, p. 48.

[2] J. S. Dines, *Fourth Report on Wind Structure.* Advisory Committee for Aeronautics.

[3] J. Fairgrieve, 'On the relation between the velocity of the gradient-wind and that of the observed wind at certain M. O. stations,' *M. O. Geophysical Memoirs*, No. 9. M. O. 210 i, 1914.

FIG. 17 (II, 1). GEOSTROPHIC AND SURFACE-WINDS AT PYRTON HILL.

Geostrophic wind-direction	WNW	NW	NNW	N	NNE	NE	ENE	(1)
Deviation α	42°	45°	34°	29°	24°	11°	13°	(2)
W/G, calculated % ...	7	0	27	39	50	79	75	(3)
W/G, observed %	49	44	40	39	38	48	48	(4)

Mean velocity 42·6 per cent of geostrophic wind.

Mean deviation of wind from geostrophic 32°.

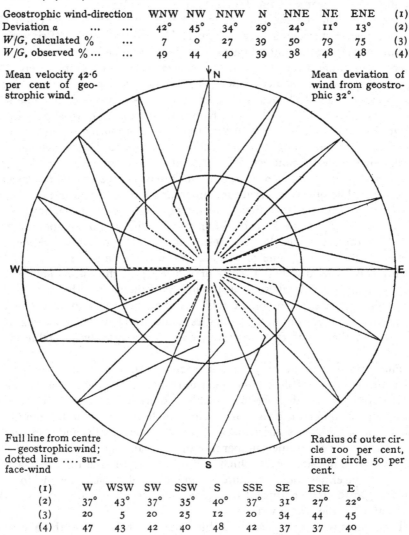

Full line from centre — geostrophic wind; dotted line .... surface-wind

Radius of outer circle 100 per cent, inner circle 50 per cent.

(1)	W	WSW	SW	SSW	S	SSE	SE	ESE	E
(2)	37°	43°	37°	35°	40°	37°	31°	27°	22°
(3)	20	5	20	25	12	20	34	44	45
(4)	47	43	42	40	48	42	37	37	40

The velocity of the geostrophic wind is set out as a circle, the radii of which show the orientation of the geostrophic wind. The percentage which the surface-wind for each orientation bears to the geostrophic wind is indicated by a dotted line from the centre to a point within the circle. The point is placed so as to indicate the deviation of the surface-wind from the direction of the isobar. The extreme points of the dotted line are connected with the geostrophic winds, with which they correspond, by lines which represent the vector-differences of the two. The diagram may be called a geostrophic wind-rose.

FIG. 18 (II, 2). GEOSTROPHIC AND SURFACE-WINDS AT SOUTHPORT

Geostrophic wind-direction	WNW	NW	NNW	N	NNE	NE	ENE	(1)
Deviation α	46°	45°	41°	36°	31°	30°	29°	(2)
W/G, calculated % ...	—	—	10	22	34	37	39	(3)
W/G, observed % ...	46	45	41	69	60	53	49	(4)

Mean velocity 51·6 per cent of geostrophic wind.

Mean deviation of wind from geostrophic 44°.

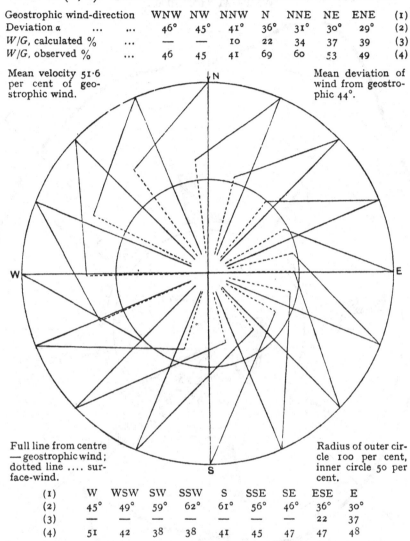

Full line from centre — geostrophic wind; dotted line .... surface-wind.

Radius of outer circle 100 per cent, inner circle 50 per cent.

(1)	W	WSW	SW	SSW	S	SSE	SE	ESE	E
(2)	45°	49°	59°	62°	61°	56°	46°	36°	30°
(3)	—	—	—	—	—	—	—	22	37
(4)	51	42	38	38	41	45	47	47	48

Note. The calculation of the ratio W/G, of the surface-wind to the geostrophic wind, from the deviation α is given in chap. IV of this volume. It fails altogether when the deviation is more than 45°. It is noteworthy that the deviation of 45° is equalled or surpassed at Southport for the orientations on the S and W side of the line, from SE to NW, approximately transverse to the coast-line. The maximum deviations are for geostrophic winds between S and SW. The results point to refraction of the isobars, most pronounced with winds along the coast from the open sea as shown on p. 119. Similar conditions favour the oscillations shown in fig. 4, p. 31.

FIG. 19 (II, 3). RELATION BETWEEN GEOSTROPHIC AND OBSERVED SURFACE-WINDS AT HOLYHEAD.

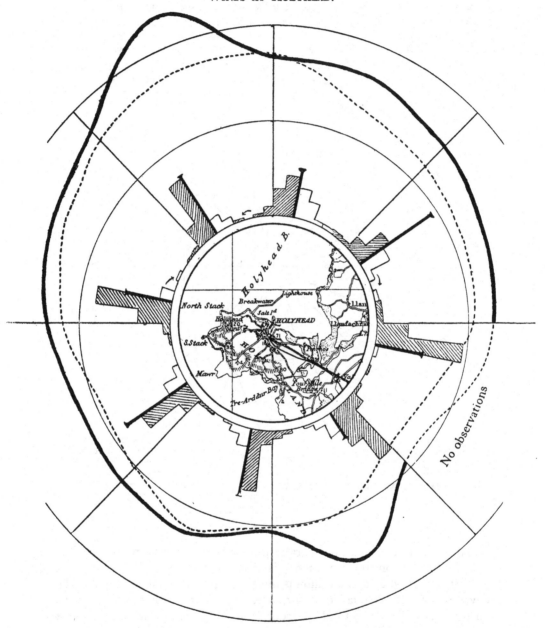

The outer circle represents 100 per cent. of the geostrophic wind, and the inner circle 50 per cent. The full line represents the ratio W/G for winds of force 6 on the Beaufort Scale; the dotted line the ratio for winds of force 4. For the explanation of the columns representing the deviation of the direction from the isobar, see p. 105.

Radius of Map = 4 miles.

The diagrams are necessarily somewhat complicated because so many items have to be expressed. A simpler form of geostrophic wind-rose showing only the ratio of W/G without any indication of the deviation of direction is used for representing the results for Falmouth and Pendennis Castle in the work referred to, and also by J. Fairgrieve in his Report. In these cases a continuous line is drawn connecting the points representing the percentage of the geostrophic wind for the several orientations of the geostrophic wind which shows at once the comparative values of the relation for the orientations of the site.

In the work on the rest of the stations included in the table endeavour has been made to present all the facts to the reader by giving the percentage of the geostrophic wind for different orientations in the form of a continuous line drawn within a circle representing the geostrophic wind and surrounding a reproduction on a reduced scale of a circular portion of the ordnance map of the district on the scale of four miles to an inch, which gives the reader an idea of the features of the geographical relief of the land in the immediate neighbourhood of the observing stations. Information as to the deviation of the surface-wind from the geostrophic wind is given in the form of columns representing percentage frequencies of deviations of given amount arranged in groups, for steps of two points, on either side of a middle group embracing four points, namely two points of veer and two of back. The percentage number of cases in this central group is indicated by the length of a pin-shaped mark on the scale of one inch of length (shown by the distance between two consecutive circles in the figure) to 50 per cent. The percentage frequency of the other groups is shown on the same scale by the length of the respective columns, the shaded columns indicating the surface-winds which are *backed* from the geostrophic wind and the unshaded columns those which are *veered*. The combination of a veer and back of two points within the central group may be held to give an artificial prominence to that group, but the classification in these cases with estimated winds is not very precise and the veering of the surface-wind from the geostrophic wind can only be looked upon as due to some local peculiarity in the determination of the wind. Hence the portion of the diagram on the right-hand side of the central pin may be regarded mainly as an indication of the peculiarity of the site in that respect. In these frequency diagrams winds of all forces are included.

The percentage relation of the velocity to the geostrophic wind, for surface-wind estimated as force 4, is given by a dotted line and the corresponding results when the surface-wind is estimated as force 6 (which are not included in Table I) by a full line. The diagrams for Holyhead and Yarmouth are reproduced here. Regarding for the moment the results for force 4 the former shows very notable difference between the small percentages for winds from the south-eastern region coming over the Welsh mountains a long way off, and the winds from the open sea in the opposite quarter. At Yarmouth the same kind of feature is characteristic of the results but in this case it is the easterly side which gives winds with a closer approximation to the gradient as compared with the winds in the western quadrants which come overland.

FIG. 20 (II, 4). RELATION BETWEEN GEOSTROPHIC AND OBSERVED SURFACE-
WINDS AT GREAT YARMOUTH.

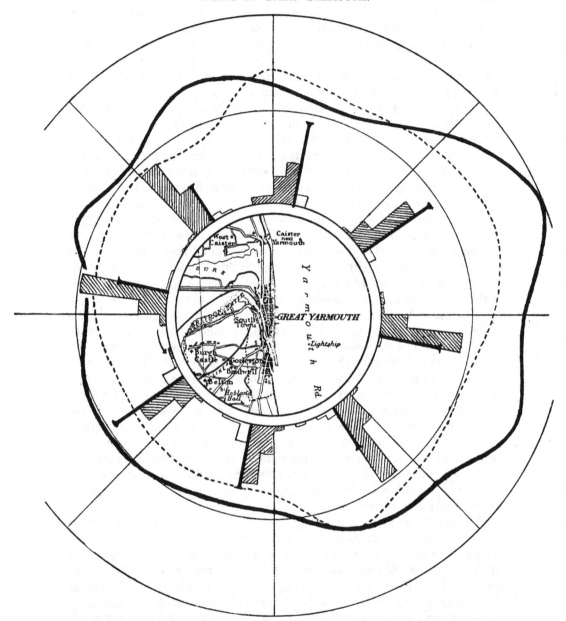

The outer circle represents 100 per cent. of the geostrophic wind, and the inner circle
50 per cent. The full line represents the ratio W/G for winds of force 6 on the Beaufort scale;
the dotted line the ratio for winds of force 4. For the explanation of the columns representing
the deviation of the direction from the isobar, see p. 105.

    Radius of Map = 4 miles.

The results for force 6 are rather disturbing. At Holyhead, contrary to expectation, with a single exception they show a greater percentage of the gradient than those of force 4. At Yarmouth the excess is less marked though it exists for many directions. There are not sufficient examples of force 6 at Holyhead to complete the circle and the numbers of observations for other orientations are not enough for a satisfactory conclusion; but when the diagrams for the other stations are examined there is more unanimity of opinion in favour of a higher percentage with force 6 compared with 4 than can be explained by lack of observations. The explanation probably lies in a tendency on the part of the observers to over-estimate the stronger winds and this tendency seems also to explain the high estimate of wind from the NNW at Holyhead which goes actually beyond the equivalent of the gradient.

We have already pointed out that in dealing with the relation of surface-wind to the gradient we have to rely upon inadequate measurements for lack of better. As the material of observation improves, by the introduction of suitably exposed anemometers and in other ways, it will become possible to revise the crude results which at present are all that we have to offer. [A contribution to the subject has been made already by S. N. Sen[1], who discusses the relation to the gradient of the records at Deerness, Holyhead, Great Yarmouth and Scilly.] At the worst we are better off than our colleagues who have to deal with the winds in the interior of continents, for which the distribution of pressure as plotted on the map refers to a hypothetical region at sea-level far below the position where the observations of wind are made.

In general, with regard to the relation between the surface-wind and the gradient over the land we may conclude from the considerations set out in this chapter that there is no ground for surprise if, on any occasion, the surface-wind when expressed as a fraction of the geostrophic wind be found to range a long way from unity; and large divergence in direction is equally possible while the upper wind still keeps within the first law of isobaric motion. The effect of friction makes no provision for surface-wind velocities in excess of the geostrophic wind. The suggestion may be put forward that some at least of the cases that occur are those in which the katabatic effect adds to the velocity of the surface-wind by a gravitational flow in the direction of the isobar, and this possibility should be borne in mind because it may prove to be an explanation of certain local winds of considerable violence such as the bora of the Adriatic, the violent winds in the fiords of Norway and Iceland which may be noticed sometimes in the charts of the *Daily Weather Report*; also the winds on the east and west coasts of Greenland[2] and the extremely violent gales experienced on the shores of the Antarctic by the expedition under Sir Douglas Mawson[3]. (Vol. II, p. 255.)

[1] *Geophysical Memoirs*, No. 25, M.O.254e, London, 1925.
[2] W. H. Hobbs, 'The Rôle of the Glacial Anticyclone in the Air Circulation of the Globe,' *Proc. Amer. Phil. Soc.*, vol. LIV, No. 218, 1915.
[3] Sir Douglas Mawson, D.Sc., B.E., *The Home of the Blizzard*, chap. VII. Heinemann, 1915.

THE RELATION OF THE SURFACE-WIND TO THE GRADIENT
OVER THE SEA

The available observations at sea are exclusively estimates according to the Beaufort scale and very few direct comparisons have been made between the wind and the corresponding gradient. The best prospective material for the purpose is perhaps that contained in two of the publications of the Meteorological Office, namely, the Synchronous Weather Charts of the North Atlantic and adjacent continents, 1 August, 1882, to 3 September, 1883, which were published in 1886, and the Synchronous Charts of the North Atlantic illustrating the stormy period of the winter of 1898–99, which were published in 1900 but are now out of print[1]. An inspection of these charts, as of any other synchronous charts of any part of the ocean, confirms the view that close isobars are accompanied by strong winds, but so far as is known no direct comparison has been made between the numerical values of the wind-velocity and the gradient. It should be noticed that in these charts, as in all the charts issued by the Office before 1911, the values of the pressure are not corrected for the variation of latitude; but this makes little difference for the middle latitudes of the North Atlantic, which offer the best examples for comparison, because they are not far from the datum latitude of 45°.

The maps of the regions of the British Isles which are included in the *Daily Weather Report* and from which geostrophic winds for certain regions are regularly computed do not afford a satisfactory measure of the gradient over the sea on the west, because the isobars cannot be carried beyond the western shores[2]. The land-locked seas, the North Sea, the Irish Sea and the English Channel can be bridged by isobars drawn according to the pressures on either side. In those cases however the span of the bridge is rather long and observations from those seas are comparatively rare. A series of such observations for the North Sea has, however, recently become available. They are made according to ships' time in accordance with the rules for observations at sea and consequently refer to 8 a.m., noon, 4 p.m. and so on, whereas the maps to which they must be referred are for 7 h or 13 h or 18 h. Hence the comparison is at the best somewhat vague. Accordingly, in dealing with the comparison of observations with computed gradients, the method of frequencies has been employed and the *mode* or group of maximum frequency has been taken as giving the most probable value of the relation of surface-wind to gradient.

In the examination of the comparison at the Meteorological Office[3] the winds of different strength have been considered separately, light winds with

---

[1] The publication of Daily Synchronous Charts of the Atlantic was begun by Captain Hoffmeyer, Director of the Danish Meteorological Institute, in 1873 and has been continued by that Institute in co-operation with the Deutsche Seewarte, Hamburg.

[2] Since this paragraph was written the transmission of wireless reports from ships in the Atlantic has been reorganised and isobars are now regularly carried beyond the western shores, but not in sufficient detail as a rule to check the relation of wind to the distribution of pressure.

[3] *Report to the Hydrographer*, M. O. No. 4977, 13 April, 1918.

geostrophic velocity below 8·5 m/sec, moderate and strong—with geostrophic velocity between 8·5 m/sec and 18 m/sec, and very strong when the geostrophic velocity exceeded 18 m/sec. The observations have also been grouped separately for direction and velocity. For direction the groups have been made according to the number of "points" in the veer of the geostrophic wind from the surface-wind, a point being 11¼°; and for velocity according to the ratio of the surface-wind to the geostrophic wind, the limits selected for the ratios being ·24 to ·36, ·36 to ·48, ·48 to ·60, ·60 to ·72, ·72 to ·84, ·84 to ·96 and ·96 to 1·08, so that the mean values for each group range approximately about ·3, ·4, ·55, ·65, ·8, ·9 and 1. The winds for the different quadrants have also been dealt with separately. The results for the moderate and strong winds are given in the following table.

RELATION OF SURFACE-WIND TO THE GEOSTROPHIC WIND (BETWEEN 8·5 m/sec AND 18 m/sec) OVER THE NORTH SEA[1].

A. Direction: Percentage frequency of points in the veer of the geostrophic wind from the surface-wind.

Deviation in points ...	−4	−3	−2	−1	0	+1	+2	+3	+4	+5	+6	+7	
	%	%	%	%	%	%	%	%	%	%	%	%	
NW quadrant	0	1	0	2	21	32	**33**	14	1	0	0	0	(104)
SW ,, ...	0	2	4	0	23	**30**	19	14	7	2	2	0	(103)
SE ,, ...	2	2	2	2	16	**29**	16	19	9	1	1	0	(99)
NE ,, ...	0	2	2	0	2	**31**	27	22	3	0	3	3	(95)

B. Velocity: Percentage frequency of the ratios of surface-wind to geostrophic wind within assigned limits.

Limits of ratio	·24–·36	·36–·48	·48–·60	·60–·72	·72–·84	·84–·96	·96–1·08	
NW quadrant	1	5	16	**38**	29	10	0	(99)
SW ,,	8	16	**34**	29	12	2	1	(102)
SE ,,	13	19	**26**	18	18	2	4	(100)
NE ,,	9	20	9	**29**	20	6	9	(102)

Note. The differences between the total numbers in each row and 100 arise from the casting up of fractional percentages in the several cases. The "modes" are shown by thick type.

In considering these results we have to remember the limitations of the comparison, the difference of the time of observation, the liability to error in the estimation of the direction and force of the wind at sea, the uncertainty in the determination of the local gradient and the absence of any allowance for the curvature of the isobar. To these causes we may attribute the great width of the range of the relation both in direction and velocity. We may remark that we are prepared to accept positive deviation from the isobars up to 8 points or 90° but we have no reason to give for negative values of the deviation of the geostrophic wind from the surface-wind and further inquiry is needed before

[1] See H. Jeffreys, *Proc. Roy. Soc.* A, vol. XCVI, 1920, p. 233.

a definite opinion can be formed of their reality. Nor can we, at present, give any satisfactory reason for surface-winds which are estimated to be of the full velocity of the geostrophic wind or beyond it, of which the easterly quadrants seem to afford a substantial number of examples. We find everywhere a tendency for observers to form high estimates of winds in the north-east quadrant, so the occurrence of higher estimates in that quadrant is not surprising, but there is not sufficient material to form a satisfactory opinion as to the true meaning of that experience. We know that the force of gravity upon any body moving westward is greater than that upon the same body moving eastward because the centrifugal effect due to the rotation of the earth is diminished in the one case and increased in the other, but the difference is hardly large enough to show in a rough investigation of this kind.

But taking the figures in the table as we find them we may note that for the winds in the north-west quadrant the frequencies group themselves with a fair approach to symmetry round a mode of about ·67 of the geostrophic wind for velocity, with a veer of about $1\frac{1}{2}$ points or, say, 18°; and a similar statement holds for the winds in the south-west quadrant with a mode of about ·55 of the geostrophic wind and a veer of one point or 11°. No such approach to symmetry of arrangement is shown for the two eastern quadrants. The winds in the north-east quadrant show two distinct modes with ratios of about ·4 and ·7 and perhaps a third about unity, and the deviations are divided between one point, two points and three points. Those in the southeast quadrant show a mode with a ratio of about ·55 with the definite suggestion of another at ·8 and a hint of a third about unity, while in the table for direction there are modes at one point (11°) and three points (34°).

There is not enough material here upon which to found a theory but it is not out of place to remark that the conclusions which the tables suggest are what we might fairly expect from the considerations which have been set out in our discussion of the influence of surface-temperature upon the relation of the wind to the gradient over the land. We have seen that when the surface is relatively cold and is therefore absorbing heat from the air which passes over it, as in the night-hours, the ratio of the surface-wind to the geostrophic wind is diminished; whereas, on the contrary, when the surface is relatively warm and is therefore supplying heat to the air which is passing over it, as in the day-hours, the ratio of the surface-wind to the geostrophic wind is increased. In the open sea there is no appreciable diurnal variation of temperature of the water which forms the surface and consequently no diurnal variation in the relation of the wind to the gradient, but instead of that we have more or less permanent differences of temperature between the water surface and the body of air which flows over it which must have their effect upon the relation of the surface-wind to the geostrophic wind.

Looking at the distribution of isotherms of the water in the North Sea[1] we may conclude as a general rule that winds in the north-west quadrant generally pass from colder to warmer water and winds in the south-west quadrant from

---

[1] *The Weather of the British Coasts*, M. O. Publication, No. 230, chap. XII, 1918.

warmer to colder water, and hence the winds of the north-west quadrant ought to approach the geostrophic wind more nearly than the winds of the south-west quadrant. This may account for the difference of the mode of ·67 of the geostrophic wind for the north-west and ·55 for the south-west quadrants, though the difference between the deviations from the isobars being 18° for the north-west and 11° for the south-west would still require explanation.

The same ideas give some clue to the apparently erratic behaviour of the winds in the eastern quadrants. Winds coming from the north-east have also to pass over the water with its variations of temperature, as likewise have the winds which come from the south-east, but the great land area of the globe lies immediately to the east and south-east of the North Sea and the temperature of the body of air which passes over the North Sea from the eastward is controlled by the land, whereas the body of air which comes from the westward is very little affected by the land if its course is from the north-west, and even if it comes from the south-west it is not so much affected as that which comes from eastern quarters.

The vicissitudes of the latter are known from the experiences of our climate to be extremely varied. From the east we get the hot spells of summer and the cold spells of winter and in all continental countries there is a great range of temperature between day and night. These vicissitudes will have left their mark on the air that is launched from the eastern shores of the North Sea, and hence the relation of the temperature of the air to the temperature of the sea over which it passes will vary on different occasions between the extremes of much warmer and much colder. The relation to the gradient will therefore naturally show, according to the occasion, the one or the other of the modes which we have recognised for the winds from the north and the south in the western quadrants.

Thus we cannot look for complete simplicity of relation of the surface-wind to the gradient even over the sea though in the open ocean the causes of variation may be greatly simplified because the régime of the temperature is free from the complications which affect our land-locked seas.

To complete the information which has been derived from the observations of wind and gradient over the North Sea we may add some notes on the more detailed tables in which the winds of different strengths are treated separately. They are as follows: In the case of the north-west quadrant the winds are at a veer of one or two points and for velocity uniformly in the group about ·65. For the south-west quadrant the groups of different strength all show a veer of one point; as regards velocity the lighter winds have a mode about ·65 and the stronger about ·55; there is one secondary mode for moderate winds about ·4.

In the south-east quadrant the veer is mostly one point, but for strong winds there is a secondary mode at three points; for velocity there are many secondary modes which in three instances are at two groups apart. In the north-east quadrant, for direction the veer increases with the velocity of the wind, for velocity the distribution of the modes is very irregular.

These notes show some suggestion of a closer agreement between the surface-wind and the geostrophic winds with lower velocities as the theory to be discussed later would indicate, but more material is needed before a satisfactory opinion can be formed. An apology is indeed owed to the reader for dealing with so important a subject as the relation of surface-wind to gradient over the sea in so inadequate a manner, but the importance of the subject is itself the excuse for making what use is possible of the material that is at hand in the hope that more may be forthcoming.

SOME PECULIARITIES OF SURFACE ISOBARS

Pressure-distribution at sea-level and wind at the 500-metre level.

For some years it was the practice of the Meteorological Office to notify the direction and velocity of the geostrophic wind at the surface as the best available estimate of the wind at the level of 500 metres or 1500 feet. Subsequently the result of the computation was notified in slightly different form as the probable wind at from 1500 to 2000 feet. The practice originated with a request from the Ordnance Committee for an estimate of the direction and velocity of the wind in the upper air for experimental work at Shoeburyness. Either by accident or design, because wind is only a disturbing element in gunnery, the times selected for the experiments were generally marked by the absence of any notable wind or gradient at the surface. After some years of practice word came that the estimates of wind in the upper air were no longer required. No reason was given, but there is ground for believing that the Ordnance Committee came to the conclusion that the method of determining the wind-velocity by the gradient had been tried and found to be not more satisfactory than guessing the upper wind by the traditional practice of applying a formula to the wind observed at the surface.

It is a striking example of the manner in which scientific inquiry should not be conducted because there was no exchange of views or of experience between those who asked the question and those who gave the answer. The question whether in ordinary circumstances the sea-level gradient is or is not a guide to the wind at the level of 500 metres or thereabouts remained exactly where it was, and a good deal of time was spent to no useful purpose because the opportunity for using the scientific method of checking calculations by results and examining the occasions of serious discrepancy was lost.

The observations with pilot-balloons, which have been set on foot at many stations since the inquiry referred to, enabled us to take up the question again in much more favourable conditions, and the inquiry is still necessary because with all the experience of the intervening ten years the geostrophic velocity is still in ordinary circumstances the best suggestion that the Meteorological Office has to offer for the wind at 1500 to 2000 feet for any locality where a direct observation of a pilot-balloon with two theodolites or its equivalent is not available.

We propose accordingly to give some attention to this particular question.

Many desultory comparisons of the wind at 1000 ft or 2000 ft, or at 500 metres, with the geostrophic wind at sea-level have been made in the Office with the general conclusion that while the agreement is good on the whole, it is better if the mean gradient over a considerable area is estimated and the means of a number of observations with pilot-balloons at the selected levels are taken for the comparison. The difference is illustrated by the following figures supplied by J. S. Dines and E. V. Newnham:

Correlation between gradient-wind at sea-level and observed wind	at 2000 ft	at 3000 ft
For single station (E. V. N.) (72 observations in the winter of 1916–17)	0·67	0·72
For three or more stations (J. S. D.)	0·76	0·82

The ratio of the mean of the winds, at the levels specified, to the mean of geostrophic winds of the isobars in Newnham's comparison is ·90 in each case; an allowance for the curvature of the isobars would make the ratio still nearer to unity. But even with this modification there are occasions on which the observed winds are not in agreement with the computed winds. To furnish a reply to the definite question whether the winds as observed by pilot-balloons at certain meteorological stations near the coast were in accord with the gradient, C. J. P. Cave made an investigation for six stations of which one is on the east coast, one on the south coast of England, two are near the east coast of Scotland and two near the coast of north-east England. The results are as follows:

COMPARISON OF THE WINDS OBSERVED AT 1000 FT AND 2000 FT WITH THE GEOSTROPHIC WIND AT SEA LEVEL: MEAN RESULTS FOR SIX STATIONS ON THE EAST OF GREAT BRITAIN FOR JANUARY 1917.

Direction. Deviation from the geostrophic wind in "points."

	Backing >4	3 or 4	1 or 2	Less than 1	1 or 2	3 or 4	Veering >4
Percentage frequency of observations at 1000 ft	4·5	24	36	13	14·5	6	2
Percentage frequency of observations at 2000 ft	1·5	12	39	19	18	7	3·5

There is an obvious mode for deviation in the direction of backing of the actual wind through one or two points which is more pronounced at 2000 ft than at 1000 ft.

Velocity. Ratio of the observed wind to the geostrophic wind at sea-level.

	<·7	·71 to ·80	·81 to ·90	·91 to 1·0 1·0 to 1·10	1·11 to 1·20	1·21 to 1·30	>1·3
Percentage frequency of observations at 1000 ft	29	18	14	18	8	4	9
Percentage frequency of observations at 2000 ft	14	11	15·5	26	8	4	21·5

The modes in this case are less pronounced than they are in the case of direction and the high values for the frequency of ratios less than ·7 at 1000 ft and greater than 1·3 at 2000 ft show that the magnitudes included in the comparison were not in strictly comparable form.

Considering that we are dealing with winds in the free air the agreement as regards direction is not good and the irregularities as regards velocity are so notable that no generalisation is possible. The results for individual stations are no more satisfactory than the means for the six.

Irregularities at the coast line

The results draw attention to the fact that there are special causes of irregularity at stations near the coast and the selection of these stations might have been made for the purpose of exhibiting the range of irregularities to which the comparison is liable. It would have been better in the first instance to choose some place of observation which is not affected by these special complications. There is no place on shore which is free from objection. It was thought that the best available selection would be an inland station like Upavon where at least the differences depending upon orientation are less marked than they are near the coast. Of the stations which were available the one which seemed likely to give the best observations was South Farnborough and in consequence the observations for a year at that station were examined in the Meteorological Office by H. Jeffreys, but it was not found possible to classify the observations in such a way as to obtain a satisfactory clue to the régime of the structure of the layer within the first five hundred metres by the statistical arrangements which suggested themselves. The observations used were those for the early morning about 7 h, a time of day when the curve of variation of wind with height as represented in fig. 23, p. 124, is in process of changing and perhaps the vertical component may be irregular. The next step in the inquiry seems to be to try a comparison for the simpler conditions of the open sea as soon as suitable observations can be obtained.

The results for the land-stations are tantalising. There is sufficient evidence of relationship to invite endeavours to reach precision and yet, from causes of which proper account cannot be taken at present by those who have before them only the barometric gradient and certain surface conditions to guide them, there are many individual cases for which the departures are so large that some means of discriminating between occasions of regularity and irregularity is desirable.

Let us now consider some of the reasons for the discrepancies which are thus observed and from which we may feel that observations in the upper air might be free. We need not revert again to the casual uncertainties of the observations of wind, and we do not intend to say more about the determination of the gradient except that, for obvious reasons, the computation of the wind from the gradient as shown on a map has little meaning when the gradient is very slight and irregular or when a barograph with an open scale shows notable embroidery.

We must however remind the reader that the comparison is made between individual estimates of the gradient at the surface and the corresponding observations of a pilot-balloon with a single theodolite and these latter are affected by any vertical component that may happen to be in the wind at the time of observation. From Table II of chap. VI it appears that vertical components are specially frequent within the first kilometre of height at Pyrton Hill. Near the sea-coast irregularities arising from this cause are probably still more frequent. It is also probably on this account that J. S. Dines's comparison with the mean value of a group of neighbouring observations gives a higher correlation ratio than the comparison for single stations.

We start from the position that the geostrophic wind can only be expected to be in agreement with the undisturbed wind at its own level and at the time of the map from which the wind is computed. For comparison with the observed wind at 500 metres the geostrophic wind as deduced from the distribution of pressure at the surface requires correction. The correction is by no means negligible. It depends upon the distribution of temperature and, using the formulae which have been developed in a subsequent chapter (VII), we find that near the surface a horizontal gradient of temperature of $1°$ F per 60 nautical miles or 0·5tt per 100 kilometres alters the computed geostrophic velocity by one mile per hour within 1000 feet of height. A gradient of temperature of that amount from south to north will add one mile per hour to the component velocity from the west for 1000 feet of elevation and the same temperature-gradient from west to east will make the same addition for the same height to the component of the geostrophic wind *from the north*. In the commoner units of this book the effect of horizontal gradient of temperature upon the geostrophic wind is as follows:

Horizontal gradient of temperature		*Increase* upward of geostrophic wind-velocity
1tt per 100 kilometres	...	0·3 m/sec per 100 metres
From W to E ...	...	Component from N
From S to N ...	...	Component from W

These changes in the components of velocity will appear as large percentages of the geostrophic wind computed for the surface if that wind itself is light. If the gradient of temperature has the same orientation as the gradient of pressure and is large enough to give a steeper slope to the isothermal surface than that of the isobaric surface, the geostrophic wind-velocity will *increase* with height without change of direction and, on the other hand, if the orientation of the gradient of temperature is opposite to that of pressure, the geostrophic wind will *decrease* with height without any change of direction; but if the orientation of the gradient of temperature is inclined to the gradient of pressure at any finite angle, the direction as well as the velocity of the gradient-wind will change with height.

8-2

The extent to which this correction might affect the comparison between the observed wind at 500 metres and the gradient-wind is not easily determined. A fair estimate for stations within the general area of the British Isles may be formed from the maps of the normal distribution of maximum and minimum temperature over the area in the *Book of Normals*[1]. In January there is an average gradient of maximum temperature from Scilly to Cromer in Norfolk of $\frac{3}{4}°$ F in 60 nautical miles, rather steeper at the south-western end of the line; for the minimum, or night temperature, the gradient is irregular in the midland and eastern counties but over Wales there is a gradient of about 12° F in 60 nautical miles which would mean a correction of 12 miles an hour to the computed value of the geostrophic wind. In July, on the other hand, the gradients of temperature at night are generally slight and irregular over the inland counties with a fringe of steep gradient from the coast inland amounting to more than 15° F per 60 nautical miles on the east coast; and in middle day, according to the distribution of maximum temperature, there is a large area of high temperature over the hinterland of England fringed by steep gradients towards the coast-lines which is again most pronounced on the eastern side and amounts to more than 15° F per 60 nautical miles on the Norfolk and Kent coasts.

The coastal regions, especially those on the eastern side, are thus indicated as peculiarly liable to large corrections to the computed value of the gradient-wind and this applies to the belt ten or twenty miles broad along the coast as represented on a map drawn to a small scale. If we go into detail, the gradient would certainly be found to be very pronounced along the extreme fringe of the coast. The steepness of the gradient in some cases is indicated by a diagram in the Meteorological Office representing a section drawn from west to east across the British Isles showing temperature and sunshine during cloudless weather, 24–30 March, 1907[2]. It shows an average difference of mean maximum of 13° F, and an extreme difference of 17°, between Geldeston near Beccles, six miles from the coast, and Yarmouth, which would probably work out to give a gradient from Geldeston to the coast at the rate of 130° F or 170° F per 60 nautical miles. We are not entitled to suppose that these differences exist throughout the vertical range of 500 metres; if they did, the correction to the computed geostrophic wind at a station on or near the coast would be prodigious. But its extension in a modified degree upwards will easily account for a considerable difference between the barometric gradient at the surface and the gradient with which the wind obtained by observations of a pilot-balloon ought to be compared.

These large local differences of temperature are usually called upon to account for sea-breezes which are not amenable to the surface-gradient, but it will be seen that in accepting an explanation in that general form we should be leaving the local details of the gradient out of account.

---

[1] Specimens of these maps for January and July are given in *The Weather Map*, M. O Publication, No. 225 i, 1917, p. 88, and maps for each month in M. O. 236 (§ III).
[2] *Forecasting Weather*, 1911, p. 285.

The "refraction" of isobars

There is moreover another reason why a station on the coast presents a complication in the relation of observed wind to gradient which may be operative in windy weather when the local gradient of temperature is not very marked. This second reason is the dynamical effect upon the stream of air due to the sudden transition between a surface with a comparatively low co-efficient of eddy-viscosity, such as the sea, and one with a comparatively high coefficient, such as a land-surface, particularly a hilly or rugged land-surface. This change must probably be represented by a sudden transition of pressure in the surface layers which produces a "refraction" of the isobaric lines on crossing the coast. We have already referred to it in a note on the diagram on p. 103, and we have mentioned that this view is borne out by the fact that there is a difference of pressure between coastal stations and inland stations, the inland pressure being the higher[1]. Objection has been taken to the view that an increase of pressure over the land can be due to the partial arrest of horizontal motion of the air by the land on the ground that in a "wind-channel" there is no increase of pressure upon the surface of a disc exposed to a current in its own plane. It is not clear that the objection holds in the case of wind over land. Even if the exaggeration of scale makes no difference to the argument, the relief must cause local differences of pressure and at the coast presumably an increase for on-shore winds; and the mere addition of the volume of the land to that of the air which passes over it must produce some increase of the pressure at sea-level.

A form of refraction of isobaric lines in line squalls[2] has been dealt with in papers which have come from the Meteorological Office. In these cases we see on the map a set of nearly straight isobars running in one direction connected abruptly with another set of nearly straight isobars running in another direction. On the map it has been customary to mark the junctions of the lines as simple discontinuities of direction and previously they were rounded off by smooth curves. At the junction there is really a curious dislocation corresponding with the crochet d'orage as represented in the carefully drawn diagrams of the papers referred to, some of which are reproduced in *Forecasting Weather*[3]. In these cases, which correspond with single refraction, the velocity normal to the line of separation of the two fields is maintained though the velocities on either side of the boundary are different. The boundary itself is a locality of great vertical motion of air generally represented by a squall and a shower of rain or hail at the surface.

The discontinuity of direction of isobars is an important feature in the Norwegian analysis of the phenomena of cyclonic depressions. A number of

[1] See a paper by Gold, 'Comparison of Ship's Barometer Readings, etc.,' *Q. J. Roy. Meteor. Soc.*, vol. XXXIV, 1908, p. 97.

[2] R. G. K. Lempfert, 'The Line Squall of Feb. 8, 1906,' *Q. J. Roy. Meteor. Soc.*, vol. XXXII, p. 259. R. G. K. Lempfert and R. Corless, 'Line Squalls and Associated Phenomena,' *ibid.*, vol. XXXVI, 1910, p. 135.

[3] *Loc. cit.*, 1911, pp. 239–243.

examples are given in an exposition of the method of analysis in the Meteorological Office by J. Bjerknes, which is now published as a *Geophysical Memoir*[1].

The discontinuity in the isobaric lines is associated with sudden change of wind and corresponding change in the temperature of the air, the sequence of transition being generally from warm air to cold air.

The advancing cold air is introduced by a cold front at the surface of discontinuity. Sometimes the transition is by a single step; but not infrequently by a succession of secondary steps. The memoir is illustrated by a series of maps and diagrams which explains the attitude adopted.

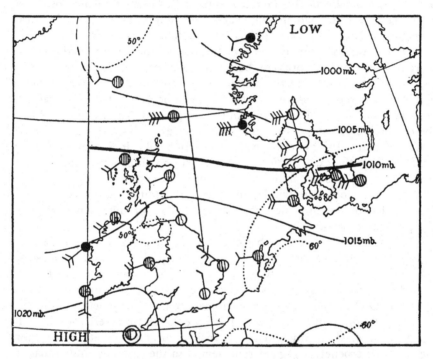

Fig. 21 (VIII, 2). Chart of Isobars, Winds, and Isotherms for 7 h from the *Daily Weather Report* for 16 August, 1918.

To illustrate irregularities in the relation between the local wind at coastal stations and the distribution of sea-level pressure for which explanation may be sought in local configuration or other disturbance of dynamic condition. See p. 120.

The full lines are isobars, the dotted lines isotherms.

The arrows fly with the wind. The number of feathers indicates the force on the Beaufort Scale. The small circles represent the positions of the stations: the number of cross bars within the circles indicates the number of quarters of the sky covered by cloud. A black dot shows that rain is falling.

The refraction which takes place at the coast-line is different in that there is, as we may suppose, no permanent change in the direction of the isobars or

[1] *Geophysical Memoirs*, No. 50, M. O. 307j, London, 1930.

their separation. We may suppose the effect to be a dislocation of the isobars giving a figure (fig. 21) somewhat similar to that representing the refraction of light by a plate. What happens in the area represented by the belt of coast is at present undetermined, but certainly some complications of the flow of air result: the cliff-eddy is in certain cases one of them and in hilly country other eddies are recognised.

This is doubtless a highly speculative method of treating the difficult question of the effect of the coast-line on the geostrophic wind but there are certain facts which seem to be in favour of it.

There is the high degree of divergence of the wind from the isobar at Southport noticed on p. 103 which would be accounted for by a deviation of the isobars themselves through a suitable angle. Suitable angles for different orientations are represented in fig. 22, here. If that is the true explanation of the phenomenon, the effect would seem to be most marked at Southport when the wind is along the coast and the land is on the right. That peculiarity may be due in that instance to the special shape of the Lancashire coast. Readers will be familiar with the manner in which the lines of the coast on either side of the English Channel are marked by cumulus clouds on days when a westerly wind blows along the Channel, bringing with it probably air which is tending towards instability before it has to meet with the mechanical interference of the coast-lines.

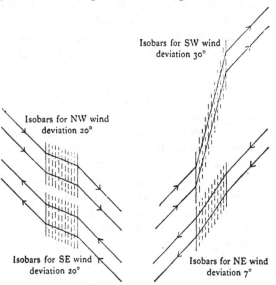

Fig. 22 (VIII, 1). Refraction of isobars at Southport computed from the data represented in fig. 18, chap. III, p. 103.

The transition from land to sea must have its counterpart in the transition from sea to land, but whereas it is easy to think of thermal effects such as the formation of clouds by the piling up of air in consequence of the arrest of its motion on coming from sea to land, no such easy expression of the transition from land to sea occurs to us. In 1918 information came from the north-east coast of a "barrage" which made the manœuvring of airships difficult or even impossible in a westerly wind leaving the land, but the information is not in sufficient detail to indicate how far the barrage may be accounted for by the mechanical effect of the refraction of the isobars crossing the coast nor what part of the effect is due to the enhancement of the eddy-motion by hills in the immediate neighbourhood and the consequent increase in the effective resistance opposed by the air to the travel of an airship. So far as we know no experiments have been made to compare the forces exerted upon an obstacle by an eddy-stream with those of a stream of the same mean velocity without eddies.

We are not yet in a position to apply numerical corrections to the computed geostrophic wind in individual cases on account of the local temperature-gradient or the refraction of the isobars by the coast-line, but both causes must combine in any actual case, and there seems no reason to consider that the discrepancies actually noted between the observed wind at 500 metres and the geostrophic wind at the surface are due to any real finite difference between the wind in the free air and the geostrophic wind for the same level.

Even when all allowances are made there remains the difficulty of drawing isobars with sufficient precision. We give (fig. 21) an example of the map for 7 h on 16 August, 1918, in which the adjustment of isobars to winds is almost inexplicable as it stands and may well be offered as an exercise for the student.

In passing we may, however, remark that in what precedes it has been usual to assume that the observed wind might be expected always to be less than that corresponding with the gradient. An observation showing wind in excess of the gradient would have been taken as due either to katabatic influence or possibly to an error of observation. If, however, we take the view of the development of pressure-distribution which is set out in chap. XI, a wind in excess of the gradient becomes the first stage in the creation of a new gradient. In a note by Miss L. D. Sawyer we find examples[1] of wind at 2000 ft (derived from pilot-balloons) more than 50 per cent beyond the geostrophic limit, on seven occasions in 1918.

The whole subject depends very largely upon the question of the manner of creating wind-velocity, with regard to which at the moment, apart from recognising the importance of radiation and consequent entropy, water-vapour, slope and the rotation of the earth, we have no suggestion to make.

[1] *Daily Weather Report*, 1918, 2 April, 16 June, 4 July, 9 and 22 August, 21 November, 11 December.

CHAPTER IV

THE CALCULUS OF SURFACE TURBULENCE

Diffusion according to the law of viscosity is typical of the vertical adjustment caused by wind in the horizontal flow of air, in the distribution of entropy (or potential temperature), and of water-vapour.

THE VARIATION OF WIND WITH HEIGHT IN THE SURFACE LAYERS

By way of giving a practical basis to the ideas which we propose to develop in this chapter let us set ourselves to face the question "Is it possible, from a measure of the wind-velocity by a fixed anemometer or by an instrument held in the hand, or from an estimate on the Beaufort scale on land or sea, to obtain working measures of the wind-velocity in the layers immediately above the observer and if so, up to what heights may the calculation be extended with reasonable accuracy?"

Considering the last part of the question we may recall the conclusion drawn in volume II, p. 389, from W. H. Dines's table of correlation between deviations from the normal pressure and temperature in the upper air, namely, that at all levels from three kilometres to nine kilometres in all seasons of the year the correlation between the deviations of pressure and temperature is very close, generally above ·75 and always positive, whereas at the level of two kilometres the correlation, though still equally high for the winter half-year from October to March, is only of the order of ·5 for the summer half-year. Below that level at one kilometre it is of the order of ·5 for the winter half-year and ·3 for the summer half; and at the surface there is no correlation at all. Anticipating to some extent what follows we may attribute the interference with the direct correlation, in part at least, to the turbulence of the motion in the surface layers caused by the friction with the land or water over which the air passes. The effect of the turbulence, which causes mixture of the layers above with those beneath, diffuses upwards at all seasons of the year, but especially in the summer half when the dynamical effect is exaggerated by the transference of heat from the ground to the air in contact with it. The region indicated by the expression "the surface layers" may therefore be regarded roughly as having a thickness between one and two kilometres in the winter half-year and between two and three kilometres in the summer half-year. Above these levels we may contemplate a separate régime of winds to be treated provisionally as independent of the turbulence due to the surface. We have already seen that the air at the top of the Eiffel Tower is affected by turbulence in the winter and still more so in the summer, so that we must not be surprised to find the influence extending to a kilometre or more.

In endeavouring to form a mental picture of the régime of winds in the lower layers on these lines the natural order would be to start from the undisturbed wind and consider the surface-wind to be connected therewith by a formula which can be represented by a diagram showing the relation of wind with height. It is fair to assume that in such a diagram the wind would have its least velocity at the surface and increase upwards until it reached the undisturbed wind; if the undisturbed wind be assumed to be the geostrophic wind for its own level, further changes at higher levels would correspond with changes in the geostrophic wind at successive levels and such changes, depending on the gradual change in the gradient due to the distribution of density of the air at successive levels, would necessarily be very gradual except in extraordinary meteorological circumstances.

We have already seen that for the same undisturbed or geostrophic wind in the upper levels the corresponding surface-wind will certainly have different values for different times of the day, for different seasons of the year, for different localities in respect of geographical relief, and for different azimuths; consequently a series of curves will be required to connect the undisturbed wind with the surface-wind. The portions of the curves which we can plot from observations in the lowest layers are the tail-ends of curves the special forms of which are dependent upon the particular conditions of the time. It is practically hopeless to suppose that any general formula can be devised to be applied in all cases to give the variation from the surface up to the undisturbed wind and consequently the approach to the solution of the question by defining empirically the relation of the surface-wind to the wind in the successive layers immediately above the surface must necessarily consist of the numerical equivalents of a series of diagrams which cannot be immediately coordinated.

The question is further complicated by the variations in the direction of the wind at different levels which are associated with the observed changes in velocity. On the general principle that the retardation of any layer of moving air destroys the power of that layer to maintain its balance with the distribution of pressure, it is clear that the retarded layers of air near the surface will be deviated from the path of the undisturbed current above them by yielding to the pressure-gradient which their velocity cannot balance, and rearranging themselves with a component of motion towards the side of lower pressure which will be greater the greater the departure of the velocity from the measure required for balance. In ordinary circumstances there is a deviation of some 20° to 30° between the direction of the surface-wind and that of the geostrophic wind, or that of the lower clouds, which is generally in close agreement with the line of the isobars; but in this matter again the results are dependent upon meteorological and local conditions and in some cases the deviation of the surface-wind from the line of the isobars is much greater[1].

At the anemometer

We must look to some general theory, if we can find one, to give us the clue to the co-ordination of all these variations, and in the meantime we will place on record the results of observation as a basis and test of future theory. For

[1] J. S. Dines, *Fourth Report on Wind Structure*, Advisory Committee for Aeronautics. Report No. 92, 1913, p. 19.

the present we will confine our consideration to the variations of speed of the wind with height. We should be glad to begin with the variations of wind near the surface of the sea, but the only direct observations that we can recall are those of G. I. Taylor[1] on the *Scotia*, who found a very irregular ratio of the velocity at 70 ft (21·5 m) to that at 45 ft (13·7 m) with a probable mean of 1·07, which is in close agreement with the scale for anemometers up to 30 metres quoted below.

It is usual to express the variation of wind-velocity at different levels as a fraction of the value at one of the levels. Thus the Meteorological Office[2] gives the following for the wind at various heights above open grass-land up to 30 metres as a fraction of the velocity at 10 metres which is taken as representing the normal height and exposure of the vane of a tube-anemometer.

Height in metres	...	0·5	1	2	3	4	5	10	15	20	25	30
Ratio of velocity of wind to that at 10 m	...	·50	·59	·73	·80	·85	·89	1·00	1·07	1·13	1·17	1·20

And for the layers quite close to the ground observations by G. I. Taylor and C. J. P. Cave (R. and M. No. 296, A. C. Ae. *Report*, 1916–17) gave the following result:

	Height in feet	1·25	4	6
Ratio of velocity to the velocity at 6 ft	over grass 1 in–6 in high	·65	·86	1·00
	over short cropped grass 2½ in high	·71	·87	1·00
	over pond with small waves	·82	·90	1·00

The ratio of decrease appears to be independent of the magnitude of the velocity but dependent upon the nature of the surface. When the ground is flat, the projections being blades of grass or small plants, the decrease is also independent of the direction of the wind, but in the case of ploughed fields or trenches, where furrows run in one direction only, the direction of the wind will make a difference.

In discussing the variations of wind with height obtained in the series of observations with kites at the station maintained for the University of Manchester at Glossop Moor, Miss Margaret White came to the conclusion that the variations could be represented better by invariable additions to the wind recorded at the surface than by an increase proportional to the velocity at the datum level.

The best statement of the actual position is a polygraph of pilot-balloons such as that given for Salisbury Plain by G. M. B. Dobson in fig. 35, p. 162.

The observations on the Eiffel Tower at 305 m as compared with those on the tower of the Bureau Central Météorologique at 21 m above ground give the following results. The velocities are given in metres per second.

	Jan.	Feb.	Mar.	Apr.	May	June	July	Aug.	Sept.	Oct.	Nov.	Dec.	Year
Eiffel Tower	10·2	9·8	9·4	8·7	8·3	7·6	7·4	8·0	8·2	9·3	9·2	9·9	8·83
Bureau	2·4	2·5	2·5	2·4	2·2	2·2	2·0	2·0	1·8	1·8	2·0	2·2	2·17

[1] *Scotia Report*, 1914, p. 65.   [2] *Annual Summary of the Monthly Report*, 1916.

Hellmann[1] has compiled observations over flat meadow-land at Nauen at heights 2 m, 16 m and 32 m above ground from which he confirmed an empirical formula $v = kh^{\frac{1}{5}}$. This is not very different from the formula $v = kh^{\frac{1}{4}}$ suggested by Archibald[2], which was found to be in accord with Vettin's results[3] for the motion of clouds at much greater heights and confirmed by observations with kites. The formula is in good agreement with J. S. Dines's observations, to be referred to later, for moderate winds in the middle of the day; but the increase of velocity is not sufficiently rapid for cloudy weather or for the early morning; and, from the nature of the case, it is probable that a logarithmic formula is more likely to be applicable than one depending on a single power of the height.

Above the anemometer

Passing on to observations for levels accessible only by aircraft or pilot-balloon the most instructive observations for our immediate purpose are those of J. S. Dines[4] carried out at South Farnborough, in October and November, 1912, with which are included other observations made previously at Pyrton Hill. Two theodolites were used in almost all cases. His conclusions are best represented by the diagrams which accompany his report and are accordingly reproduced here.

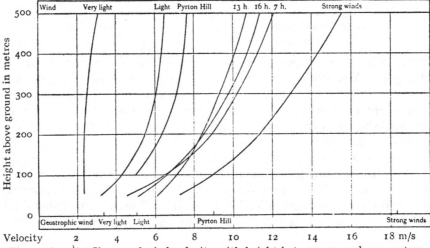

Fig. 23 (IV, 1). Change of wind-velocity with height between 50 and 500 metres.

The numbers of observations upon which the curves are based are not numerous but they show some noteworthy results. The first diagram represents the variations of wind-velocity with height up to 500 metres for different groups of winds selected according to their velocity at 500 metres, namely very light winds when the velocity at that level was 4 m/sec or less, light winds more than 4 m/sec and not more than 10 m/sec and strong winds with

[1] *Meteor. Zeitschr.*, 1915.　　　　　　　　[2] *Nature*, vol. XXVII, 1883, p. 243.
[3] Vettin, *M. Z.*, vol. XVII, 1882, p. 267.　　[4] *Fourth Report on Wind Structure.*

a velocity greater than 10 m/sec at that level. Another curve is added representing the combination of ascents at Pyrton Hill. The geostrophic wind is marked for each of the groups. The shape of the curve for very light winds is peculiar but it tends to confirm the suggestion put forward in the preceding chapter that light winds at the surface tend to show closer accordance with the gradient than stronger winds.

With the original diagram for these several groups of winds we have incorporated the curves of another diagram in the same report which shows the result of grouping the observations according to the time of day for the purpose of disclosing the diurnal variation of wind-velocity. The effect is quite conspicuous. In the early morning at 7 h the velocity is least, but the curve shows greater increase near the ground, so that at less than 100 metres the morning wind becomes weaker than the afternoon wind, marked as 16 h; the midday wind for 13 h, though much the strongest of the three at 50 metres, has a smaller fractional increase with height and becomes the least strong at less than 200 metres. At what level precisely this reversal takes place on any particular occasion is not disclosed, but the reversal is a characteristic phenomenon which is exhibited by the maximum in the night and minimum in the day shown by the records of wind at all high-level stations.

In the same report the variations of wind-velocity with height on cloudy and cloudless days are compared though only for a few occasions. The results show that the velocity increases from the surface more rapidly on cloudy days than on cloudless days and hence cloudiness as compared with freedom from cloud has the same kind of effect as the early morning compared with midday, and both owe their influence to the difference in the conditions of the warmth of the surface relatively to the air above it which affects the turbulence of the surface layer.

These different types of curve representing the variation of the velocity of wind with height should be borne in mind as carrying the key to differences between the various kinds of exposure, and hence to the differences to be expected at different stations.

We shall see later that theoretically the effect of turbulence on geostrophic wind at an upper level is represented by the addition of a component the point of which follows a spiral, so that the change in the velocity of the wind is logarithmic.

In the early days of pilot-ballooning, as recorded in Cave's *Structure of the Atmosphere in Clear Weather*, we thought it to be linear and assumed a uniform rate of increase from a zero value at the ground to the geostrophic wind at about 500 metres. Since then many observations have been co-ordinated[1] showing that the variation is more nearly expressed by geometrical progression than arithmetical.

The precise formula to be adopted is however not a matter of great

[1] E. H. Chapman, 'The variation of wind-velocity with height,' *M. O. Professional Notes*, No. 6, London, 1919; W. J. Humphreys, 'The way of the wind,' *Journal of the Franklin Institute*, 1925, pp. 279–304.

practical importance. A glance at the anemometric records in which we have propounded the problem of dynamical meteorology will be sufficient to show that in consequence of turbulence and the gustiness which expresses it, the air-flow within a hundred metres of the ground is a theoretical ideal not easily related to the actual experience of wind which an object near the ground may have to face.

The formula of velocity proportional to height above sea-level (until the geostrophic velocity is reached) was put forward as a rough working rule in the writer's Report on Wind Structure[1]. A linear formula is however too empirical to find a permanent place in scientific literature when an exponential formula is available, and from what has been said previously in this chapter it is clear that no formula can hold for all hours of the day. One important principle emerged from the study of the approach to the geostrophic wind in different localities, namely that for those localities where the exposure is so good that the surface-wind is near to the geostrophic wind the rate of increase of velocity with height will be slow, whereas in those localities where the surface-wind is a small fraction of the geostrophic wind the rate of increase of the wind from the surface upwards will be rapid. Thus at a station at low level on the coast the wind should increase rapidly from the surface value and attain the geostrophic value at less elevation above the surface than at a high inland station where the rate of approach to the geostrophic wind from the surface will be much slower.

Observations with pilot-balloons in the Isles of Scilly by C. J. P. Cave[2] and J. S. Dines in 1911 show the geostrophic wind at less elevation above the ground than at Ditcham Park on the crest of the South Downs in Hampshire.

Nothing has yet been said about the cases, which will be noticed in any collection of soundings with pilot-balloons[3], in which the velocity of the wind decreases with the height from the surface or from a level not far from it. They are not likely to occur when the surface-wind is in normal relation with the distribution of pressure, but whenever the surface-wind is to be classed as katabatic or anabatic, controlled by gravitational effects at the surface rather than by the general distribution of pressure, the surface-wind may have no relation to the gradient. Such conditions are most likely to occur when the distribution of pressure is very uniform and there is no general gradient to exercise control over the surface-air, and it is to be expected that in these circumstances wind will diminish with elevation and perhaps be reversed at no great height.

On the occasion of a kite competition at Worthing[4] held by the Aeronautical Society of Great Britain on a threatening day in July 1903 the distri-

[1] Advisory Committee for Aeronautics, *Reports and Memoranda*, No. 9.

[2] 'Soundings with pilot-balloons in the Isles of Scilly, November and December, 1911,' *Geophysical Memoirs*, No. 14, M. O. 220 d, London, 1920.

[3] An instructive collection of diagrams representing a series of soundings with pilot-balloons is given in C. J. P. Cave's *Structure of the Atmosphere in Clear Weather* (Cambridge University Press, 1912).

[4] *The Aeronautical Journal*, January, 1904.

bution of pressure afforded very little prospect of wind sufficient to raise kites, but fortunately for the competition a sea-breeze set in which gave a surface-wind sufficient to get the kites off the ground. In ordinary circumstances, once off the ground, the kite's future is assured because the wind increases aloft, but on this occasion when a certain level was reached the kites lay on the surface-current as if it had been a cushion and travelled along it without making elevation, and on that account a very limited height of ascent was possible, hardly exceeding 1800 ft.

It is a well-known experience of meteorologists that winds are sometimes reversed aloft even when they are true to the surface-gradient, and in particular easterly or north-easterly winds (sometimes also southerly or northerly winds, though seldom or never westerly ones) fall off in the upper air and are replaced by winds from an opposite direction or nearly so. The reversal in these cases however seldom takes place below the level of 500 metres, which has been taken as the limit of the surface layers for the purposes of this chapter. It is quite possible that the influence of the surface may extend beyond the level of 500 metres, but it is less dominant there.

Reversals at higher levels are controlled by the regime of temperature and their consideration belongs to chap. VII.

THE THEORY OF EDDY-MOTION IN THE ATMOSPHERE

G. I. Taylor's analysis

It will be apparent from what precedes that, even if the exceptional occasions such as those of katabatic winds are left out of account, the variation of wind-velocity with height requires for its representation a whole series of curves all of which, except that representing the case of very light winds, have the same general character and may perhaps belong to the same family. Such a family of curves might be obtained, for example, from an empirical formula for the variation of wind with height such as that of Archibald or of Hellmann by assigning different values to a constant according to the strength of the wind at a standard height, the time of day at different seasons of the year and so on. In any case it is idle to suppose that any empirical formula can hold for winds of all directions at all heights because the régime of winds may change entirely at levels where the distribution of pressure is different from that at the surface. We will, therefore, in this chapter limit our view to the winds which belong to the system indicated by the distribution of pressure at the surface. The lowest kilometre may be taken as a rough indication of the thickness of the layer, though the selection of that limit is at present arbitrary.

A rational basis of explanation of the variation of the velocity and direction of wind with height in the lowest layers is to be found in the theory of atmospheric turbulence to which we have already referred. Here we will follow the development of the theory given by G. I. Taylor[1] in a paper to which reference

[1] 'Phenomena connected with Turbulence in the Lower Atmosphere,' *Proc. Roy. Soc.* A, vol. XCIV, 1918, 137.

has already been made. It is based upon a previous paper on eddy-motion[1], the application of which is used in the discussion of the conditions of formation of fog on the banks of Newfoundland in the report of the work carried out by the s.s. *Scotia*, 1913[2]. It is not too much to say that these contributions by their method of treatment of eddy-motion have opened up the field of meteorological investigation in a very remarkable manner. They enable us to obtain a clear insight into the complicated phenomena of the lower layers of the atmosphere, which have not yielded to the ordinary procedure by the method of means.

The fundamental conception is concerned with the distribution of potential temperature in the vertical. In consequence of the turbulence or eddy-motion of the atmosphere heat diffuses downward towards a cold surface according to a law similar to the ordinary law of diffusion of heat in a solid but with a greatly increased coefficient. It may be recalled (p. 8) that the fundamental equation for the diffusion of heat in a solid[3], viz.

$$\frac{d\theta}{dt} = \mu \frac{d^2\theta}{dz^2},$$

is the expression of the fact that the rate at which temperature is communicated to the solid at a particular point is proportional to the change in the gradient of temperature at that point. That equation when applied to find the distribution of temperature in the atmosphere, on account of the difference of potential temperature between an undisturbed upper layer and a cold surface, similarly expresses the law that the rate at which potential temperature is increased at any point is proportional to the rate of change in the vertical of the variation of potential temperature with height; but when turbulence exists the constant of the proportion is many times that which would be appropriate if the air were solid or restrained in some other way from using its mobility to help towards equalising its potential temperature.

The direct expression for the variation of the potential temperature θ, of unit volume of the air, in terms of the variation of height z, and time t, on the hypothesis set out, is the equation

$$\rho\sigma\delta\theta/\delta t = \frac{\delta}{\delta z}(k_\theta \rho\sigma\delta\theta/\delta z),$$

where ρ is the density and σ the specific heat of the air; k_θ, the coefficient of eddy-diffusion[4], is dependent upon the state of the air as regards turbulence.

This equation is reduced to a more simple form as a first approximation by regarding ρ and k_θ as independent of the height. The specific heat of air is

---

[1] G. I. Taylor, 'On Eddy-Motion in the Atmosphere,' *Phil. Trans.* A, vol. CCXV, 1915, p. 1.

[2] 'Ice Observation, Meteorology and Oceanography in the North Atlantic Ocean,' *Report...s.s. Scotia*, 1913, *to the Board of Trade*.

[3] Kelvin, *Encyclopaedia Britannica*, ninth edition, Art. 'Heat.'

[4] In the course of the discussion we shall require a number of coefficients which may or may not be numerically the same, we therefore indicate the family relationship by using k for each and identify the individuals by suffixes: thus k_θ is the coefficient for the diffusion of potential temperature θ, k_{mm} for momentum.

known to be practically constant at all pressures and temperatures. Consequently we get the equation

$$\delta\theta/\delta t = k_\theta \delta^2 \theta/\delta z^2 \qquad \ldots\ldots(1),$$

of which a solution is given by

$$\theta = A e^{-bz} \sin(2\pi t/\tau - bz) \qquad \ldots\ldots(2),$$

where

$$b^2 = \pi/(\tau k_\theta),$$

whence if A_1 and A_2 are the amplitudes at heights z and $z + h$ we can compute k_θ from the equation $bh = \log_e A_1 - \log_e A_2$.

As an example of the application of the theory we may take from the Report of the voyage of the *Scotia* the account of the distribution of temperature and humidity over the sea as disclosed by a sounding with a captive balloon at 7 p.m. on 4 August, 1913, represented by the curves in fig. 33, p. 151. Taylor traces the course of the air which formed this current from the shores of Labrador along a devious course over the ocean and attributes the counterlapse of temperature which reaches from the surface to a height of nearly 400 metres to the loss of heat at the surface through the direct effect of the eddy-motion during three days while the air was passing over water colder than itself. The next section of the curve up to 750 metres shows an approximation to the adiabatic lapse for dry air, the result of the passage of air over warmer water as it came down from north to south between 30 July and 2 August, while the top portion of the curve shows the remains of the counterlapse of temperature formed while the air was passing from the coast of Labrador northward over cold water before 30 July; the dryness of the air, at the top, itself suggests its origin from over land. Thus he traces in the details of the shape of the curves representing the present distribution of temperature and humidity, the past history of the air and explains the formation of fog in the lowest layer, about 120 metres thick, as a consequence of the mixing of the cold air of the surface with the warmer upper air caused by the eddy-motion. It should be noticed that the cooling effect of eddy-motion has reached nearly 400 metres but the fog only extends about one-third of the way up the line of counterlapse of temperature which is continuous without change of slope through the fog and beyond it. Also that some days previously the effect of the cold surface had extended upwards beyond a kilometre.

As to the height to which the effect of the eddy-motion extends Taylor gives a formula $z^2 = 4k_\theta t$, where z is the height affected by the eddy-motion when the air has been moving over a surface of lower temperature for a time t, k_θ being the "eddy-conductivity." The curves of distribution of temperature deduced from that equation as applied to a current of air with an isentropic lapse rate passing over cold water are represented in fig. 24 which is also taken from the Report of the *Scotia*. In the diagram the diagonal line represents the supposed original isentropic lapse of temperature with height. The several curves represent the distribution of temperature in the vertical after certain periods of passage over cooling water. The curve marked 1 shows the height-temperature curve when the travel has been over water which cools the surface-air by 1tt in 30 miles run for a period of $1\frac{3}{8}$ hours. That marked 2 shows the effect of a travel of $5\frac{1}{2}$ hours in the same circumstances. For the other three curves a rate of fall of temperature along the surface of 1tt in 60 miles is assumed and the first of them numbered 3 corresponds with a travel of $5\frac{1}{2}$ hours. For No. 4 22 hours are supposed to have elapsed and for No. 5 87 hours. Though the curves do not show the sharp angles that are drawn on

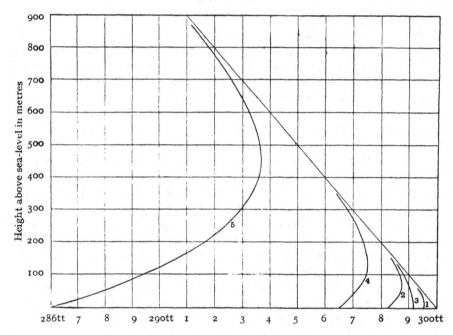

Fig. 24 (IV, 2). Curves showing the distribution of temperature set up in an atmosphere originally isentropic, with surface-temperature 300tt, by wind of 4·5 metres per second (10 miles per hour) moving over a track along which the surface-temperature decreases, according to the theory of eddy-motion, with coefficient of eddy-conductivity 3×10^3 c, g, s units.

Scale of temperature changes.

Curve		hrs.								
1.	Distribution after	$1\frac{3}{8}$ (	14 miles fetch) when temperature decreases 1tt in 30 miles
2.	,,	,, $5\frac{1}{2}$ (55	,, ,,)	,,	,,	,,	,,	,,
3.	,,	,, $5\frac{1}{2}$ (55	,, ,,)	,,	,,	,,	1tt in 60	,,
4.	,,	,, 22 (220	,, ,,)	,,	,,	,,	,,	,,
5.	,,	,, 87 (870	,, ,,)	,,	,,	,,	,,	,,

p. 151 and other diagrams representing the variation of temperature with height, their shapes are very suggestive of many of the figures that are obtained for the variation of temperature with height from soundings of the upper air. Whenever we have a cold anticyclonic spell of weather in winter, whether there is a fog at the surface or not, the variation of temperature with height is similar to one or other of the five curves of fig. 24, the particular example being determined by the length of time that the surface has been absorbing heat from the air above it with the aid of the mixing due to eddy-motion.

The value of k_θ is notably different in different situations, and this we might anticipate from the differences to be expected in the turbulence of the air arising from differences in the nature of the surface which offers resistance to the motion of the air. An order of progressive values of k_θ may be supposed, beginning with the surface of the sea, or a perfectly smooth surface of land, through the roughness of grass to that of the irregularities of trees or town-buildings the various effects of which are represented in Table I (p. 100) which gives the different ratios of W/G and are illustrated in figs. 3 to 9 of chap. 1. Some of the values of k found by Taylor for different situations are shown in the following table:

COEFFICIENTS OF EDDY-DIFFUSION

	c, g, s units
At sea over the Grand Banks (from the distribution of temperature)	0.3×10^4
Over Salisbury Plain (from the distribution of velocity)	5×10^4
Eiffel Tower (from the daily range of temperature at different levels)	
Lowest stage (18 m to 123 m) 9×10^4 in Oct. to 24×10^4 in Mar.	Mean 15×10^4
Highest stage (197 m to 302 m) 1.6×10^4 in Feb. to 30.1×10^4 in July	Mean 11×10^4
Whole range (18 m to 302 m) 4.3×10^4 in Jan. to 18.3×10^4 in June	Mean 10×10^4
Eiffel Tower (from wind measurements) Åkerblom	7.6×10^4

The small value for air over the Grand Banks is to be associated with the counterlapse of temperature there which gives the air a certain "resilience" against the frictional forces, and restrains the diffusion.

In like manner as the surface may act as a boundary at which heat is absorbed, so it may act as a boundary at which momentum is absorbed; and in that case momentum also diffuses downward from the "undisturbed" current of air in the upper regions to be lost at the surface, and the distribution of velocity with regard to height and time will follow the law of the equation

$$\rho \delta u / \delta t = k_{mm} \rho \delta^2 u / \delta z^2 \qquad \qquad \ldots\ldots(3),$$

where k_{mm} has the same value as k_θ in the thermal equation because the diffusion both of heat and momentum is governed by the eddies which cause the mixing of the layers.

"Roughly k may be taken as $\frac{1}{2}wd$ where w represents the mean vertical component of velocity due to the turbulence and d represents roughly the mean vertical distance through which any portion of the atmosphere is raised or lowered while it forms part of an eddy till the time when it breaks off from it and mixes with the surroundings. This may be taken to be roughly equal to the diameter of a circular eddy."

For the steady state under these conditions, assuming that the system has been established long enough for the initial conditions to have died away, Taylor has obtained a formula for the relation of the surface-wind to the undisturbed wind (taken as equivalent to the geostrophic wind) in terms of the angle α of deviation between the geostrophic wind G and the surface-wind W, which takes the form

$$W/G = \cos \alpha - \sin \alpha \qquad \qquad \ldots\ldots(4).$$

The equation is tested by comparison of the calculated results with some observations of the surface velocity and its angular deviation from the gradient, by G. M. B. Dobson[1] over a very suitable exposure on Salisbury Plain; the comparison is as follows:

			Light winds	Moderate winds	Strong winds
Observed value of W/G	...		·72	·65	·61
α observed	...	...	$13°$	$21\frac{1}{2}°$	$20°$
α calculated	...	...	$14°$	$18°$	$20°$

The closeness of the agreement is remarkable and incidentally we may note that in this case also the lighter winds show closer agreement with the gradient both as regards direction and velocity. The comparison is, moreover, supported by the examination of a conclusion arrived at by Dobson from his observations, namely that the geostrophic velocity is attained at a height considerably below that at which the direction of the geostrophic wind is reached. The difference is usually as much as that between 300 metres for the velocity, and 800 metres for the direction, and this is shown to be a direct consequence of the theory.

From the application of this theory Taylor arrives at the further conclusion that for a given geostrophic wind G, which is reduced by surface-friction to such an extent that the surface-wind is inclined at an angle α to the undisturbed wind, the rate of loss of momentum to the surface, that is the force of surface friction F, is

$$2k_{mm}\rho G \sin \alpha / B \qquad \qquad \ldots\ldots(5),$$

where B is equal to $\sqrt{\omega \sin \phi / k_{mm}}$, ω is the angular velocity of the earth's rotation, and ϕ is the latitude. And in another paper[2] he obtains for the frictional force of air over the grassy land of Salisbury Plain the value

$$F = 0\cdot0023\rho W^2,$$

whence $\qquad\qquad\qquad 0\cdot0023\,W^2 = 2k_{mm}G \sin \alpha / B.$

Substituting numerical values of ω and ϕ, and remembering that W/G is equal to $\cos \alpha - \sin \alpha$, we get

$$\frac{1}{BG} = \frac{20\cdot4}{\sin \alpha}(\cos \alpha - \sin \alpha)^2.$$

In deducing equation (4) the following equations were obtained[3] for the components of wind-velocity u along the isobar and v perpendicular to the isobar:

$$\left.\begin{aligned} u &= G - A_2 e^{-Bz} \cos Bz + A_4 e^{-Bz} \sin Bz \\ v &= + A_2 e^{-Bz} \sin Bz + A_4 e^{-Bz} \cos Bz \end{aligned}\right\} \qquad \ldots\ldots(6),$$

[1] *Q. J. Roy. Meteor. Soc.*, vol. XL, 1914, p. 123.
[2] *Proc. Roy. Soc.* A, vol. XCII, 1916, p. 198.
[3] 'On Eddy-Motion in the Atmosphere,' *loc. cit.*, p. 15.

where $\qquad A_2 = G \tan \alpha \, (1 + \tan \alpha)/(1 + \tan^2 \alpha),$

and $\qquad A_4 = G \tan \alpha \, (1 - \tan \alpha)/(1 + \tan^2 \alpha),$

from which we obtain

$$\left. \begin{aligned} (G - u)/G &= e^{-Bz} \sin \alpha \, \{\cos (Bz - \alpha) - \sin (Bz - \alpha)\} \\ v/G &= e^{-Bz} \sin \alpha \, \{\cos (Bz - \alpha) + \sin (Bz - \alpha)\} \end{aligned} \right\} \quad \ldots\ldots(7),$$

as an equation for the relation between velocity and height. Since

$$k_{mm} = \omega \sin \phi / B^2,$$

the relation will be different for different values of k_{mm}, and if k_{mm} is subject to diurnal variation there will be a corresponding diurnal variation in the curves which represent the relation of velocity and height.

The corresponding values of α, B, and k_{mm} for values of G to be taken at will, are given in the following table.

CONDITIONS OF EDDY-VISCOSITY FOR GIVEN DEVIATIONS

α	BG	k_{mm}/G^2
°	c, g, s units	c, g, s units
4	·0040	3·54
6	·0065	1·35
8	·0094	0·635
10	·0129	0·338
12	·0171	0·192
14	·0223	0·116
16	·0286	0·069
18	·0366	0·042
20	·0456	0·027
22	·0599	0·0156
24	·0775	0·0094
26	·101	0·0055
28	·135	0·0031
30	·181	0·0017
32	·270	0·00085
34	·384	0·00038
36	·588	0·00016

With these values of the constants Taylor has constructed the curves which are represented in fig 25, the shapes of the curves being determined by selected values for α, at the surface, marked against them. The abscissae are the ratios of the wind-velocity to the geostrophic wind and the ordinates are the ratios of the numerics of the height and the geostrophic wind, so that, for example, when the geostrophic wind is 10 m/sec the figures at the side will represent heights in dekametres, and those at the base velocities in dekametres per second.

It is interesting to note the difference in the curves which would be suitable for representing the variation of wind with height under different conditions as to turbulence. If we take from p. 131 the three values of the coefficient k_{mm}, 3×10^3 appropriate for the sea, 5×10^4 for Salisbury Plain and 10×10^4 for Paris respectively, we see that to give a value of α equal to 20° (requiring a

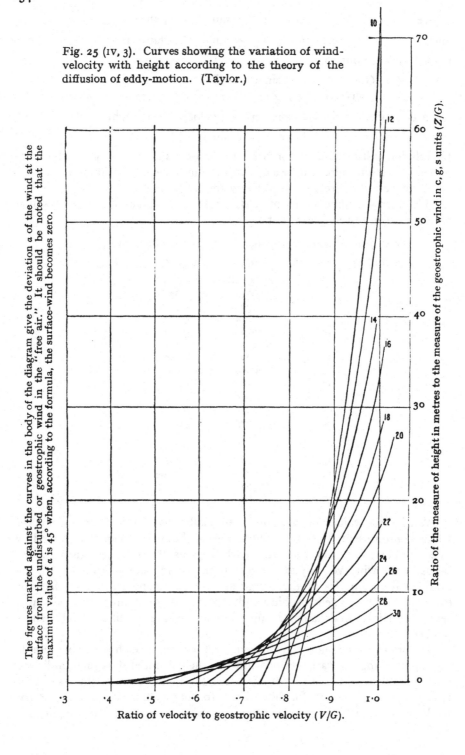

Fig. 25 (IV, 3). Curves showing the variation of wind-velocity with height according to the theory of the diffusion of eddy-motion. (Taylor.)

The figures marked against the curves in the body of the diagram give the deviation α of the wind at the surface from the undisturbed or geostrophic wind in the "free air." It should be noted that the maximum value of α is 45° when, according to the formula, the surface-wind becomes zero.

Ratio of the measure of height in metres to the measure of the geostrophic wind in c.g.s units (Z/G).

Ratio of velocity to geostrophic velocity (V/G).

quotient 0·027 for k_{mm}/G^2) the geostrophic winds would have to be about 3 m/sec, 12 m/sec and 18 m/sec respectively. Hence different curves are appropriate for the same geostrophic wind if on account of the circumstances of the site, the time of day or the season of the year, the coefficient k_{mm} which defines the turbulence is different, a conclusion which we have already seen to be in accord with experience.

This adjustment to the different circumstances of the site or time is a strong point in favour of the theory as a dynamical explanation of the variation of wind with height; and the evidence is strengthened by the further application of the theory, in the paper from which we are now quoting, to calculate, from a transformation of the curves which we have reproduced, the heights at which the day maximum of wind-velocity with which we are familiar at the surface gives place to the day minimum which was represented in the results for the Eiffel Tower and is characteristic of all mountain stations and, as we have already seen, is disclosed in the results obtained with pilot-balloons at South Farnborough. It appears as a result of these calculations that a variation in k_{mm} by an amount which fits in well with all the other known data concerning the turbulent motion of the air near the ground is sufficient to explain both quantitatively and qualitatively all the facts concerning the daily variation of wind-velocity at different heights above the ground which are brought to light by Hellmann's observations.

The spiral of turbulence

The lines representing the wind at successive levels, drawn *from* the point at which the wind is measured, are a series of vectors the extremities of which lie on an equiangular spiral with its pole at the extremity of the vector representing the geostrophic wind.

This result was obtained by Hesselberg and Sverdrup[1] in 1915 and has naturally attracted much attention.

In a paper in the *Quarterly Journal* D. Brunt[2] gives the analysis. He assumes that the coefficient does not vary with height and that the geostrophic wind is the same at all the levels to which the calculation extends.

We use vector notation which enables us to treat the two components at right angles by a single equation in which the parts referring to the two components are indicated by prefixing $\sqrt{-1}$ represented by i to one of them. The sign of addition $(+)$ then indicates that the y-component behind the i is to be combined with the x component by the process of geometrical addition.

The use may be derived by considering that geometrically $- a$ means a distance a drawn opposite to a and may be reached by operating $\sqrt{-1}$ twice to turn a through a right angle each time. Turning a through an angle θ in this way is by operating $a (\cos \theta + \sqrt{-1} \sin \theta)$ which is equivalent to $ae^{i\theta}$.

Thus the velocity V at any level z may be expressed as $V = u + iv$ where u and v

[1] 'Die Reibung in der Atmosphäre,' *Veröff. d. Geophys. Inst. d. Univ. Leipzig*, Heft 10, 1915; 'Die Windänderung mit der Höhe vom Erdboden bis etwa 3000 m Höhe,' *Beitr. Phys. frei. Atmosph.* Bd. VII, 1917, p. 156.

[2] *Q. J. Roy. Meteor. Soc.*, vol. XLVI, 1920, p. 175. See also F. J. W. Whipple, *ibid.*, p. 39.

are the components along and perpendicular to the isobar and the frictional force per unit volume R can be represented vectorially by $k_{mm}\rho\partial^2V/\partial z^2$.

In the figure below let the axes Ox, Oy be drawn parallel to the isobar and perpendicular to the isobar, respectively, the point O being at any height z above the ground. Let G be the magnitude of the gradient-wind there, directed along Ox, and let V be represented by OV. At the ground OV will make an angle a with Ox. Let the frictional force R on unit volume of air at O be represented by OR, making an angle β with the reversed wind-direction. Draw OC perpendicular to OV.

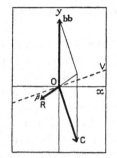

The pressure-gradient bb acting along Oy is $2\omega\rho\sin\phi.G$. The rotation of the earth produces a deviating force $2\omega\rho\sin\phi.V$ along OC. The forces in the plane xy acting on unit volume of air at O may therefore be written

$$2\omega\rho\sin\phi.G \text{ along } Oy, \quad 2\omega\rho\sin\phi.V \text{ along OC,} $$
$$R \text{ along OR} \quad\quad ......(8).$$

Fig. 26. The equilibrium of the frictional force of turbulence R with the pressure-gradient bb at O and the centrifugal force of the earth's rotation for the velocity V.

These may be written vectorially in the form

$$2\omega\rho\sin\phi.iG, \quad -2\omega\rho\sin\phi.iV, \quad k_{mm}\rho\partial^2V/\partial z^2.$$

For air in steady motion these forces must be in equilibrium, and their algebraic sum must be zero. This condition reduces to $k_{mm}\rho\partial^2V/\partial z^2 - 2\omega\rho\sin\phi.i(V-G) = 0$.

In Taylor's notation substituting $B^2 = \omega\sin\phi/k_{mm}$ we may write this equation in the form

$$\frac{\partial^2V}{\partial z^2} - 2iB^2(V-G) = 0 \quad\text{or}\quad \frac{\partial^2V}{\partial z^2} - (1+i)^2B^2(V-G) = 0 \quad ......(9).$$

It should be noted that the only assumption made hitherto is that R can be represented by $k_{mm}\rho\partial^2V/\partial z^2$. In deriving equation (9) no assumption has been made as to the constancy of B or of G. In order to reduce it to an integrable form, we shall assume that B is constant and G either constant or a linear function of the height z. The solution of equation (9) is then

$$V-G = C_1e^{(1+i)Bz} + C_2e^{-(1+i)Bz} \quad ......(10).$$

Since we cannot admit infinite values of V for infinite values of z, we must have $C_1 = 0$,

$$V-G = C_2e^{-(1+i)Bz} = Ce^{-(1+i)Bz+i\gamma} \quad ......(11),$$

where C and γ are both real constants, whose values must be determined from the boundary conditions. We assume with Taylor that the boundary condition at the ground is that the direction of slipping is in the direction of strain. In terms of equation (11) this means that for $z = 0$, $\partial V/\partial z$ is in the same direction as V. Let a be the angle between the surface-wind and the gradient-wind. Then for $z = 0$, $\partial V/\partial z$ must make an angle a with the gradient-wind. Differentiating equation (11) we find at $z = 0$,

$$\frac{\partial V}{\partial z} = -CB(1+i)e^{i\gamma} = \sqrt{2}\,CBe^{i(\gamma+5\pi/4)} \quad ......(12),$$

since

$$1+i = \sqrt{2}e^{i\pi/4} \text{ and } -1 = e^{i\pi}.$$

But since $\partial V/\partial z$ makes an angle a with the real axis, for $z = 0$ it is possible to write

$$\frac{\partial V}{\partial z} = De^{ia} \quad ......(13).$$

Equations (12) and (13) are identical, and so $\gamma + 5\pi/4 = a$ or $\gamma = a - 5\pi/4$.

Substituting in (11), $\quad V-G = Ce^{-Bz+i(a-5\pi/4-Bz)} \quad ......(14).$

To evaluate C we use the condition that for $z = 0$, V makes an angle a with the axis of x.

Since $V = G - C \cos (a - \pi/4) - i\,C \sin (a - \pi/4)$ must be identical with $V = M \cos a + i\,M \sin a$; the condition reduces to

$$\frac{-G + C \cos (a - \pi/4)}{\cos a} = \frac{C \sin (a - \pi/4)}{\sin a} \text{ or } C = \sqrt{2}\,G \sin a.$$

Equation (14) therefore becomes

$$V = G + \sqrt{2}\,G \sin a . e^{-Bz + i\,(a + 3\pi/4 - Bz)} \qquad \ldots\ldots(15).$$

Thus the wind V at any height is equivalent to the gradient-wind together with an added component of magnitude $\sqrt{2}G \sin a . e^{-Bz}$ acting at an angle $a + 3\pi/4 - Bz$ with the direction of the gradient-wind. This second component decreases with height according to the exponential law, and its direction rotates uniformly counter-clockwise with increasing height. It follows that if in the figure previously described we draw vectors QV to denote the wind at different heights, these vectors will sweep out an equiangular spiral, if G is constant.

If we return to our original notation and write $V = u + iv$, we find from equation (15)

$$\begin{aligned}
u - G &= \sqrt{2}\,G \sin a . e^{-Bz} \cos (a + 3\pi/4 - Bz) \\
&= -\sqrt{2}\,G \sin a . e^{-Bz} \cos (a - \pi/4 - Bz) \\
v &= \sqrt{2}\,G \sin a . e^{-Bz} \sin (a + 3\pi/4 - Bz) \\
&= -\sqrt{2}\,G \sin a . e^{-Bz} \sin (a - \pi/4 - Bz)
\end{aligned} \right\} \qquad \ldots\ldots(16).$$

These equations are equivalent to those derived by Taylor (Equation (7)).

Further, since $R = k_{mm}\,\rho\,\dfrac{\partial^2 V}{\partial z^2}$ differentiating equation (15) and remembering that $1 + i = \sqrt{2}e^{i\pi/4}$ we find

$$R = 2\sqrt{2}\,k_{mm}\,B^2 \rho\,G \sin a . e^{-Bz + i\,(a + 5\pi/4 - Bz)}$$

$$= 2\sqrt{2}\,\rho\omega \sin \phi . G \sin a . e^{-Bz + i\,(a + 5\pi/4 - Bz)}.$$

From the fact that $R = -\,\partial F/\partial z$, where F is the frictional force on unit surface, Brunt further deduces that R at the ground must act at an angle $\pi/4$ with F, and hence with the reversed wind-direction.

This analysis refers to cases where B and hence k_{mm} do not vary with height. In a note added to the original paper Brunt discusses cases where k varies either inversely as the height or inversely as the square of the height. The problem of k varying as a linear function of the height is also considered by S. Takaya[1] in a paper "On the coefficient of eddy-viscosity in the lower atmosphere." The solution in these cases involves complicated mathematical analysis into which we do not wish to enter here. Readers who are interested in the details may refer to the original papers.

Y. Isimaru has developed a slightly different aspect of the subject in two papers to which we have already referred on p. 72. Following Fujiwhara, instead of representing the eddy-resistance by a single coefficient k he considers it as composed of two components due to longitudinal eddy-viscosity and transverse eddy-viscosity. By analysis similar to Brunt's he develops a relation between the surface-wind and the gradient-wind in the form $W = G\,(\cos a - \sin a \tan \delta)$, where δ is the angle which the resistance makes with the reversed wind-direction and is such that $\tan 2\delta$ equals the ratio of the coefficients of longitudinal and transverse eddy-viscosity. The equation contains Taylor's result as a special case. Isimaru compares the values of δ

[1] *Memoirs of the Imperial Marine Observatory, Kobé, Japan*, vol. IV, No. 1, 1930.

KAIKIAS IN MODERN DRESS

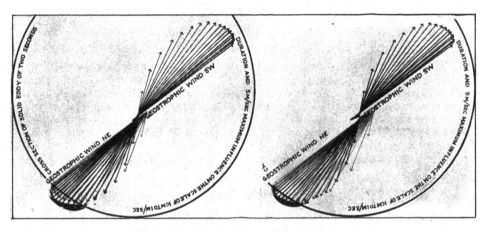

(a) View from the sky looking downwards.

Glass plates, in three groups of twenty each, showing the effect of friction in reducing the mean velocity of a north-east wind, 10 m/sec at 500 metres, and the reversal of the wind from north-east at 500 metres to south-west at 1500 metres.

Successive plates represent steps of 25 metres each. The third from the bottom (labelled LAND-WIND) represents approximately the normal effect of the friction of the land upon the north-east wind. The fifth (labelled SEA-WIND) represents roughly the effect of a sea-surface upon the same wind.

For an explanation of the circle and of its scale relation to the anemometer at the foot of diagram (b) see p. 144.

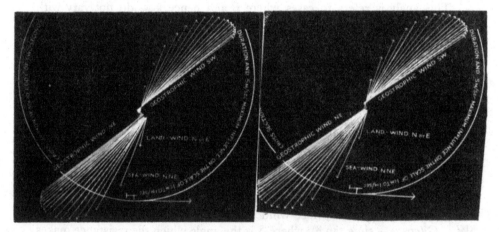

(b) View from the ground looking upwards.

FIG. 27. Transition from no wind at the surface through north-east geostrophic wind at half a kilometre to the complete reversal at one and a half kilometres of the gradient and the wind.

derived from the theory with those obtained from observations and finds good agreement.

The spiral obtained from Brunt's analysis is different according to the measures of the surface-wind and one can suppose the spiral extended downward until a point is reached where the friction has reduced the surface-wind to zero.

This case is represented in the model which is shown in stereo in fig. 27. The whole model is intended to exhibit complex but real phenomena. First of all, the increase from the surface-wind over the land or over the sea until a NE geostrophic wind is reached at half a kilometre.

Between that height and $1\frac{1}{2}$ km another change is operative; the wind changes from the NE geostrophic wind of the surface-pressure to that of a reversed gradient, viz. for SW, at $1\frac{1}{2}$ km. This is the kind of transition which can be observed in the region of the trade-winds and counter-trades either in the northern or southern hemisphere. The transition in the northern is from north-east near the ground to south-west up above, a transition of which the Greeks were aware 25 centuries ago.

The double transition is complicated but undoubtedly interesting and it must correspond more or less accurately with every case of the relative motion of superjacent layers of air if the motion is sufficiently strong to cause turbulence.

The case pictured may not agree exactly with the analysis because in the analysis we assume the barometer-gradient to be operative throughout the vertical layer affected, and in the model the two gradients at top and bottom of the transition are equal and opposite and there is no history of what happens between them.

Another point to which attention should be directed is the meaning of the stratum of no wind. According to the theory of turbulence it is not a region of calm but a region in which the energy of flow has been converted into the energy of turbulence without any effective flow. Hence an anemometer placed in the stratum of no wind ought to be expected to show gustiness but no resultant wind.

It follows from these considerations that the simplicity of the spiral, which appears so real in the mathematical analysis, may in actual practice be sadly marred by the gustiness which is an essential part of its creation. The unsophisticated reader must not expect to be able to put up a pilot-balloon and express its motion as an equiangular spiral; but if he would perform the experiment and say what actually is to be seen in circumstances that ought to give a spiral, and do give a spiral for the "flow" of the wind, he will add to our knowledge of the structure of the atmosphere. If he wishes to find a recognisable spiral he had better employ mean values of a number of ascents—they ought at least to make some recognition of the flow as distinguished from the eddies of a single ascent[1].

[1] A comparison of the results derived from theory and observation is given by Hesselberg and Sverdrup in papers already referred to, by W. J. Humphreys, in 'The way of thewind,' and by Takaya.

Since the advent of the spiral in the dynamics of the atmosphere a great deal of attention has been paid to the measurement of the coefficient k_{mm} upon which it is based.

The k which we have included in the formula, with G. I. Taylor, is only one of many "constants" that may be the subject of investigation. We have seen that Isimaru considers it as made up of two components and on p. 71 we quote four from Rossby. Here we give a series of values from L. F. Richardson to which we have added Takaya's values.

We must remark however that the equation with k is modified by Richardson to give $\frac{\partial \theta}{\partial t} = \frac{\partial}{\partial z}\left(\rho k \frac{\partial \theta}{\partial z}\right)$ instead of $\frac{\partial \theta}{\partial t} = k \frac{\partial^2 \theta}{\partial z^2}$ as in G. I. Taylor. So we get a coefficient, ρk or ξ, which may be described as the coefficient of eddy-viscosity multiplied by the density of air under investigation. The values of k quoted in the table on p. 131 are at least 800 times those of the coefficients set out here.

The reader may if he wishes attempt to adjust the proper value of ρ to the circumstances; the author would be more than shy of doing so.

TABLE OF OBSERVED VALUES OF EDDY-COEFFICIENTS

These quantities are obtained by equations on the analogy of the diffusion equation in which the quantity is referred to unit mass of the working substance. If they are referred to unit volume in the same way as the coefficients of kinematic viscosity they must be divided by the density of air, the values of which vary notoriously with height.

Coefficient	Height	Locality and season	Type of data and notes		Author
g/sec cm	m	EDDY-VISCOSITY, $k_{mm}\rho$, *from variation of wind with height*			
83	21 to 305	Paris, winter	Eiffel Tower winds. Allowance is made for		Åkerblom
113	,,	,, summer	variation of barometric gradient		
70	0 ,, 1700	Salisbury Plain	Wind at {9·5 m/sec No allowance is made		Taylor
56	0 ,, 1700	,, ,,	30 m {5·9 ,, for variation of baro-		,,
32	0 ,, 1700	,, ,,	{3·3 ,, metric gradient		,,
0·9 to 8	0 ,, 200	Newfoundland sea	Kite ascents from ss. *Scotia* ,,		,,
0·9 to 4·8	0 ,, 200	Sea	Kite ascent, 2 Aug. 1913, 11 h ,,		,,
0·9	0 ,, 9	(Stevenson and Hellmann)	Variation of barometric gradient with height		Hesselberg
40	9 ,, 209	Lindenberg	is allowed for except for the range 0 to		and Sverdrup
50	109 ,, 309	,,	3000 m		,,
50	209 ,, 409	,,			,,
60	309 ,, 509	,,			,,
50	9 ,, 3000	,,			,,
140	0 ,, 1200	N. Atlantic: Trade	Allowance is made for variation of baro-		Sverdrup
260	1200 ,, 1800	,, Intermediate	metric gradient		,,
90	1800 ,, 2000	,, Anti-trade			,,
				$\sigma \times 10^3$	
30·5	Surface	Spring: morning	Dobson's ascents at Upavon as-	0·598	Takaya
172·3	,,	mid-day	suming that $k_{mm}\rho$ varies with	0·248	,,
38·9	,,	Summer: morning	height and that the ratio of the	−1·174	,,
271·6	,,	mid-day	value at height z to the surface	−1·028	,,
31·5	,,	Light winds	value is $(1+\sigma z)$	2·155	,,
241·5	,,	Moderate winds		−1·192	,,
138·9	,,	Strong winds		−0·1301	,,
16·8	,,	Winter	Pilot-balloon data at Kasumi-	−0·981	,,
52·8	,,	Spring	gaura, 1927	−2·751	,,
		Other methods			
2·6	9		Hellmann anemometers, assuming that near		Schmidt
1·3	2		the ground the vertical flux of momentum		,,
1·1	1		is independent of height		,,
15	400	Lindenberg along wind	By integrating the loss of momentum with		Richardson
110	,,	,, across wind	respect to height		,,
1030	0 to 2000	Land, mid-day	Speed of ascent of cumuli		,,

Coefficient	Height	Locality and season	Type of data and notes	Author
g/sec cm	m	EDDY-CONDUCTIVITY $k\rho$: (a) *dust, smoke and other floating solids*		
6	1000	Hills 150 m high, Feb.	Irregularities in flight of free balloon	Richardson
120	250	Argonne forest, Aug.	Dispersal of smoke	,,
6	9	Moor	,,	,,
0·16	3·4	Flat fields	,,	,,
25	5	1920 Sept. 10	,,	,,
7	2	,, ,,	Firework smoke	
		(b) *Water-content*		
61	2 to 302	Paris, Dec. to Feb. 9 h to 12	Evaporation at Bureau Central and difference	Schmidt
21	2 ,, 302	21 h to 24	of water per mass between Parc St Maur and	,,
84	2 ,, 302	Paris, July 3 h to 6	summit of Eiffel Tower	,,
60	2 ,, 302	21 h to 24		,,
·006 to ·38	8500	Whole globe	Comparison of precipitation with up-grade of	Richardson
120	500	,,	water per mass of atmosphere	,,
≤0·8	0·5	,,		,,
		(c) *Carbon dioxide*		
? 0·001	10,000		Distribution of carbon dioxide	Schmidt
		(d) $k_\theta \rho$, *potential temperature*		
		Sea, Newfoundland banks:		
1·6	0 to 170	July 29	Life-history of air-currents and distribution	Taylor
3·0	0 ,, 200	Aug. 2, 11 h	of temperature with height, July 29, Aug. 2	,,
3·1	0 ,, 370	Aug 4 19 h	and 4, 1913	,,
4·1	370 ,, 770	,, Year ,,		,,
180	18 ,, 123	Paris, Year	Eiffel Tower temperatures, neglecting radia-	Taylor
20	197 ,, 302	,, February	tion	,,
364	197 ,, 302	,, July	,, ,,	,,
54	18 ,, 302	,, January	,, ,,	,,
221	18 ,, 302	,, June	,, ,,	,,
31·2	197 ,, 302	,, Spring	Eiffel Tower temperatures, allowing for radia-	Schmidt
46·5	197 ,, 302	,, Summer	tion	,,
10·6	197 ,, 302	,, Autumn	,, ,,	,,
9·9	197 ,, 302	,, Winter	,, ,,	,,
9·2	2 ,, 123	,, Year	,, ,,	,,
0·6	1·2 ,, 1·4	Allahabad, cold	Temperatures allowing for radiation	,,
0·7	1·2 ,, 1·4	,, hot	,, ,,	,,
2·8	1·2 ,, 1·4	,, rainy	,, ,,	,,
4·6	31·7 ,, 50·6	,, cold	,, ,,	,,
3·3	31·7 ,, 50·6	,, hot	,, ,,	,,
7·4	31·7 ,, 50·6	,, rainy	,, ,,	,,

The differences in the values of the coefficients shown in the table are remarkable. If we include Richardson's estimate from the speed of ascent of cumuli, the range of the variation of the "constant" is from 1 to 1000. Sverdrup's coefficients for maritime air, determined from the changes of velocity at various levels in the North Atlantic trade-wind with a foundation of polar air southward bound, range from 90 to 260, while Taylor's figure for equatorial air northward bound over the Grand Banks is less than 1.

It is not easy to form a definite idea of the details of the motion which has to be interpreted by these great differences. So far as the surface-wind is concerned we have an effective representation of what is happening in the gustiness exhibited by the South Kensington exposure (fig. 9), and the corresponding information for Pendennis Castle (fig. 3). The inclusion of the effect of cumulus as turbulence, in the same table with the restricted effect of air travel over an inversion (which accounts for the low value over the Grand Banks), suggests a wide range of possibilities.

It is to be regretted that we have not now time or opportunity to review some details of the subject in the light of certain meteorological principles which have emerged in the course of the work on these volumes.

In vol. III and again in this volume we have laid stress upon the fact that the stratification of atmospheric movement is ordered by isentropic surfaces, not necessarily level surfaces. Whence it follows that alterations of entropy imply potential changes in the vertical which in a general formula must count as turbulence. We should like to compare the measures of turbulence with the vertical distances necessary for unit increase of entropy.

And the whole of the work of this volume has led to the conclusion, referred to in many places but not set out before chap. xi, that as regards horizontal motion in the upper air the rotation of the earth is always bringing the distribution of pressure automatically into harmony with the wind-velocity, and thereby the slopes of the isobaric surfaces are subject to continuous adjustment. Except in cases of penetrative convection they cannot show much departure from the equilibrium position.

We should anticipate therefore that in the upper air the variation of velocity in the space between a lower layer moving from north-east and an upper layer moving from south-west is mainly controlled by change of gradient rather than by turbulence. The photographs of fig. 27 show the results of the hypothesis of turbulence applied equally to the zero velocity at ground-level and to a zero velocity in the free air. We wonder if that is reasonable. We wish it were possible to have, as Moltchanoff[1] has suggested, an anemometer record for the free air to correspond with what we have for a position near the ground; in its absence we are left to guess whether the two records would be as nearly identical as the hypothetical evaluations of the mean wind.

The coefficient is determined by the use of a formula of assumed type to account for an observed variation of wind with height; the material employed consists of two readings of wind-velocity and a typical formula. For the wind-velocities a good deal must depend upon what allowance is made for the variation of barometric gradient, which might, according to our conclusions, account practically for everything. And something also depends on the details of the formula. Taylor based his "over sea" coefficients on the variation of potential temperature; Brunt's examination[2] of the physical process shows that absolute temperature is required, not potential temperature, though in the circumstances the numerical difference may not be very large.

And, moreover, Richardson has pointed out that in problems of diffusion of any entity by turbulence we are concerned with the flow across a given area in a given time. The mean value per unit area per unit of time does not of necessity have a value independent of the size of the area or of the time over which the mean is taken. In other words, we ought to use for each problem we discuss a value of k appropriate to the size of the parcel of air we are discussing. Richardson has shown that the diffusion equation is appropriate to the evaluation of the flow across a surface moving with the mean motion of the air; and in an atmosphere containing eddies of varying sizes, the mean motion depends on the size of the area over which the mean is taken.

[1] *Beitr. Phys. frei. Atmosph.*, Bd. xiv, Heft 1/2, 1928, p. 43.
[2] *Proc. Roy. Soc.* A, vol. 124, 1929, p. 201.

CHAPTER V

GUSTINESS AND CLOUD-SHEETS

On Bacon's principle that "words are but the images of matter" and Ernest Barker's aphorism that "the first necessity of argument is the use of clean words which are always used to denote the same things and connote the same attributes" it might be well to examine the relation of an anemometer record to the traditional use of words like shift, gust, bump and squall, as well as breeze, gale, blizzard and hurricane, and their international equivalents.

THE theory which has been adduced in explanation of the various contingencies in the relation of the surface-wind to the gradient receives strong support of an incidental character from what we have learned about the gustiness and shiftiness of ordinary winds. When the tube-anemometer, devised by W. H. Dines in 1890, was set in operation the wind was seen from the records to consist of a series of rapid alternations of velocity, and when a direction-recorder was subsequently added the alternations in velocity were found to be accompanied by corresponding alternations in direction.

We have stated the general problem of the meteorological calculus as being the interpretation of the record of a pressure-tube anemometer with the understanding that the incidents of history recorded in the trace, in so far as they can be regarded as referring to entities with a certain vitality, are probably the result of some analogy in the local motion of the air to the rotation of a solid. We may recognise something which bears out this suggestion in shiftiness, in gustiness, in squalls, whirlwinds and tornadoes, and the suggestions of revolving fluid to be found in the isobars of weather-maps.

Our first business is with the gustiness and shiftiness which is the common characteristic of anemometer records when there is a reasonable flow.

It is agreed that the gustiness so recorded is due to turbulence: that turbulence is eddy-motion treated statistically, quite unrestricted as to the three dimensions, horizontal and vertical; and that an eddy represents the effort of spin to preserve the identity of the parcel of air which has been made to spin with some analogy to a revolving solid by the interference of an obstacle of some sort with the steady flow of the current.

The illustrations of anemograph records which we have given in chap. 1 make it quite clear that the effect of the eddies which are expressed statistically as turbulence is of the same order of magnitude as the flow, sometimes annihilating it or even reversing it and sometimes doubling its speed.

In this chapter we propose to examine further this common effect of turbulence in the form of gustiness and shiftiness. So long as it is treated statistically it is perfectly manageable with a constant which is allowed unlimited variation according to the variation of the flow with height; but it is important for us to find out, if we can, something of the motion which becomes personified by spin.

As an introduction to this part of the subject we may refer to the model of the double spiral of wind-velocity (fig. 27) on p. 138.

(143)

We have put on it a circle which indicates the relation of the magnitude of one of the possible individual eddies to the dimensions of other things with which they are associated.

We take for example the scale of the model (in its original form) representing wind-velocity on the scale of 1 cm to 1 m/sec. That gives a gradient-wind of 10 cm representing 10 m/sec. If we assume that a flow of 10 m/sec is affected by a gust which will reduce the flow transiently to one-half, and we attribute the effect to air rotating temporarily like a solid, we can get the dimension of the hypothetical solid by the fact that it takes a certain time for the gust to pass the anemometer. If we allow two seconds for the revolving mass to be carried past the anemometer, its effect will be spread over 20 metres. The axis of spin may be in any direction, but with the dimension figured in the model a 10 m/sec wind carrying a 20-metre disc of air rotating about a vertical axis and passing over the anemometer will produce an effect equivalent to a 5 m/sec gust.

If we could imagine the actual wind to be made up of a flow of 10 m/sec filled with rotating masses of air like that represented in the figure, we should get the statistical result which is amenable to the mathematical treatment of chap. IV by supposing the eddies to succeed one another in the effect on the anemometer by passing along at different levels. In fig. 27 the effect of the eddy in the position sketched would be little different from 5 m/sec against wind during the short period of its action, but a similar eddy with its centre at the level of the anemometer would simply add a north component during half its transit and a south component in the other half.

This will go to explain that the quasi-solid masses of air which exhibit their activities as gusts are larger than one would be disposed to think them. These are the gusts which we now propose to discuss.

The number in a given time and the range of consecutive fluctuations which make up the alternations of velocity and direction in an ordinary wind are quite irregular. We have become accustomed to refer to them as the "gustiness" of the wind. The word is not very appropriate because besides these rapid fluctuations of velocity there are other marked increases in the velocity of many winds lasting for some minutes, which we call squalls, and the transition between the normal fluctuations in the wind and the occasional squall is not well expressed by it.

G. W. Walker gave the following figures for the frequency of gusts in relation to the mean velocity of wind at Eskdalemuir. The figures give the number of reversals of the motion of the pen per minute taken over a period of 15 minutes.

Wind velocity mi/hr	37	24	22	22	22	15	15	14	14	9
No. of reversals per min.	21	17	17	17	16	12	11	8	14	9

If we regard the disturbance as carried along with the wind the figures indicate a radius of from 25 to 45 metres for the disturbance.

It has been shown[1] that at a height of 19 metres over grass land the linear

[1] F. J. Scrase, *M. O. Geophysical Memoirs*, No. 52, M. O. 331 b, 1930.

dimensions of eddies are most frequently from 2 to 8 metres but may reach 50 to 70 metres. "One frequently finds disturbances which take a few minutes to pass and indicate eddies or other disturbances of the order of 3 miles in horizontal dimensions."

The gustiness of the anemometer ribbon

The nature of ordinary gustiness will be best understood by the study of the illustrations of the records obtained from an anemometer with its vane 98 ft above the ground at Pyrton Hill by J. S. Dines[1] and reproduced here. They show the separate records of velocity and direction for two occasions with the ordinary time-scale (2 cm equivalent to 5 hrs in the reproduction) and above the regular record in each case is an inset representing the velocity during "quick runs" on the time-scale of 10 cm to 8 minutes.

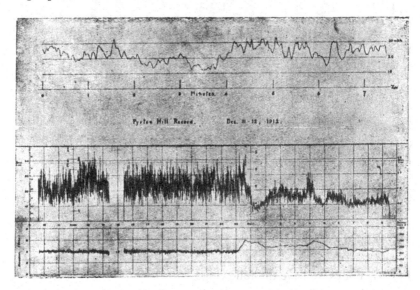

Fig. 28 (v, 1). Record of wind at Pyrton Hill, 11–12 Dec., 1912, with an analysis of the gusts during a quick run for seven minutes between 14 h 30 m and 15 h 20 m.

We may note that in the quick runs the trace of the velocity is a very irregular line and on the closer time-scale the trace appears as an irregular ribbon; the width of this ribbon represents the range of the gusts of which the wind is composed. It will be noticed that speaking in general terms the width of the ribbon varies with the mean velocity of the wind and is roughly speaking proportional to it. It may thus be taken as a measure of the gustiness of the wind which passes the anemometer, and as a numerical measure applicable to winds of different strengths we may use the ratio of the width of the ribbon in a short interval of time to the mean velocity for that interval. Using

[1] Advisory Committee for Aeronautics, *Report*, 1913. *Fourth Report on Wind Structure*, fig. 1 and fig. 2. Blocks lent by H.M. Stationery Office.

this measure we find that the gustiness of winds is different for different situations and is, moreover, different for different orientations in the same situation; in some cases, as, for example, at Shoeburyness, where there is a land-exposure on the one side and a sea-exposure on the other, the difference of gustiness for different orientations is very marked.

Judged by this standard we find that the most gusty exposure for the stations with tube-anemometers which report to the Meteorological Office is that of Dr J. E. Crombie of Dyce, near Aberdeen, where the mast of the ane-mometer projects fifteen feet above surrounding tree tops; the coefficient of gustiness in this case is 1·3. The result caused some surprise as it was thought a belt of trees would reduce turbulence rather than increase it.

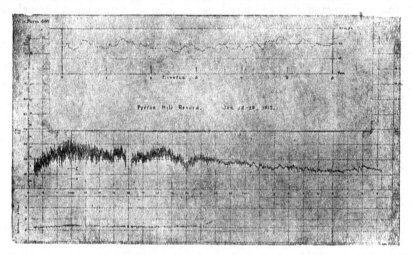

Fig. 29 (v, 2). Record of wind at Pyrton Hill, 28–29 Jan., 1913, with an analysis of the gusts for six minutes at 16 h.

For other sites we have the following:

Marshside, Southport	Coefficient of gustiness				·3
St Mary's, Scilly	,,	,,	,,		·5
Shoeburyness, ENE wind ...	,,	,,	,,		·3
,, W wind	,,	,,	,,		·8
Holyhead (Salt Island)	,,	,,	,,		·5
Pendennis Castle (Falmouth), S wind	,.	,,	,,		·25
,, ,, ,, W wind	,,	,,	,,		·5
Aberdeen (Roof of King's College)	,,	,,	,.		1
Alnwick (Roof of Schoolhouse) ...	,,	,,	,.		·8
Richmond (Roof of Kew Observatory)	,,	,,	,,		1
Lerwick, Shetland	,,	,,	,,		·3
South Kensington (fig. 9)	,,	,,	,,		1·6

The gustiness which is represented by these figures may be regarded as the direct result of the eddy-motion or turbulence due to the obstacles which are presented to the direct flow of the air by the surface and its irregularities, the eddies being more pronounced the greater the resistance presented by the obstacles. Where the surface is nearly flat, as over the sea or a flat spit of land, the turbulence is less marked, but when the air has to make its way over trees or groups of buildings the eddies are larger and more pronounced and the turbulence produces greater effect on the record of the anemometer[1].

A useful piece of evidence confirming this view comes from Eskdalemuir where the anemometer is mounted on the central block of the Observatory. It stands between two other buildings on a hill-side which slopes towards the east. Above the central block on the west is the Superintendent's house, below it on the east is the Assistants' house. Nothing particular is remarked about the easterly winds which pass over the lower structure; but whenever the wind veers from south to west the passage of the wind-shadow of the house at the higher level across the central block is always marked on the record by an increase of the gustiness shown in the traces both of velocity and direction.

J. S. Dines[2] has put together some information about the variation of gustiness with height. He found for two anemometers, one with its vane at 36 ft and the other at 98 ft, that the gustiness factor of the lower (estimated as the ratio of the range of velocity during an hour to the mean velocity for the hour) was 137 per cent of that of the higher, and from observations of the variation of the pull of a kite-wire he found the gustiness to vary with height very differently on different occasions with the general average of a factor of gustiness (range of pull on the kite ÷ mean pull) of 2·5 for the step 0 to 500 ft and 1·5 for the step 500 to 1000 ft. Four ascents with a westerly wind gave for the lower step irregular values 2·6, 3·7, 3·5 and 1·3 and for a south-westerly wind 2·0, 4·5, 2·0 and 1·4, while for seven ascents when the wind had an easterly component the factors were all high but more uniform, namely 2·7, 3·0, 7·3, 2·7, 3·2, 3·5, 2·5. The station is situated on the western slope of the Chiltern Hills. Easterly winds come over the hills.

[1] In a paper on 'Gustiness in particular cases,' *Q. J. Roy. Meteor. Soc.*, vol. LI, 1925, p. 357, A. H. R. Goldie notes the difference in the type of gustiness at Falmouth (Pendennis Castle) in the transition between what might be called equatorial and polar winds, that is to say, between S or SW winds from "over the sea" and W or NW winds from "over the land." In the former the eddies were of uniform type, in the latter there was as a rule a succession of minor squalls and the range of gusts varied very irregularly, but was, on the whole, decidedly less in actual magnitude. If we are agreed about the use of the words employed, the conclusion would suggest that the contrast exhibited in fig. 3 requires some explanation. We gather from what has been said about Eskdalemuir that when an anemometer is mounted on a building the gustiness may be quite sensitive to small differences of orientation such as that between the SE of the upper record of fig. 3 and the SW of Goldie's note. Apart from any question of buildings, at Pendennis SE points to the French coast, SW to the open Atlantic across 12 miles of land, NW across 13 miles of land to Ireland. Open sea lies between 110° and 190°.

There are other curiosities about gustiness. A strong wind with exceptionally small gustiness is recorded for Leafield, an open moorland site near Oxford, by N. K. Johnson in the *Quarterly Journal* for 1928, p. 179; and in *Meteor. Mag.*, vol. LXV, 1930, p. 154, C. S. Durst gives pictures of gustiness at Cardington greater when there was a counterlapse of temperature than in intervals when there was none—contrary to the inference drawn from the study of turbulence.

[2] Advisory Committee for Aeronautics, *Report*, 1911–12. *Third Report on Wind Structure*, p. 219. 1910–11, *Second Report on Wind Structure*. Reports and Memoranda, No. 36.

These factors of gustiness appear to be directly related to the constant k of Taylor's formula and the whole question of gustiness seems to be merely a phase of the general question of eddy-motion.

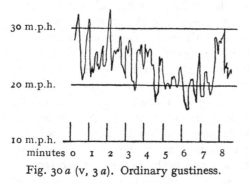

Fig. 30 a (v, 3 a). Ordinary gustiness.

An investigation of the details of the motion represented by gustiness has been made by J. S. Dines[1]. He has pointed out the difference between the record of an anemometer with an open time-scale, represented by fig. 30 a, and the sudden changes in wind-velocity, represented by the examples of fig. 30 b, which last for a minute or more and would be appreciable by an aeroplane as a definite "bump" in consequence of the almost instantaneous transitions between phases of the

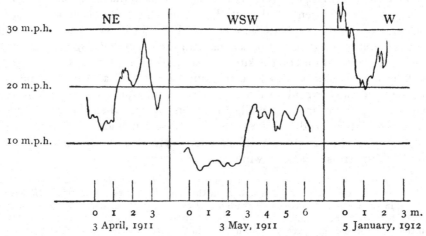

Fig. 30 b (v, 3 b). Sudden changes of wind-velocity of comparatively long duration.

relative motion which last long enough to alter the lift, whereas the fluctuations of ordinary gustiness are alternations which are complete in a few seconds.

An endeavour made by J. S. Dines[2] at the writer's suggestion to analyse the motion of eddies by arranging an anemometer to give a vector-diagram representing the direction and speed of the wind at each moment by a vector-radius drawn from a centre, has failed to indicate any simplification of the idea of turbulent motion in the layers of air near the ground. The diagram

[1] Advisory Committee for Aeronautics. *Third Report on Wind Structure*, p. 216, 1912.
[2] Advisory Committee for Aeronautics. *Second Report on Wind Structure*, 1911, Plate 5, fig. 6.

(fig. 31 *a*) reproduced here shows the record, lasting one minute, of some exceptionally variable winds from the north-east at 36 ft. A corresponding diagram (fig. 31 *b*) with wind of similar range of velocity from the west at 98 ft shows much less violence of oscillation in direction but it must be noted that the large range of the oscillations shown in the diagram may be due to the momentum of the vane which may carry the writing pen beyond the true position of the wind, and in this connexion it may be remarked that with the

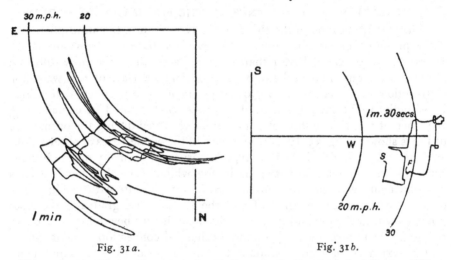

Fig. 31 *a*.   Fig. 31 *b*.

Fig. 31 (v, 4). Vector diagrams of variation of velocity of wind at Pyrton Hill.

view of reducing oscillations of this character a new form of vane has been designed and is now in operation.

J. S. Dines has also taken records of the variations in altitude of a small balloon tethered by a thread 100 ft long to the top of the anemometer pole

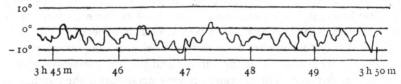

Fig. 32 (v, 5). Theodolite record of altitude of a tethered balloon.

98 ft high at Pyrton Hill. One of these records for 19 January, 1912, is represented in fig. 32, and is reproduced in order to show that the vertical fluctuations of short period in an air-current are of the same kind as the fluctuations in the velocity and direction of the horizontal motion as represented in the trace of an anemograph. It follows that the effect of the imperfect eddies due to the turbulence of the flowing air is to produce the same kind of alterations in the horizontal and vertical directions; and, consequently, the change in direction, the change in the horizontal speed of the wind and the superposed

vertical fluctuations may all three be regarded as aspects of the same physical or dynamical process.

G. I. Taylor has carried the theory of eddy-motion so far as to show that there is a direct numerical relation between the mean variation in the horizontal direction of the wind and that in its speed[1]. He has, moreover, obtained evidence to show that the effect of the turbulence spreads equally in all directions round any selected point[2].

THE RELATION OF TURBULENCE TO THE FORMATION OF CLOUDS

The first application of the theory of eddy-motion in the atmosphere was the explanation of the formation of fog over the banks of Newfoundland[3]. It was shown by G. I. Taylor that by taking account of the eddy-diffusion between a current of air in the upper regions and a surface of cold water the distribution of temperature in the lowest layers of the atmosphere determined by soundings with kites from the deck of s.s. *Scotia* could be explained. A specimen of this distribution of temperature observed during the occurrence of fog is shown in fig. 33. It is quite typical of the whole series of observations and the general application of the type is borne out by many other observations of the temperature of the air in fog which show that fog is always associated with a counterlapse or increase of temperature with height[4]. The coldest stratum is at the ground and the temperature gradually increases upwards within the fog and for some additional height beyond it. The cloud of fog is the immediate effect of the mechanical convection of cold caused by the mixing of consecutive strata in the turbulence of the eddy-motion and in spite of the fact that the coldness of the lowest layers makes for stability and restrains convection. The formation of the cloud extends as far upward as the reduction of temperature is sufficient to produce a mixture, due to the mechanical process of the eddies, which has a temperature below its dew-point. When condensation has begun the further effect of the turbulence is to mix the fog-laden layers as well as to extend their upward boundary so that the water condensed at any particular level has not all to be borne by the air at that level, and the thickness of the cloud is more uniform than the ascertained variation of temperature would lead us to expect.

We must therefore picture to ourselves a set of rolling eddies producing cloud by mixture beginning at the surface and gradually extending upwards as the current flows on. The operation is very persistent if the conditions are maintained, yet it is very self-contained because the coldness of the surface layer always tends towards stability and therefore towards limiting the operation to those layers which are directly affected by the eddies set up at the

[1] Advisory Committee for Aeronautics, *Report*, 1917–18, vol. I, p. 26, London, 1921.

[2] 'Turbulence,' *Q. J. Roy. Meteor. Soc.*, vol. LIII, 1927, pp. 201–11.

[3] *Report of s.s. Scotia to Board of Trade*, 1914.

[4] The reader may refer to the results of observations with kites (particularly those at Brighton by S. H. R. Salmon), *Weekly Weather Report*, 1906 to 1911, and *Geophysical Journal*, M. O. No. 209 d, and to more recent observations by captive-balloon (L. H. G. Dines, *Meteor. Mag.*, vol. LXV, p. 277), and on a wireless-mast (G. S. P. Heywood, *Q. J. Roy. Meteor. Soc.*, vol. LVII, 1931, p. 97).

surface and gradually developed in higher levels, but only by gradual incorporation of the next higher layer with those already affected.

This view of the formation of fog being accepted we have next to note that many, if not all, of the varieties of stratus cloud are also marked by a counter-

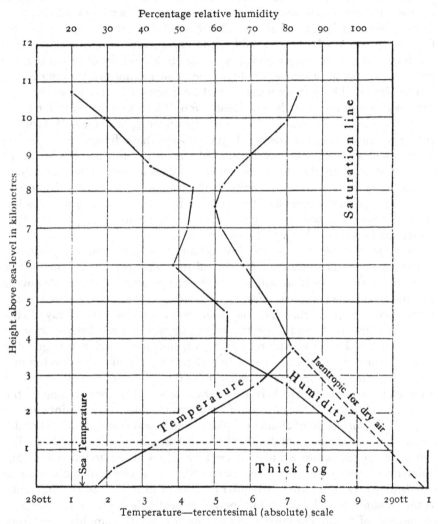

Fig. 33 (v, 6). Captive balloon ascent from s.s. *Scotia*, east of Newfoundland, 4 August, 1913, 19 h. Wind 5 miles per hour (2·2 m/s) at all heights from SE½S (140°).

lapse of temperature in their mass, extending beyond their upper limits. This has been noted on many occasions during kite ascents. Not all clouds have that characteristic but only clouds of certain types such as stratus and strato-cumulus. Cumulus clouds have no inversion of temperature at their tops.

In explanation of this peculiar separation of clouds into two different classes let us consider what would happen if a current, such as that which produces fog gradually extending upwards from a cold surface, passed over a surface which was not colder than but as warm as or warmer than the lowest layer of the moving air. The turbulence due to the eddy-motion would of course exist in that case also, and its coefficient of diffusion upward would be greater than that for a cold surface. The effect of the turbulence would therefore be to cause the layers affected by it to approach the isentropic condition for dry air, that is, to make a greater approach to uniformity of distribution of potential temperature by the gradual process of mixing or churning in the layers affected. This implies a sacrifice of temperature in the upper layers to the advantage of the lower layers. Hence would arise a reduction of temperature of the upper boundary of the stratum affected, as compared with the layer just above it, and consequently just beyond the influence of the turbulence. In other words a counterlapse would be formed marking the boundary of the stratum affected by turbulence just as it does in the case of a stratum in which fog is formed; in which, indeed, the process is exactly similar in this respect, viz., that the change of temperature from the surface upwards is less abrupt from cold to warm and therefore is nearer to the isentropic lapse than if there had been no mixing of the lower layers.

But it is clear that if the air in eddy-motion at any point can be represented as having a vertical component for part of its course there will come a time when the vertical elevation will reduce the temperature of the air below its dew point; cloud will therefore form and the top of the stratum affected by the turbulence will be marked by a layer of clouds or cloudlets, always being formed so long as the eddies persist, travelling with the wind while they are in existence and always being reformed with sufficient regularity to give the idea of a permanent drifting layer. Such clouds are not likely to develop into rain as a general rule because above them by the process of their formation is a layer of counterlapse which is a guarantee of stability; but so long as the surface conditions and the current of air above the surface are maintained so long will the formation of cloud take place. And it will give a very level under surface because the condensation will always take place under exactly similar conditions of temperature which range themselves in horizontal layers; but different samples of air may well have different amounts of moisture and clouds may therefore form in irregular patches or in rolls. And the difference in the amount of condensation may occasion local differences in the thickness of the layer of cloud. The appearance of a layer of strato-cumulus cloud from below is familiar enough. The development of the art of flying and of photography in connexion therewith placed a new method of observation at our disposal. We reproduce, with remarks based upon the original notes, a number of examples of photographs of clouds from above which were supplied by Captain C. K. M. Douglas, R.A.F., in 1918, through the courtesy of the commandant of the Meteorological Section of the Royal Engineers in France.

EDDY CLOUDS: PLATES I, II AND III

PHOTOGRAPHS OF CLOUD-SHEETS FROM AEROPLANES

WITH REMARKS BASED UPON NOTES CONTRIBUTED BY

CAPTAIN C. K. M. DOUGLAS, R.A.F.

Nos. 1, 2, 3, 4, *Sheets of strato-cumulus cloud in ripples, waves or rolls.*

Examples 1, 2 and 4 were taken on the same day, 15 August, 1918; the first, at 1700 feet, in the early morning about 7 h, and the other two in the evening at 18 h when the cloud-sheet of the morning had worked upwards and developed a much more turbulent appearance. The tops of the rolls were then at 5000 feet. The clouds were formed in a light northerly wind on the eastern side of an area of high pressure lying over the English Channel.

The third example shows a cloud-sheet with tops at about 4000 feet formed in a fresh westerly wind at 7 h 30 m on 17 August, 1918.

"These clouds are accompanied by eddy-motion within and below them which keeps up a supply of water-vapour from below. It is the expansion and cooling of this water-vapour as the air carrying it gets diffused up by eddies that causes the clouds. Similar cloud-sheets are common at all heights up to 20,000 ft. The turbulence reduces the temperature at the cloud-level and there is often a rise of temperature above it." It was only 1° F in the first example but in the later examples of the same day when the cloud was several thousand feet higher it had increased to 8° F.

Nos. 5 and 6. *Low Clouds of Lenticular Type.*

"These clouds are of interest as they represent a type rather similar to those shown in Plate XIV of the series recently published by the Meteorological Office[1]. They were however at a much lower altitude, about 2500 to 3000 feet at their upper surface. They were accompanied by very little turbulence and occurred in a stable layer, the temperature being 53° F at 2000 feet and 50° F at 4000 feet." The normal lapse of temperature between these levels is between 5° and 6° F. The peculiarity of the lenticular clouds which have the smooth, gently rounded form so well imitated by the long wreaths or "sastrugi" of examples 5 and 6 is that they seem to be the loci where cloudlets are persistently formed and move independently of the general motion of the cloud-bank. The banks in this case lay in north and south lines and moved from SSW. Comparing their forms with the diagrams of wave-motion in figure 52 (chapter VIII) it is impossible to resist the suggestion that these long banks may represent waves, which are stationary or nearly so, across the current of air which forms the wind. On this occasion there was a wind at 7000 feet from WSW, while the surface-wind was from the south. Northern France was under the northern margin of an extended anticyclone and there was a stationary "low" centred off the Hebrides. [7 h 30 m, 8 Aug., 1918.]

[1] *Cloud Forms according to the International Classification*, M. O. Publication, No. 233, 1918.

236

1. Flat sheet of cloud at about 1700 feet, in rolls or waves of strato-cumulus advancing towards the observer. 15 August, 1918, about 7 h.

237

2. The same layer of clouds as No. 1 at a greater height in the evening of the same day. Strato-cumulus in rolls and hummocks with tops at 5000 feet advancing from the north (obliquely from the right towards the observer facing north-west) with sunlight from the west. The sun is out of the picture on the observer's left. 15 August, 1918, 18 h.

Note the clearer belt beyond the strato-cumulus, over the English Channel, and the bank of clouds on the horizon over England.

238

3. Strato-cumulus clouds about 4000 feet with high clouds in another layer above, probably alto-stratus which is generally at 10,000 feet or higher. 17 August, 1918, 7 h 30 m.

239

4. Another view of the cloud-sheet represented in No. 2 on the opposite page looking east with the sun behind the observer. The lines of cloud are moving obliquely away from the observer from the foreground on the left to the background on the right. 15 August, 1918, 18 h.

Note the more turbulent appearance of the evening clouds represented in Nos. 2 and 4 as compared with the morning clouds represented in Nos. 1 and 3.

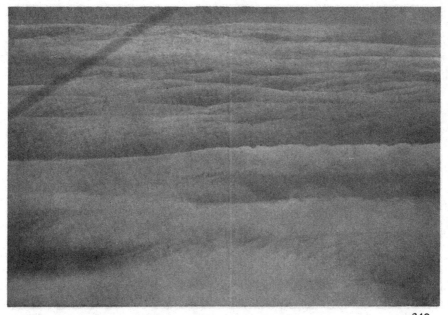

240

5. A vast field of long strips or bands of cloud nearly parallel, with smooth and lightly rounded upper surfaces in the form of "hogs-backs," like wreaths of drifted snow (sastrugi).

241

6. Another part of the cloud-sheet, No. 5, viewed to NE against the light showing an eruption of cumulus heads in the middle distance. Overhead is another cloud-sheet at 7500 feet, of which the margin is seen at the top of the picture. 8 August, 1918, 7 h 30 m.

The bands of cloud lay roughly north and south. The wind at their level was from SSW. The form suggests wave-motion across the bands from W to E.

There are a number of well-known phenomena which may be accounted for on this hypothesis of cloud marking the boundary of the stratum affected by the turbulence of the surface. There is first of all the well-known circumstance that the sky over our Islands may be covered by a layer of cloud the whole day through without allowing a break for the sun on the one hand or developing into rain on the other. The adjustment which is necessary to cause condensation without rain is a very delicate one. It could hardly be attempted in a physical laboratory by any simple process of reduction of air pressure, but if arrangement could be made for a certain limited amount of rarefaction to be superposed upon a gradual diminution of temperature with height the necessary adjustment might be made and such an arrangement seems to be secured by the effect of eddy-motion upon a flowing current when the surface is as warm as, or warmer than the air above it. The explanation is even more *à propos* if we consider that layers of cloud such as are here spoken of have a marked tendency to disappear over land stations towards or after sunset when the level reached by the turbulent motion becomes lower. We have seen that the height to which turbulence extends has a marked diurnal variation, being increased by the accumulated warmth of the day. D. Brunt[1] has pointed out that there is a diurnal variation of cloud at Richmond (Kew Observatory), which is a very good indicator of the diurnal variation of the thickness of the layer affected by turbulence. (Vol. II, p. 164, see also chap. V of vol. III.)

Clouds of this character are often formed in easterly winds, which at the surface are dry winds. They are indeed so characteristic of anticyclonic weather in winter, that they have received from W. H. Dines the name of anticyclonic gloom. It should be noted that these clouds are to be found in the current which brings the surface-wind. They do not belong, as one is apt to think, to the transition between the surface-wind and upper winds in the opposite direction coming from a warmer quarter with a larger supply of water, or to any other process of mixing of currents from different sources in the upper air; they are developed in the body of the surface-current itself[2].

Another example which was frequently noticed in the course of the four years of war is the heavy cloud in winds from the north which have a long

---

[1] *M. O. Professional Notes*, No. 1.

[2] In the course of correspondence W. H. Dines expressed the opinion that loss of heat by radiation is the most probable cause of the cloud in the still air of a winter anticyclone and other clouds of stratus-type. The suggestion deserves more careful consideration than is possible at this stage. The relation between the ultimate effects of radiation and the formation of cloud is a very complicated thermodynamical process. The immediate effect is thermal: the next step is the dynamical process of the adjustment of level according to the entropy or potential temperature. The final result of local loss of heat by radiation may be the "warming" of the air which has lost its heat. See Shaw, 'La Lune mange les Nuages,' *Q. J. Roy. Meteor. Soc.*, vol. XXVIII, 1902, p. 95, reproduced in *Forecasting Weather*, 2nd edn. p. 245. It is doubtful whether any hypothesis can rely upon air being still for dynamical purposes. There is always a slow drift even in the densest fog as mentioned elsewhere. G. I. Taylor has represented that the pattern of the eddy-motion of the atmosphere is independent of the mean velocity of the air-current to which the eddy-velocity is proportional. Hence the motion in an anticyclone may be only a slow model of the same pattern in a stronger current which in the end causes the same thermal effect.

"fetch" over the North Sea. The cloud gets heavier in the southern part of its course and not infrequently develops into a persistent drizzle of rain over Flanders and northern France. Winds from the same quarter often give clear weather in the central and southern parts of England with cloud on the eastern coasts. We may perhaps attribute the difference to the enhancement of the coefficient of eddy-conductivity by the warmth of the water and to the increase of water vapour by evaporation from the sea.

A third example is to be found in the clouds of the trade-winds which take the form of strings of cumulus at a uniform level. According to Piazzi Smyth the level of clouds in the north-east trade-wind at Teneriffe was 5000 ft, and the thickness of the trade-wind itself was about 10,000 ft, so that the clouds were in the middle height of the great current of air, not in its upper margin.

At St Helena in the heart of the south-east trade-wind the mechanical effect of the Island itself, as an obstacle to the current, increases the amount of condensation; and to such an extent that near the top of the Island at 2000 ft above sea-level the air preserves an almost uniform humidity of 90 per cent at 9 a.m. and the mean amount of cloud, the year through, at that hour is 85 per cent.

The particular type of cloud which is formed by the process which we have described depends upon the lapse rate of temperature in the region in which the condensation begins. If the lapse rate approaches that of the adiabatic for saturated air the initial condensation may give rise to cumulus heads, or even develop into a shower, whereas if the lapse of temperature is not near to that of the adiabatic for saturated air the cloud layer must remain thin as the condensation will be dependent upon the forced vertical motion due to the eddies.

It seems possible that the scud which is often to be seen drifting in mid-air under a nimbus cloud after heavy rain is similarly due to the eddy-motion of the lowest layers operating upon the saturated air close to the ground. Eddy-motion is operative even with light winds, and when the lowest layers are completely saturated very small amounts of forced elevation due to turbulence would be sufficient to produce condensation where the eddies have some upward movement. It is noteworthy that the fragments of cloud here spoken of tend to arrange themselves at a definite level and it is possible that the remarkably level line along the slopes shown by morning clouds on a mountain side may be due to the regularity of the mixing of the surface layers in the eddy-motion of the slowly moving air. We have to remember that there can be no permanent cloud in perfectly still air. The lightest fog would settle if it were enclosed and kept still. Fog and cloud are always in a state of motion, sometimes only moving slowly but never still, and it would appear that the turbulence of the motion is necessary to keep the cloud in suspension.

In summarising this section let us remark that it has hitherto been usual to regard cloud as being associated with cyclonic weather and the upward convection of columns or limited masses of air. From the consideration of the inevitable effect of the churning of successive layers of air by the eddies which constitute the turbulence of currents of air moving over sea or land it becomes evident that if the process goes on unaltered for a sufficient length of time the formation of cloud must occur, immediately at the surface if that is cold enough to give a mixture below the dew-point, in the upper layers if the surface is not colder than the air which flows over it; and in that way we may account for the formation of the many clouds which have been shown to have a counterlapse of temperature within them and above them, and also of certain detached clouds formed like scud in exceptional positions.

Thus we may regard cloud in some frequent forms as being associated with currents of air of long fetch[1] whether they belong to cyclones or anticyclones. Over the land the diurnal variation of temperature by affecting the eddy-motion introduces a corresponding diurnal variation of the conditions for the formation of cloud; over the sea the long travel of wind must end in cloud of some sort, unless the diffusion upward of the eddy-motion is restrained by some lid of counterlapse of temperature which confines the effect of the churning of the surface layers to the production of a shallow isentropic atmosphere in the lowest layers, and the moisture of the surface layer is not sufficient to cause saturation within the stratum of turbulent air. Such is probably the daily experience of the air over the hot deserts of the tropical regions. The effect is represented in fig. 81 of vol. III.

In continuation of this subject the following notes are apposite.

1. L. H. G. Dines, of Valencia Observatory, Cahirciveen, has written that at Valencia with a north-west wind which travels over successively warmer water, the weather is almost always of a violent squally type with vigorous convection to a height of 10,000 ft or so and violent showers, the air between the showers being comparatively dry.

2. It may be remarked that the cooling of the air in the higher levels indicated by the formation of cloud, in consequence of the mixture of layers by eddy-motion, must have as its counterpart the warming of the air near the surface, and we may thus account for the relative warmth of the eastern side of Britain in a persistent WSW wind which was remarked upon in *Nature* by H. Harries, 9 Jan. and W. H. Dines, 16 Jan., 1919. The dynamical effect was estimated quantitatively by Lieut. John Logie, R.A.F.

Föhn and chinook winds

There is another class of phenomena of weather for which the aid of the theory of eddy-motion must be invoked; these are the warm, dry, oppressive winds in the valleys on the lee-side of mountain ranges. They came up for discussion under the heading of convection in vol. III. On the northern side of the Alps such winds are well known as *föhn* winds and in the prairie country to the east of the Canadian Rocky Mountains as the *chinook*. It is a common practice to explain the hotness and simultaneous dryness of these winds by tracing the history of the air from low levels on the windward side through a period of rarefaction and reduction of temperature, as it is driven up the slope, culminating in the condensation of vapour and the formation of rain about the ridge, during which the great store of latent heat of evaporation is set free. By that time the air has acquired a greatly enhanced potential temperature, or an increased entropy as we may regard it, and the operations

[1] M. A. Giblett, 'Some problems connected with evaporation from large expanses of water,' *Proc. Roy. Soc.* A, vol. XCIX, 1921, p. 472.

terminate in the realisation of this entropy in the raised temperature of the air when it reaches the lower levels again with its water taken away and a store of heat left in its stead.

This simple life-history is not quite satisfactory because there is no sufficient reason adduced for the air of the valleys on the windward side to climb up to the ridge, nor for it to get down again from the ridge to the valleys on the leeward side. Various reasons may be urged for some amount of climbing due to mechanical forces on the windward side; but there is nothing to be said in favour of the air at the ridge with its great store of entropy finding its way into the valleys as a warm dry wind except that the eddy-motion in the current which passes the ridge will gradually excavate the cold air from the valleys on the lee-side and ultimately the régime on the lee-side of the ridge will be a flow of isentropic air extending in thickness from above the ridge even to the deeper valleys. The cooling effect of the surface as the warm air reaches it will produce a certain amount of stability in the lowest layers. But the process is not completed all at once. Some interesting details of its circumstances are given in von Ficker's and Frl. Lammert's papers[1].

Eddies and the general turbulence of the atmosphere

Before bringing our consideration of eddy-motion to a close we will call attention to two examples that will enable the reader to carry in his mind a general idea of the state of turbulence which exists in the atmosphere wherever an air-current passes along a boundary surface with a density different from its own, or, indeed, presumably wherever a discontinuity of velocity is associated with a discontinuity of density, for it is not the particular difference of density between water and air, or between the ground and the air above, that causes waves in the water and eddies in the air, but the existence of a discontinuity in the density, which may be called infinite when the boundary is solid, very large when the boundary is water, and very small, though still operative, when the boundary is a distinctly heavier gas. The process of the formation of eddies in such cases is most easily seen in the case of a liquid passing an obstacle. This case is represented in an illustration (fig. 34) taken from a report of the Advisory Committee for Aeronautics[2]. It will be seen that, as the current passes the plate which forms the obstacle, eddies—in reality unfinished and therefore unprotected vortices—are formed at regular intervals in the current. They travel along with the current and gradually disintegrate, filling that portion of the stream with irregular eddy-motion. A succession of obstacles would imply a corresponding series of disintegrating eddies having some recognisable form near the obstacle which causes them; in the further distance

[1] Heinz von Ficker, *Innsbrucker Föhn-studien*, Wien. *Denkschr. Ak. Wiss.*, vol. LXXVIII, 1905, p. 83. L. Lammert, *Veröff. d. Geophys. Inst. d. Univ. Leipzig*, Bd. 2, Heft 7, 1920; see also R. Streiff-Becker, 'Altes und Neues über den Glarner-Föhn,' *Glarus*, ' *Mitt.* 1930' *der Natf. Ges. des Kantons Glarus*, 1930.

[2] *Report*, 1909–10. Reports and Memoranda, No. 31, fig. 7. Block lent by H.M. Stationery Office.

they have no defined form but merely fill the current with irregular turbulence. As the velocity of the stream is increased the eddies are formed at shorter intervals until a practically permanent eddy is formed at the obstacle itself.

A permanent eddy of this kind is formed at the edges of cliffs or the ridges of houses or walls in all strong winds. When the wind blows upon the steep face of a cliff an observer standing at the edge may be effectually screened from the direct wind by the deflexion of the current upward in the formation of the eddy and he may feel only the return current which comes towards the edge of the cliff completing the circulation in the interior of the eddy. The writer recalls a remarkable example of a westerly gale at Dover in the spring of 1889 when the top of the Admiralty pier, apparently exposed to the full force of the wind, was the only place in Dover where it was possible to walk without discomfort on account of the violence of the wind. The protection was quite absent from the part of the pier where it joined the land and where the wind could travel up the slope of the beach without forming an eddy of the same size as that due to the nearly vertical wall

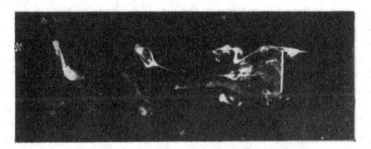

Fig. 34 (v, 7). A succession of coloured eddies formed in a current of water passing a plane obstacle. (National Physical Laboratory.)

When the air passes out to sea along a level surface at the top of a cliff a well-marked permanent eddy is formed on the face of the cliff. Photographs illustrating the course of a balloon in an eddy thus formed are included in the collection of photographs at the Meteorological Office.

The reader can make experiments for himself, simply with an empty match box or even his own hat, in the eddy formed by a strong wind blowing upon a nearly vertical cliff. A most remarkable example of a cliff-eddy can be found at the Rock of Gibraltar when a strong levanter blows on the steep eastern face of the Rock. Its effect upon the tube-anemometer which was maintained at the signal station on the Rock was very remarkable. When the velocity of the wind reached a certain limit it passed the opening of the anemometer in a direction nearly vertical and the effect was a reduction of pressure in the recording float. A limit is thus fixed to the velocity which the instrument can record and gusts of greater velocity appear on the record as entirely fictitious lulls, due to the withdrawal of the pen to the zero line by the "suction" of the air passing the anemometer. The sheet of air which forms the eddy in this case goes upward for some hundreds of feet. The phenomena were investigated by H. Harries of the Meteorological Office by means of small balloons and balls of cotton wool during a visit to Gibraltar. A description was contributed to the Royal Meteorological Society and to the discussion of a paper before the Aeronautical Society[1] in January, 1914.

Except in strong cliff-eddies it is not easy to trace the progress of a part of the air which forms the eddy by floating an object in it. In the ordinary way the motion in the eddy is superposed on the flow of air and should appear to the observer as the displacement of a few metres from the line of flow which itself is not necessarily horizontal. Reference has been made in vol. III to G. M. B. Dobson's endeavour to trace eddy-motion by "no lift" balloons in the flow of air on a sunny day over Salisbury Plain. It eventuated in the balloon making an upward travel of 1000 feet which we have supposed might be the indication of an isentropic surface.

The reader may form a good idea of what happens on a smaller scale by watching the smoke which issues from a tall chimney in a strong wind[2]. It is obviously in a state of turmoil but gradually spreads out laterally and vertically by the action of the eddy-motion in the air. If we imagine the process represented by the trail of smoke to be continued for a distance of some hundreds of miles we can form an effective idea of the result of the spreading upwards of eddy-motion due to turbulence, and we may also find some instruction in the superficial analogy between the trail of smoke from a factory chimney and the trace of a tube-anemometer. If the reader will imagine a factory chimney with its top in the trace of fig. 29 of this chapter at the point indicated by "midt" he will find the trace on the left very suggestive of a smoke trail. We will leave him to think out for himself how far the superficial analogy has a real significance. At first sight it looks as though the chimney was an essential factor in the production of the eddies which appear to spread from it, but the eddies are already in the atmosphere and the smoke only makes their effect visible. They are not due to the motion past the chimney. A. Mallock[3] has usefully pointed out that the trail of smoke left by a steamer travelling through still air is not disturbed by eddies. There is ample relative motion of the funnel with respect to the air but there are no eddies in the surrounding air.

On a somewhat larger scale interesting effects of eddy-motion can be observed on the lee side of a big ship in a steady cross wind. Gulls can keep their flight without any flap of the wings, and light objects like thin paper are carried smoothly to great heights. A cross wind is rather better for observing than no wind at all; but even in that the upward run of the air in the eddy in the wake of the ship is sufficient to carry the sea-bird, with little or no effort on its part except the adjustment of the wings to the fuselage. The same kind

---

[1] Harries, *Q. J. Roy. Meteor. Soc.*, vol. XL, 1914, p. 13; Shaw, 'Wind Gusts and the Structure of Aerial Disturbances,' *Aero. Soc. Jour.*, 1914, p. 172.

[2] A. E. M. Geddes, 'Some notes on atmospheric turbulence,' *Q. J. Roy. Meteor. Soc.*, vol. XLIX, 1923, p. 1; L. F. Richardson, *Phil. Trans.* A, vol. CCXXI, 1921, p. 1.

[3] Advisory Committee for Aeronautics, *Report*, 1912–13, p. 329.

of atmospheric condition is one of the means indicated in chap. VIII by which a gliding aeroplane supports itself.

The analysis of air-motion

So far our information has been derived from anemometers or aeroplanes, or computed from the theory of turbulence. A different method has been adopted by W. Schmidt in a memoir on the structure of winds[1]. He explains that an anemometer gives a record for a single point only of the atmospheric structure. In order to study the variation in the velocity of wind in different parts of an advancing front he places in the face of the wind a number of rings of wire covered with silk tulle suspended in 50-cm frames from axes supported on masts by brackets or stretched wires. The strength of the wind is indicated by the deflexion of each indicator and the simultaneous deflexion of each is obtained by kinematographic photography with one or more exposures per second. In that way an endeavour is made to analyse the advancing front of air into laminar or vortical motion, with the result that vortical motion is found to be rare. Diagrams representing the structure are given for a number of localities which show a complication difficult to express in words.

Another point of detail about atmospheric structure is the indication of vertical oscillations of pilot-balloons to which attention has been drawn by E. Fontseré[2].

Further particulars about special features of gustiness, especially in relation to aviation, are available[3].

Squalls

We have now completed our discussion of the foot of the atmospheric structure which brings us "nel mezzo del cammin" for this volume. Looking back to the statement of the problem at the end of chap. I we may note that when we have made allowance for the interference incidental to "the exposure" the general run of the record is in direct relation with the distribution of pressure. To this part of the problem we shall return in later chapters. Meanwhile we shall have something to say about the winds in that region of the troposphere which we have called the limb of the structure.

The irregularity of the wind as represented by the ribbon has been classed as gustiness and attributed to turbulence, for which a certain statistical order in disorder has been established.

With these explanations we have courage to face most of the items of the statement. Exceptions are exhibited in figs. 4, 5 and 6. Fig. 4 shows some

---

[1] *Sitzber. Akad. Wiss. Wien,* 'Mathem.-naturw. Kl. Abt. IIa,' Bd. 138, Heft 3 und 4, 1929.

[2] Commission for the Exploration of the Upper Air, *Report of the meeting at Leipzig,* M. O. 300, 1928, p. 59.

[3] 'The relation of bumpiness to lapse of temperature at El Khanka from 27 July to 3 August, 1920,' *M.O. Professional Notes* No. 20, 1921, p. 118; C. W. B. Normand, 'Meteorological conditions affecting aviation in Mesopotamia,' *Q. J. Roy. Meteor. Soc.,* vol. XLV, 1919, p. 376; G. I. Taylor, 'Turbulence,' *ibid.,* vol. LIII, 1927, p. 201; L. F. Richardson, 'Turbulence and vertical temperature difference near trees,' *Phil. Mag.,* vol. XLIX, 1925, p. 81.

remarkable oscillations in the wind at Southport on 7 Jan. 1907. The oscillations were repeated in the record of 12 Jan. 1907, and explained in the *Quarterly Journal of the Royal Meteorological Society* for 1910 as a kind of resonance arising from the configuration of the distant environment of the station. Similar oscillations are shown in a record for Eskdalemuir, 6 March, 1918 (*Forecasting Weather*, p. 355), and at Fleetwood, 4–5 February, 1927 (vol. III, fig. 12).

Figs. 5 and 6 show notable features which are attributed to squalls. Fig. 5 is an example of the severe squall which often ushers in the most active stage of a thunderstorm, and belongs to the order of line-squalls which have been noticed in vols. II and III. Fig. 6 shows a succession of squalls or strong winds of from ten to twenty minutes' duration. For these we have no explanation ready made. They might easily be represented as secondary refractions of the isobars of an advancing cold front, and are therefore also to be associated with line-squalls or the rain-squalls of fig. 11 of vol. III.

On the continent such squalls are treated as *grains* or *Böen*, and have a literature of their own, as for example Durand-Gréville's[1] 'Rubans et Couloirs de Grain'; Moltchanoff's[2] 'Die Struktur der Böen in der freien Atmosphäre'; W. Peppler's[3] 'Über starke Vertikalböen in der freien Atmosphäre.'

In taking leave of the foot of the structure and its gustiness we find it too shifty a foundation for a home of the gods who control the weather. Noisy Jupiter may find accommodation there for his electrical apparatus, but the home of the real weather gods is above the low clouds and we are not likely to understand the structure until we have mapped the sequence of changes from their point of view.

That may not be possible yet for lack of observations; but what is certainly possible if it were deliberately attempted is an occasional map of the weather in the 4 km level, which is at any rate beyond the reach of the shiftiness of the foot.

What is important for our purpose is that the foot should be regarded as extending from the ground to the first limit of the typical variation of wind with height, which may be marked as the region of motion under balanced forces where with straight isobars there is equality between the geostrophic wind and the actual wind. From the pictures of fig. 35 (p. 162) it would appear to be distinctly less than 500 metres, but that refers to a particular exposure on Salisbury Plain. From another point of view the limit of the foot may be regarded as that of diurnal variation of temperature, which on p. 112 of vol. II we have found to be about one kilometre, from that level upwards we may regard the limb as expressing the structure.

[1] *Bulletin de la Société Belge d'Astronomie*, 1906.
[2] *Beitr. Phys. frei. Atmosph.*, Bd. XVI, 1930, p. 152.
[3] *Ibid.*, p. 115.

CHAPTER VI

KINEMATICS OF THE LIMB AND TRUNK

The structure of the atmosphere has not yet been reduced to an arithmetical or algebraical formula. Graphic representation affords a means of exhibiting the relation between the elements by which the structure is expressed.

in mournful numbers
Life is but an empty dream.

VARIATION OF WIND WITH HEIGHT IN THE UPPER AIR DISCLOSED BY PILOT-BALLOONS[1]

FOR the first stage of our inquiry into the problem of the record of the wind we have devoted our attention to the relation between the observed wind and the geostrophic wind at the surface where the disturbance of the relation between the wind and the distribution of pressure is certainly very considerable, and to the variation with height of the direction and velocity of the winds in the lowest layers. We have found that the observed phenomena in relation to these matters are the natural consequences of the eddy-motion of the surface layer and that the wind gets more nearly in accord with the gradient as the effect of the eddy-motion becomes less marked. We have laid down no specification of the range of the lower layers. We have taken the first kilometre as being probably affected by the turbulence. We have given figures for variation of velocity up to 500 metres without any particular stress upon the selection of that level. The correlation coefficients between the deviations of temperature and pressure referred to on page 121 indicate 2 kilometres as the limit of disturbance for the middle of England in winter and 3 kilometres in summer.

According to the plan of this work the description of the normal structure should be included in vol. II. Our task in this chapter is to lead from that

---

[1] Observations of pilot-balloons are very numerous. The number of observations increased very largely during the war. The results referred to in this chapter are derived from the discussions of various authors for which references are given in the text. Records of additional observations are to be found in various official publications: in the *Geophysical Journal* (M. O. Publication, No. 209 d, etc.), in the official publications of the Royal Prussian Aeronautical Observatory, Lindenberg, in a special daily publication of the Italian Meteorological Institute, in the publications of the Weather Bureau of the United States, and in the monthly publications of the International Commission for Scientific Aeronautics. A summary of the observations with a single theodolite at Aberdeen by A. E. M. Geddes, is given in a publication of the University Press of Aberdeen, 1915. The observations with kites which previously formed the chief basis of our exact knowledge of the winds in the upper air up to the level of about three kilometres were discussed by E. Gold in *Barometric Gradient and Wind Force* (M. O. Publication, No. 190, 1908), and in *Geophysical Memoirs*, No. 5 (M. O. Publication, No. 210 e, 1911). The reader may legitimately complain that the illustrations in this chapter are restricted to observations in the British Isles but the structure of the atmosphere in general is so complicated that in order to keep in mind some unity of ideas it seemed best to deal in the first instance with the problem as localised by a selection of the whole number of available observations.

For additional references to data published since the close of the war see p. 187.

structure to the dynamics of the records of individual occasions. This excuses to some extent the limitation of our illustrations to the ascents for which full details were available.

From the nature of the method which has been followed the natural course would be to regard the position at which the geostrophic wind is attained, if we knew it, as marking the limit of the disturbance caused by the surface, and it has been the practice of the Meteorological Office to quote the value of the geostrophic wind determined from the isobars of the daily maps as probably representing with sufficient accuracy the actual wind at the level of about 1500 or 2000 feet or 500 metres. But at this stage we are not in a position to go further than that in defining the limit of interference of the surface with the free course of the air, because the balance which we postulate

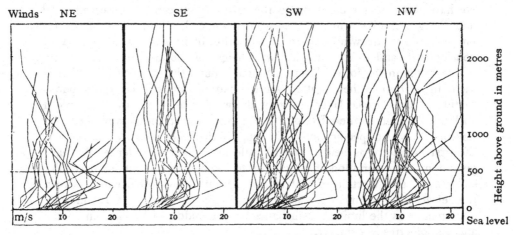

Fig. 35 (VI, 1). Change of wind-velocity with height at Upavon, between the level of 2500 metres and the surface, for winds in different quadrants.

is between the wind and the distribution of pressure at the same level. In order to make the comparison with the wind at 1500 or 2000 ft we ought to obtain the distribution of pressure at those levels. Before approaching that part of the subject it will be best to examine the variations of wind with height in the levels which we may suppose provisionally to be for practical purposes free from the disturbing influence of the turbulence due to the surface, in order to obtain some guidance as to the changes to be expected from other causes than eddy-motion.

We shall find the structure of the atmosphere from this point of view extremely complicated and we must seek the probable cause of these complications. As giving a general representation of the different cases that arise we will take the observations with pilot-balloons made at Upavon by G. M. B. Dobson[1] and discussed by him in a paper before the Royal Meteorological

[1] G. M. B. Dobson, 'Pilot Balloon Ascents at the Central Flying School, Upavon, during the year 1913,' Q. J. Roy. Meteor. Soc., vol. XL, 1914, p. 123.

Society in 1914. It gives an excellent summary of results for the layer of air from the surface to the level of nearly 2500 m (8000 ft) at a well-exposed station. This height includes the surface layer and gives also an insight into the variations through a suitable range for the next stage. The soundings are 97 in number, the great majority being followed with one theodolite only. The site of the observatory is 183 m (600 ft) above sea-level and the situation is very favourable because it gives a large expanse of nearly level plain.

From Dobson's paper we take four composite diagrams, polygraphs of the individual ascents for winds in the four quadrants, 18 in the north-east, 18 in the south-east, 33 in the south-west, and 29 in the north-west respectively. We have added to the diagrams as published in the original paper a full line at 500 m and also the line of sea-level. The line at 500 m divides the height of the ascents into two parts. Within the lower part is the rapid increase of the wind in the layer at the surface which is characteristic of all the winds and shows generally the effect of the surface-turbulence. Yet there are cases in all the quadrants, and particularly in the north-west quadrant, in which the first step from the surface shows a diminution of velocity. We shall refer to this peculiarity later, on page 171.

Looking at the appearance of the curves in the composite diagrams the direct effect of the surface seems to have come to an end before the level of 500 m was reached. We may also notice as a general feature the remarkable irregularities that are disclosed in all the quadrants. We may fairly say that the winds in the north-east quadrant fall off with height beyond 1000 m, that those in the south-east quadrant remain steady[1], those in the south-west and north-west quadrants show all variations from approximate uniformity to large increases of velocity in the upper levels, but above and against all these general statements must be written that exceptions do occur and the variations on different occasions follow no absolute rule. That is the situation which we have to face in dealing with the observations of winds in the second stage between 500 m and 2500 m.

Numerically the results are summarised as follows: as regards velocity, on the average the geostrophic wind corresponding with the surface-gradient was just reached at 915 m *with NE winds* and then the velocity *began to diminish*. *With SE winds* the velocity reached the geostrophic wind below 300 m and *kept quite close to it* from that level upwards, *with SW winds* the calculated value was reached near the 500 metre-level and thereafter the velocity *increased to* 117 *per cent* of the geostrophic wind at about 2500 m and *with NW winds* the geostrophic wind was reached below 300 m and thereafter *the velocity increased* at 8000 ft (2500 m) *to* 145 *per cent* of the calculated value. On the other hand, as regards direction, NE winds, starting with an average deviation of 27° at 50 m, only got within 6°

---

[1] This characteristic of winds in the south-east quadrant is borne out by Cave's observations at Ditcham Park which are referred to later. Cave's class of "solid current" included a large number of examples in the south-east quadrant.

of the line of the surface isobar even at the highest point tabulated; SE winds with an initial deviation of 24° veered to the isobar at 600 metres and passed beyond it. SW winds in like manner starting from a deviation of 19° passed the line of the isobar at about 800 metres and carried the veer 16° farther, and NW winds starting with a deviation of 11° kept close to the line of the surface isobar between 600 m and 1200 m and then veered 8° from it.

Grouped according to the velocity at 605 metres, light winds (below 4·5 m/sec) showed little increase of velocity with height and do not seem to have quite reached the geostrophic wind below 1000 m. Moderate winds (between 4·5 m/sec and 13 m/sec) attained the geostrophic velocity at 300 m and strong winds (above 13 m/sec) at 500 m.

For light winds the surface velocity is 87 per cent of the wind at 650 m, for moderate winds 64 per cent and for strong winds 56 per cent. Light winds also show considerably less change of direction with height than do moderate or strong winds. "It is remarkable that the moderate winds should go on veering considerably after the direction of the isobar has been reached."

These results as regards the relation of the surface-winds to the upper winds are in accordance with the conclusions which have already been reached in chap. IV, and details of the characteristics of the variation with height of the velocity and direction of strong winds have been used by Taylor, as we have seen, to verify his theory of the effect of turbulence. We have selected this particular group of observations partly on account of the favourable nature of the site and partly because they are based mainly upon the method of the single theodolite which is now in daily use at many stations. The results include any defects of the method but they present the problems which have to be faced in considering the observations from the stations where that method is employed.

As giving a summary of results obtained from seventy-three soundings made with special care, mainly by the use of two self-recording theodolites, we take from the third report of the Advisory Committee for Aeronautics[1] J. S. Dines's discussion of the results obtained from pilot-balloon ascents at Pyrton Hill, 1910–11. The site is in very open country 150 m above sea-level on the western slope of the Chiltern Hills near Watlington in Oxfordshire. But it has, on that account, some disadvantages for work with pilot-balloons. "The surrounding hills subtend an angle of 10° above the horizon from N round through E to S. From S the altitude falls off to zero at W, and from that point to N the horizon is clear." The peculiarities of the site limited the range of observation with strong winds from the western quadrants and protected the surface against strong winds from the eastern side. The observations are therefore confined mainly to days with a surface-wind below 7 m/sec. Nor are all wind-directions represented. There are no observations from south-east or south-west. Their absence is accounted for

---

[1] *Third Report on Wind Structure*, Advisory Committee for Aeronautics, *Reports and Memoranda*, No. 47, 1911–12.

partly by the site and partly by low clouds which are a common accompaniment of winds from those quarters.

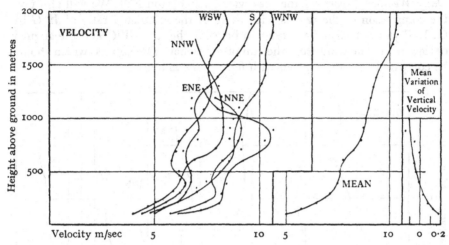

Fig. 36 a (VI, 2 a). Variation of the velocity of wind with height for soundings grouped according to the direction at the surface (J. S. Dines, 1910–11).

The results for groups arranged according to the wind-directions and also for the means of all the observations at successive levels at stated hours are represented in the diagrams of fig. 36. 36a shows the characteristic points of the results of soundings with pilot-balloons, the rapid increase of velocity near the surface with winds from any quarter, which ceases below the level of 500 m for all the examples plotted except that of the winds grouped as WNW. There is a notable falling off at about 1000 m of the velocity of the winds grouped as NNE and ENE, especially the latter, after a considerable further increase from the limit of the first increase of the surface-wind. The winds grouped as S and WSW show an irregular but generally continuous increase up to the level of 2000 m which is most pronounced in the WSW group. The inset showing mean variation of the vertical velocity is referred to on p. 167.

Variations in direction are shown in fig. 36b. The mean curve on the right of the diagram shows a veering with height amounting to nearly 30° at 1500 m. The effect is most striking with S which veers 45°, and NNW which, after a definite backing at middle heights, approximates to the same veer.

These characteristics are generally in accord with the results already quoted for Upavon on p. 162 but it must be understood that even when the observations are with two theodolites the mean values include examples which deviate a good deal in various ways from the mean. We cannot yet make any satisfactory generalisation as to the variation of wind with height above the level of 500 m that can be regarded as applicable in all individual cases.

Before passing on to other examples of the structure of the upper air disclosed by pilot-balloons we may interpolate a note as to the relation of the winds at 50 m (the lowest level of observation in the series now under consideration) with the geostrophic wind as determined from the maps of the *Daily Weather Report* for the observations at Pyrton Hill. We will also give the comparison of the results with those of the calculated ratio of W/G by G. I. Taylor's theory of the relation between the ratio W/G and the angular deviation α of the wind from the run of the isobar. The figures, which should be compared with those given with fig. 17, are as follows:

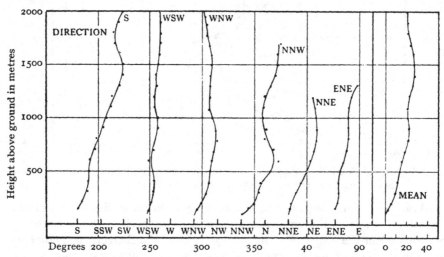

Fig. 36 *b* (VI, 2*b*). Variation of the direction of wind with height at Pyrton Hill.

TABLE I. COMPARISON OF OBSERVED AND CALCULATED VALUES FOR THE RATIO OF THE WIND AT 50 METRES TO THE GEOSTROPHIC WIND.

		S	WSW	WNW	NNW	NNE	ENE	Mean
W (100 m)	m/sec	5·0	4·25	4·75	4·25	5·75	6·0	5·0
G (from D. W. R.)	m/sec	9·5	9·5	7·75	7·75	7·5	12·5	9·0
W/G ("observed")		·55	·45	·61	·55	·77	·48	·56
α from D. W. R.		$17\frac{1}{2}°$	5°	5°	11°	30°	$22\frac{1}{2}°$	13°
$\cos\alpha - \sin\alpha$ W/G ("calculated")		·65	·91	·91	·79	·37	·54	·75

It will be seen that the agreement is fairly good for the S and ENE groups. For the others, the "calculated" value is much in excess of the "observed" value. There are various possible explanations of these differences which must be explored before a final opinion can be arrived at, and they are mentioned here in order to lay stress upon the fact that the comparison of the surface-wind with the geostrophic wind requires very close and careful observation if uncertainties of a difficult character are to be avoided. As we have already seen the diurnal variation of the wind is a matter of importance, so that it is necessary for the observed wind and the map from which the geostrophic wind is obtained to be properly synchronous and a suitable hour selected; and in the case of Pyrton Hill the peculiarities of the site may have had some influence. The south winds blow at about forty-five degrees to the line of the hills, those from WSW are climbing the hills, those from the ENE have come over them. The effects of such circumstances as these upon the direction of the wind near the surface are as yet unexplored.

The variations of vertical velocity of pilot-balloons

The great advantage of the method of two theodolites, especially when they are provided with means for recording continuously the altitude and azimuth of the balloon which is being followed, is that the observations enable the observer to determine the actual height of the balloon and thus obtain a measure of the variation of the vertical velocity. If the actual rate of ascent of the balloon in still air were exactly known, in spite of the solarisation of the balloon and other possible causes of change, the differences from the observed velocities would give the vertical component of the motion of the air at successive levels, but for dealing with small differences it is not safe to assume that the velocity of ascent in still air is sufficiently well known in ordinary cases.

In the discussion of the observations at Pyrton Hill J. S. Dines obtained the variation in the rate of ascent and the mean values of these variations are shown in a small inset in the diagram of fig. 36 a. It appears from the curve that the balloons lost on the average about 0·3 m/sec of ascensional velocity within the first kilometre and the greater part of the loss took place within the first half kilometre. In explanation of this general result, which agrees with what had been previously observed by Cave at Ditcham Park and by Hergesell at Strasbourg[1], J. S. Dines has pointed out in a note upon his discussion[2] that balloons set free at the surface will as a rule be carried by the current in the surface layer away from the region where air is descending and towards the region where air is ascending, because ascending air must be supplied by currents moving over the surface from places where it is descending. The fact that evidence for this conclusion can be detected in the mean values of a large number of soundings with pilot-balloons is very satisfactory evidence of the general precision of the measurements. Dines pursued the question of the vertical component of air-currents within the lowest two kilometres still further and found a range of vertical velocity in the lowest kilometre amounting to 3·2 m/sec on 27 June, 1911, in a sky with a few small detached cumulus clouds, and a range of 3·0 m/sec within the second kilometre on the same day. On 5 July, of the same year, also in a sky with some cumulus, he found a range of 3·6 m/sec in the second kilometre and summarising a table of results he adds, "It appears from the records obtained that vertical currents (of short duration) of 2 m/sec must be not uncommon on days with detached cumulus about, while on days of clear sky the current would not as a rule exceed ·5 m/sec."

[1] *Commission Internationale pour l'Aérostation Scientifique*, Monaco, 1909, p. 102.
[2] *Third Report on Wind Structure*, Advisory Committee for Aeronautics, *Report*, 1911–12, p. 230.

After satisfying himself about the application of a formula for the ascent of a pilot-balloon of given weight and lift[1], in a subsequent report[2] to the Advisory Committee for Aeronautics, published in 1913, J. S. Dines has discussed the information about the vertical component of the motion of the air derived from 66 soundings with pilot-balloons at Pyrton Hill which were watched with a pair of recording theodolites. He has given the vertical component at different levels in each of the soundings and from his results (which include only 89 observations in the various levels, out of a total of nearly 1000, when the vertical component is set down as zero) a table of frequencies of vertical components within certain specified limits has been compiled.

TABLE II. FREQUENCIES OF VERTICAL COMPONENTS OF WIND-VELOCITY IN METRES PER SECOND AT DIFFERENT LEVELS FROM 66 SOUNDINGS WITH PILOT-BALLOONS AT PYRTON HILL.

Level m	DOWNWARD 1·0 to 1·4	0·5 to. 0·9	0 to 0·4	UPWARD 0 to 0·4	0·5 to 0·9	1·0 to 1·4	1·5 to 1·9	2·0 to 2·4	2·5 to 2·9	3·0 to 3·4	3·5 to 3·9	4·0 to 4·4	4·5 to 4·9	5·0 to 5·4	Number of observations
100	—	2	13	27	18	1	2	2	1	—	—	—	—	—	66
150	—	4	17	23	12	5	—	3	2	—	—	—	—	—	66
200	2	5	19	22	7	3	3	2	1	2	—	—	—	—	66
250	2	8	21	16	8	3	2	4	—	1	1	—	—	—	66
300	3	8	23	16	6	2	3	2	1	1	1	—	—	—	66
400	4	11	26	9	6	4	3	—	—	2	—	1	—	—	66
500	2	10	27	15	3	4	2	—	—	2	—	—	1	—	66
600	1	8	26	16	5	3	2	1	1	—	—	—	—	1	64
700	1	4	28	15	5	2	2	—	—	—	—	1	—	—	60
800	1	3	25	13	8	2	2	1	—	—	1	—	—	—	56
900	1	2	19	19	6	—	3	—	—	—	1	—	—	—	51
1000	1	4	15	15	4	3	1	1	—	1	—	—	—	—	45
1100	—	6	9	11	8	2	—	—	1	—	—	—	—	—	37
1200	1	3	12	9	8	3	—	—	—	—	—	—	—	—	36
1300	1	5	10	10	7	—	2	—	—	—	—	—	—	—	35
1400	1	4	9	8	6	—	1	1	—	—	—	—	—	—	30
1500	—	4	8	9	5	—	—	—	—	—	—	—	—	—	26
1600	—	2	11	9	2	—	—	—	—	—	—	—	—	—	24
1700	—	1	11	6	1	—	1	—	—	—	—	—	—	—	20
1800	—	2	10	7	—	—	—	—	—	—	—	—	—	—	19
1900	—	3	7	4	—	—	1	—	—	—	—	—	—	—	15
2000	—	4	5	5	—	—	1	—	—	—	—	—	—	—	15
	21	103	351	284	125	37	31	19	7	9	4	1	2	1	995

The table shows a region of maximum frequency of vertical velocity at about 500 m. An upward vertical component of 5·1 m/sec was registered on one occasion at the level of 600 m and components between 4 m/sec and 5 m/sec three times between 500 m and 1100 m. The maximum components downward are not so large but reach a limit greater than 1 m/sec and less than 1·5 m/sec on several occasions.

The gradual change with height in the frequency of upward motion and

[1] Q. J. Roy. Meteor. Soc., vol. XXXIX, 1913, p. 101.
[2] 'Vertical motion in the free air above Pyrton Hill,' Reports and Memoranda, No. 95.

downward motion is noteworthy. At the level of 100 m there are fifty-one rising currents as against fifteen downward and the distribution gradually changes until at 500 m, the last level for which the full number of sixty-six soundings are available, there is a preponderance for descending motion of thirty-nine as against twenty-seven. But a preponderance of upward motion shows itself again at 900 m and is continued to 1500 m with a maximum at 1100 m.

The general results may be somewhat biassed in favour of the display of vertical motion because the occasions were chosen with a view to exploring cases of vertical motion and do not properly represent random sampling.

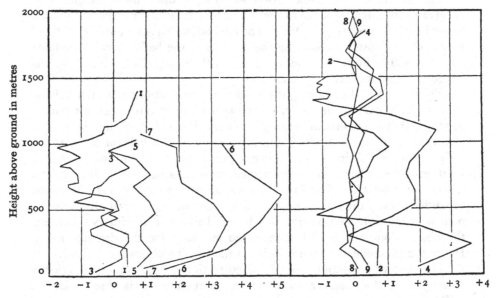

Fig. 37 (VI, 3). Vertical components of air-currents at Pyrton Hill. The velocities are marked, below the diagrams, in metres per second, upward components +, downward components −.

For comparison with the corresponding weather-maps numbers have been inserted on the diagrams to show the dates and times of the observations as follows:

1. 1911 May 4 11 h 35 m	5. 1913 May 10 12 h 10 m
2. 1912 April 30 19 12	6. 1913 May 22 10 55
3. 1912 May 9 11 51	7. 1913 May 26 11 10
4. 1912 June 22 11 12	8. 1913 Oct. 13 14 50

9. 1913 Oct. 23 15 h 40 m

It must be remembered however that the existence of an observation of a pilot-balloon implies clear weather; cases of thermal convection producing cloudy condensation are thereby excluded.

Diagrams illustrating the rapid variation of vertical velocity with height are given in fig. 37. Such currents are regarded by J. S. Dines as only transient because in those cases when a sounding was repeated after a short interval of half an hour or less they were not recorded.

The importance of the determinations of vertical components in the stratum up to the two-kilometre level lies in their bearing upon the observations with a single theodolite in the computation of which the vertical component of thĕ air-velocity is ignored; but if there is a vertical component it will affect the determination of the horizontal velocity by the single theodolite and a correction will be necessary. An ascending current will cause the computed velocity to be too small and might even cause an apparent reversal of the wind, and conversely a descending current will exaggerate the speed of the balloon away from the observer and in certain circumstances may give very inappropriate results.

This conclusion has forced itself so strongly upon meteorologists that observation by what is known as the tail-method is used for single theodolites in the British Meteorological Office. The method, which is set out in the official handbooks, consists in observing the altitude of the balloon by a theodolite and its distance by reading on a reticule in the eye-piece the length of the image of a tail attached to the balloon.

The method of determining the wind from observations of a pilot-balloon from a single point depends on assuming a uniform rate of ascent (vol. 1, p. 222) so that at any instant the height of the balloon is assumed to be known. Observations of the altitude E and azimuth A of the balloon are made at definite intervals of time from the beginning of the ascent and from these the position of the balloon can be determined, remembering that the distance east is given by $z \cot E \sin A$ and the distance north by $z \cot E \cos A$. By taking the differences of the distance east of two consecutive observations the component of the distance east travelled by the balloon in the interval between the observations can be determined and similarly for the distance north. Hence the components of the velocity in the two directions are derived.

The following computation of the correction to the computed velocity u_0 away from the observer for a vertical component w is due to Dr H. Jeffreys.

E is the angular altitude of the balloon, z its height, x its horizontal distance, A the azimuth of the balloon. The components of the velocity of the balloon are $\dot{z}$, upwards, $\dot{x}$ horizontal in the direction of the projection of the line of sight, and $x\dot{A}$, transverse to the line.

Then $\dot{z} = Z + w$, where Z is the vertical velocity of the balloon in still air.

Then
$$x = z \cot E,$$
$$u = \dot{x} = \dot{z} \cot E - z \operatorname{cosec}^2 E \, dE/dt$$
$$= (Z + w) \cot E - z \operatorname{cosec}^2 E \, dE/dt.$$

In computations from observations with a single theodolite it is usual to assume that w is zero. Thus the computed velocity is too small by $w \cot E$ along the projection of the line of sight. The height z also is in error by a calculable amount depending on the values of w during the earlier parts of the ascent. Hence, to correct for the vertical component of velocity, w, the term $w \cot E$ should be added to the computed velocity along the line of sight.

The amount of the correction depends upon the altitude of the balloon as observed in the theodolite. It requires to be borne in mind in considering the results derived from observations with a single theodolite. Thus, for example, the initial loss of velocity on leaving the ground which we have noticed in some of the individual curves included in Dobson's diagrams might possibly be attributable to ascending currents near the ground which are quite likely to be present in the daytime.

A similar explanation may be suggested for exceptional velocities, either high or low, which are sometimes obvious when the observations of pilot-balloons with a single theodolite come to be plotted on a map. Such an exceptional velocity is often shown in the observations for 1000 ft and 2000 ft from Barrow-in-Furness as compared with those at surrounding stations. It is possible that the peculiar position of the station with reference to the hills of the Lake District may be the cause of descending or ascending currents, at the level of 1000 or 2000 ft, which will be different according to the direction of the wind.

The subject of vertical velocity has been treated in a number of memoirs. It includes two aspects which are not always easy to distinguish. The one is the deviation at different levels of the velocity of a pilot-balloon from that given by a formula assumed for a single theodolite; and the other is the actual velocity of the rising air as determined by observations with two theodolites or by the tail-method.

P. Moltchanoff, in the Hergesell-Festschrift of the *Beiträge zur Physik der freien Atmosphäre*, using two theodolites, gives some observations which afford corrections to a formula.

A. J. Bamford[1], using the tail-method, gives an elaborate account of the variations in the vertical velocity shown as the result of 3500 observations in Ceylon. He had previously given some results for Palestine, using two theodolites for the lower layers. We may add that E. Kidson and H. M. Treloar, on p. 153 of the same journal (1929), give the results for the rate of ascent of pilot-balloons in Melbourne, based upon three years' observations. "The rate of ascent is very variable.... The greater the turbulence the greater the mean rate of ascent."

Meanwhile, in the *Monthly Weather Review*, August 1928, P. A. Miller notes a case of an aeroplane carried up in a current with a vertical velocity of approximately 38 m/sec as determined by the altimeter, unless there was a sudden change of pressure to account for the reading; C. E. Peebles (*Bull. Amer. Meteor. Soc.* 1930, p. 123) was whisked up 6500 ft in 60 secs, and in August 1930 the Airship R 100 rose 1500 ft without notice in less than a minute in consequence of an up-current of air when the ship was approaching Montreal. We have quoted the experience of J. Wise in a free balloon on p. 382 of vol. III.

[1] *Q. J. Roy. Meteor. Soc.* vol. LV, 1929, p. 363; *ibid.* vol. XLVI, 1920, p. 15.

Two theodolites at Aberdeen

The two examples which have been selected for illustrating the results obtained by pilot-balloons are both dependent upon inland sites and differ the one from the other merely in the fact that the station at Upavon is on a

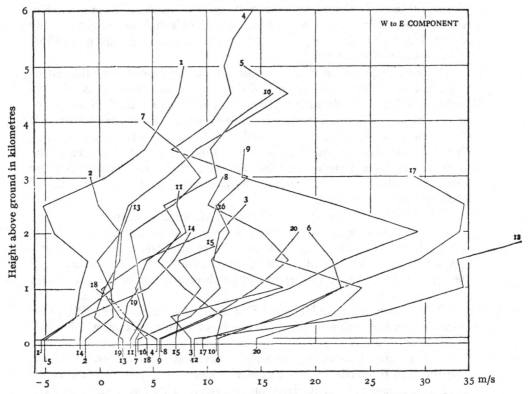

Fig. 38 *a* (VI, 4 *a*). Structure of the atmosphere below 6000 metres on the East Coast (Geddes). W to E component of velocity.

Numbers are inserted in the diagrams 4 *a*, 4 *b*, 4 *c*, to show the dates of the observations as follows:

I.	1912 May 22	6.	1913 Oct. 17	II.	1912 June 21	16.	1913 Nov. 1
2.	1912 June 26	7.	1913 Nov. 5	12.	1913 Dec. 13	17.	1913 Nov. 26
3.	1912 Nov. 22	8.	1913 Nov. 14	13.	1913 Feb. 28	18.	1912 June 7
4.	1912 Nov. 29	9.	1913 Nov. 19	14.	1913 May 16	19.	1912 June 14
5.	1913 June 13	10.	1913 Dec. 5	15.	1913 June 11	20.	1912 Nov. 20

The time of ascent was, in every case, between 11 h and 12 h.

level plateau whereas that at Pyrton Hill is on the north-west slope of a range of hills which runs from south-west to north-east. As a third example we may take the observations by A. E. M. Geddes[1] at Aberdeen because they are derived from a station on the coast. A station in such a position may be

[1] *Q. J. Roy. Meteor. Soc.*, vol. XLI, 1915, p. 123.

expected to have special characteristics as regards the structure of the first two kilometres of its atmosphere because on the one side is the land with all the disturbances due to the irregularities of relief and variations of temperature at the surface, and on the other side is the sea, the surface of which is free from those irregularities and variations. We may, therefore, expect the results

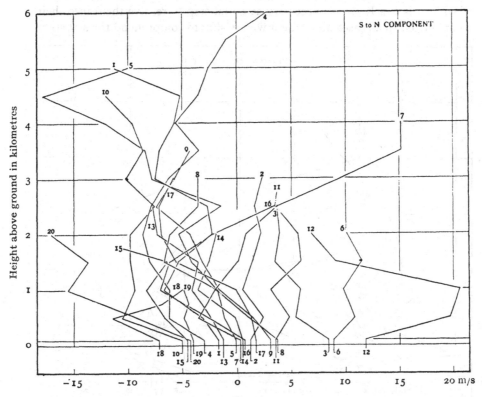

Fig. 38 *b* (VI, 4 *b*). Structure of the atmosphere below 6000 metres on the East Coast (Geddes). S to N component of velocity.

of soundings with pilot-balloons to display complications even more involved than those which are displayed at Upavon or Pyrton Hill. The expectation is certainly realised in the diagrams which represent the components of the motion of the air at different levels up to 2500 metres as taken from the tables given in Geddes's paper. Two theodolites were used for the observations and therefore the influence of the vertical component upon the velocity of ascent can be represented by the variation of the vertical velocity of the balloon which is given in the tables and is also represented in the diagrams.

Geddes groups his observations according to the fate which ultimately overtook the balloon and the examples represented in the diagram are for those lost in haze or distance and those lost in strato-cumulus cloud with three examples when stratus clouds terminated the observations.

The reader will note that in these diagrams components have been plotted instead of the resultant velocity with its direction. The use of components is desirable partly because it may simplify the examination but mainly because whenever computations have to be made with the view of combining measures of wind it is nearly always necessary as a first step to resolve them into their components. In all cases the student who wishes to comprehend the structure

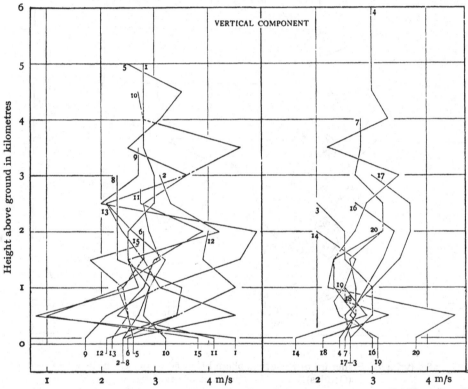

Fig. 38 c (VI, 4 c). Structure of the atmosphere below 6000 metres on the East Coast (Geddes). Vertical velocity of the balloon, showing changes with height of the vertical component of the velocity of the air.

of the atmosphere has to read simultaneously two diagrams and for any comprehensive study the use of components will ultimately be found the more satisfactory alternative. The process is also desirable in view of the difference of geographical significance between motion along parallels and motion along meridians on the earth's surface. At Aberdeen the difference must be considerable because the North Sea lies to the east and mountains to the west. Perhaps a more definite result in that particular case might have been obtained by resolving along a north-east and south-east line instead of due north and south.

The illustrations which have been adduced in this chapter are sufficient to set before us the complexity of the problem presented by the facts regarding the structure of the atmosphere between the levels of 500 and 2500 metres[1]. It is probably the layer of greatest complexity. It includes not only the varieties of structure incidental to the kinematics of the free atmosphere in its simplest form depending on the relations of its motion to the distribution of pressure and temperature, but also some disturbance due to the effect of eddy-motion wherever circumstances are such that a high value of the coefficient k of eddy-conductivity has been operative for a long period. There is also to be found the complication due to vertical motion of which the upward tendency is most marked in these strata but with which the compensating downward motion must be recognised and, with those, the consequent local disturbance of the horizontal motion.

2500 to 7500 metres

We now pass on to lay before the reader an account of another section of the atmosphere, that between 2500 and 7500 metres as representing the layers which we may be encouraged to regard as somewhat less complicated in their structure because they represent the region where according to W. H. Dines's results the deviations of pressure for any level from their normal show a close approach to proportionality to the simultaneous deviations of temperature; and in the ordinary conditions, when soundings with pilot-balloons are practicable, there is at least comparative freedom from vertical motion. These two specifications may be regarded as interdependent. If the motion of the air is confined to horizontal layers the changes in the temperature of the air will necessarily be governed by the changes in pressure, provided we may leave out of account such slow changes as are attributable to the loss or gain of heat by radiation and express themselves perhaps in the seasonal variations of temperature at the different levels.

In order to represent the structure which has been observed in the layer which extends from 2500 to 7500 metres we will make use of the results obtained by C. J. P. Cave principally at Ditcham Park which lies at a height of 167 metres, close to the ridge of the South Downs in Hampshire. In his book [2] "smoothed" results are given for 200 soundings in the years 1907 to 1910, and represent the first investigation of the upper air by means of pilot-balloons in England. For some of the soundings two theodolites were employed, for others only one. We have drawn no distinction between them in the selection made for our purpose because for exploring the layers under consideration the two methods are about equally effective[3]. From Cave's tables we have taken all the ascents which gave smoothed values for each half-kilometre from 2500

[1] The complication is even more irregular in other countries where the thermal changes and the geographical relief are more pronounced. A report on the winds of Macedonia as ascertained by soundings with pilot-balloons presented to the Advisory Committee for Aeronautics (*Report*, 1916–17, Reports and Memoranda, No. 296) illustrates this statement.

[2] C. J. P. Cave, *The Structure of the Atmosphere in Clear Weather*, Camb. Univ. Press, 1912.

[3] See some remarks upon the comparative errors of the two methods by G. M. B. Dobson, *loc. cit.*, p. 130.

to 7500 metres and have included also three which only missed the last step. There were 23 soundings in all and of these 22 have been plotted in a diagram to represent the west to east components of the wind-velocity and 18 for the south to north components. Points to the left of the zero line represent easterly or northerly winds and those to the right westerly or southerly respectively. The soundings have been omitted from the diagrams only when

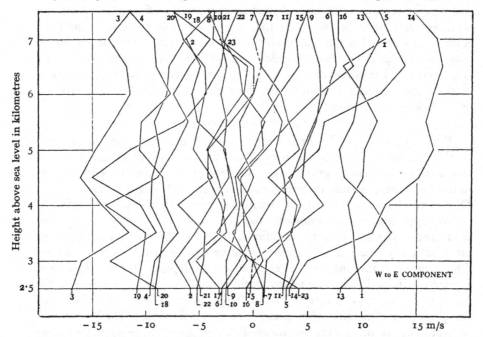

Fig. 39 *a* (VI, 5 *a*). Structure of the atmosphere from 2500 metres to 7500 metres at Ditcham Park (Cave). W to E component of velocity.

The numbers assigned to the lines in the diagrams show the dates of the observations as follows. The letters prefixed show the class as indicated on pp. 177–8.

a 1.	1908 May 18	*e* 7.	1907 Sept. 23	*e* 13.	1908 July 30	*c* 19.	1909 May	7
a 2.	1908 May 30	*e* 8.	1908 June 22	*e* 14.	1908 July 31	*c* 20.	1909 May	7
a 3.	1909 May 5	*e* 9.	1909 Feb. 18	*b* 15.	1908 Sept. 30	*a* 21.	1909 Aug.	5
a 4.	1909 May 6	*e* 10.	1908 June 3	*b* 16.	1908 Oct. 1	*c* 22.	1910 Mar.	3
d 5.	1908 July 27	*e* 11.	1908 July 28	*b* 17.	1908 Oct. 2	*b* 23.	1907 May 24	
d 6.	1908 Nov. 6	*e* 12.	1908 July 29	*a* 18.	1909 May 6			

Particulars of the soundings are given in Cave's *Structure of the Atmosphere in Clear Weather*.

practically they repeated types already shown or when they belonged to a part of the diagram that was already sufficiently filled.

In these diagrams we may begin to see some suggestion of order. In that representing the west to east components the range of velocity at each level is approximately the same but on the whole the westerly components are stronger at the higher levels. Thus at 2500 metres the range is from − 16·5 which means an east component of 16·5 m/sec to + 10 which means a west component of 10 m/sec, whereas at 7500 metres the range is from − 11·5 to 15 so that the east component has lost 5 m/sec and the west component has gained 5 m/sec at the higher level. The particular figures are perhaps accidental, but the general trend towards westerly winds at higher levels is a real pheno-menon.

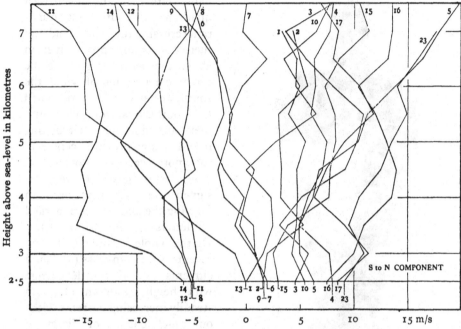

Fig. 39 b (VI, 5 b). Structure of the atmosphere from 2500 metres to 7500 metres at Ditcham Park (Cave). S to N component of velocity.

In the diagram representing the south to north components, on the other hand, we notice that the range of velocity is much wider at 7500 metres than it is at 2500 metres. It extends from − 20 which means a north component of 20 m/sec to + 20 which means a south component of the same magnitude. The range at 2500 metres is indeed very small being from only − 6 to +9. It must not be supposed that the figures quoted from the diagrams mark the practical limits of velocity of winds, or their components from north or south, east or west, at the level of 2500 metres. A more reasonable inference is that the occasions when higher wind-velocities might have been experienced at that level were not suitable for a sounding up to the higher level of 7500 metres, on account either of the formation of cloud or the difficulty of keeping a small balloon within observation in a strong wind.

Captain Cave has grouped his soundings into five classes or types of structure within the troposphere, namely (a) those which show "solid current" or little change from the direction and velocity at the surface over a large range of height, (b) those which show a considerable increase of velocity without much change of direction, (c) those which show a decrease of velocity with height, (d) those which show a reversal or great change of direction in the upper layers, and (e) those which show an upper wind (either in the north-west quadrant or south-west quadrant) crossing the lower wind and therefore coming apparently from above the central region of the cyclonic depression of which the south-westerly wind occupied the southern sector; in reality they are doubtless circulating round a low pressure centre to the north-east or north-west. The classes to which were assigned the soundings used in forming the diagrams are marked by letters in the list of dates. Of the whole number of the 23 ascents, six are in class (a) "solid currents" showing little change with height, four in class (b) steadily increasing currents without much change of direction, three are the diminishing currents of class (c), two are reversals, of class (d), and eight are in class (e), increasing cross upper currents generally from north or north-west.

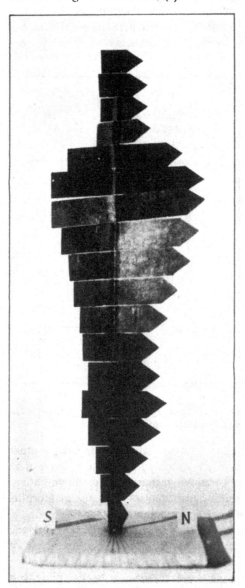

Fig. 40 (VI, 6). Gradual increase of the south component of the wind (Ditcham, 1 Oct. 1908). See fig. 50, p. 201.

This classification, which though provisional is useful as a general guide, is well represented by the various lines included in the diagrams. The diagram of the south to north components is particularly noteworthy, it shows a singular symmetry of the two sides; whether from south or north, there may be a marked increase in the components in the upper layers; one of the lines starts from − 5 and ends nearly at − 20 and another starts from +6·5 and ends at + 20.

A typical case is illustrated by a photograph of a model (fig. 41) taken from Cave's book representing the sounding which we have numbered 12 on p. 176. Each of its pointers represents the wind of a kilometre; its length expresses the speed; and the way it points, the direction of the wind. The gradual but marked increase in wind from the north in the upper air is very characteristic and has been noticed by many observers. The following note taken from an official communication to the Meteorological Office calls attention to it for 25 February, 1918. "The wind was practically constant in direction at all heights, from 15°. Its remarkable feature was its strength which increased from about 30 f/s at the ground to 160 f/s at 25,000 ft. So far as observations were obtained the wind was nearly linear, i.e. the graph of strength with height is nearly a straight line.

"Winds of this strength have now been observed at Portsmouth on three occasions (the others are 5.12.16 and 19.10.17), in each case they have been northerly winds, their other common feature is a remarkable regularity of the wind for different observations at the same height, but at slightly different times."

One of the occasions mentioned in this note, namely that of 19 October, 1917, has become historic because the northerly wind increased at high levels to about 30 m/sec and carried away a fleet of Zeppelins that ascended into it after attacking England

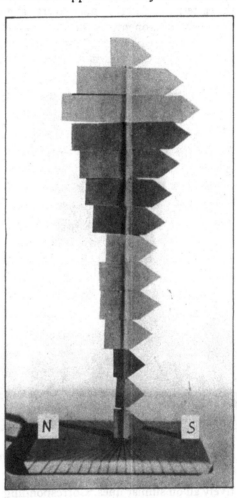

Fig. 41 (VI, 7). Gradual increase of the north component of the wind (Ditcham, 29 July, 1908). See also vol. II, fig. 60, p. 108.

and dispersed them over France where a number of them were destroyed. *Afflatus, dissipati.*

A model of the ascent numbered 16 (fig. 40), almost identical in shape, but larger in dimensions equally represents the variations in the winds of type (*b*) and shows a south wind increasing uniformly in like manner and taking on a little westerly component in the higher levels; and just as these models represent a gradual increase of southerly or northerly component of precisely similar character throughout the range from 2500 to 7500 metres, so another model (fig. 42), which is also reproduced from Cave's book, represents the sounding numbered 6 in the diagrams and shows the gradual addition of westerly component to the original east which reverses gradually the easterly wind in the upper layers. Hence there are three cases for increase of northerly component, southerly component, and westerly component respec-

a. West to east component.   b. South to north component.
Fig. 42 (VI, 8). Reversal of the air-current at the level of four kilometres
(Ditcham, 6 Nov. 1908). See fig. 50, p. 201.

tively all on similar lines. Corresponding cases of gradual increase of easterly wind for the same range of levels are certainly rare. But probably they do exist as cirrus clouds are sometimes observed moving rapidly from the east.

In order that the reader may have an additional reminder of the general problem of the variation of winds in the upper air we give also Cave's model representing the rapidly increasing westerly winds of 1 September, 1907, from 5·5 m/sec at the surface where its direction was actually from a point or two east of north to 13 m/sec from nearly west at 2500 metres. Particulars of a more striking example are to be found in the tables[1] for the sounding of 2 April,

[1] Cave, loc. cit., p. 88.

1907, which showed an increase from 3·5 m/sec at the surface to 22 m/sec at 2000 metres. These examples may be compared with those given in the diagram representing the west to east components at Aberdeen. It is instructive to speculate as to what is the dénouement of the story of which these very rapid increases with height are the beginning. So far as soundings with pilot-balloons are concerned we know that the story is generally brought to a premature conclusion by clouds or by the balloon being carried out of sight on account of distance, but there must be a development in the upper regions which is of considerable interest and may ultimately be ascertained. If the increase goes on at the same rate a velocity of the order of 100 m/sec would be reached at 9000 metres. That would be beyond any of the known measure-

ments of the velocity of clouds at any level though very high velocities were measured by means of pilot-balloons in the region of Spitsbergen. The soundings at Aberdeen, represented in fig. 38, suggest that the rapid increase of velocity is replaced by a corresponding decrease not very far up, but the mechanism of such a process is difficult to formulate. Still it must be remarked that the curves which we have put before the reader show every kind of change in every type of sounding; they differ from the extreme cases that seem almost unnatural only in degree, not in kind.

Fig. 43 (VI, 9). Rapid increase of wind-velocity with height (Ditcham, 1 Sept. 1907). See p. 201.

Above 7500 metres

For the stage next in order we select the layer between the levels of 7½ kilometres and 12½ kilometres. The range of height is specially interesting because in nearly all cases it covers the transition between the troposphere and the stratosphere. For the information we rely again upon the observations recorded in Captain Cave's book. We have put together in a pair of diagrams all the observations which extended over the range of levels mentioned. A noticeable feature of these observations is that there is as a rule very little change of direction and when a change of direction is noted it is irregular and may be called wild.

We have therefore presented the diagrams showing direction and velocity for this stage instead of the two components employed for the other layers. The results are particularly noticeable because the wind-directions group themselves very clearly about the directions NNE, SSE, SSW and NW.

Whether this grouping is fortuitous, depending on the peculiar circum-

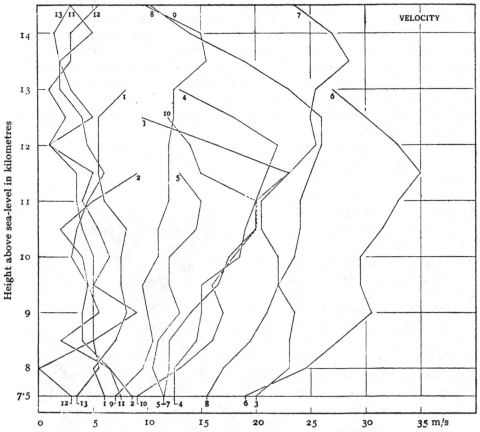

Fig. 44 a (VI, 10 a). Structure of the atmosphere above 7500 metres, at Ditcham Park (Cave). Velocity.

The numbers assigned to the lines in the diagrams give the dates of the observations as follows:

1.	1908	June 22	18 h	36 m		7.	1908	Sept. 30	16 h	31 m
2.	1908	June 3	10	19		8.	1908	Oct. 1	16	20
3.	1908	July 28	19			9.	1908	Oct. 2	16	20
4.	1908	July 29	19			10.	1909	May 6	18	25
5.	1908	July 30	19			11.	1909	May 7	18	29
6.	1908	July 31	19			12.	1909	Aug. 5	18	33

13. 1910 Mar. 3   16 h 30 m

stances when a balloon had been followed to such great heights, or whether on the other hand it indicates some general property of the air-currents at those levels it is not possible to say without further investigation.

With regard to the curves representing the variations of velocity with

height, the first point to notice is that in many cases the velocities are uniformly small. About one-half of them are within 10 m/sec This is partly to be
accounted for by the fact that small velocities make observations possible for
great heights and partly because the winds are light, even to great heights, in
many cases when the sky is free from clouds. Those winds which show an
increase of velocity from the level of $7\frac{1}{2}$ kilometres upwards generally carry the
increase to a certain point and then show a marked decrease. The level at
which the decrease commences is different on different occasions. It is at
11 kilometres on 30 July, 1908, and on 6 May, 1909, at $11\frac{1}{2}$ kilometres on

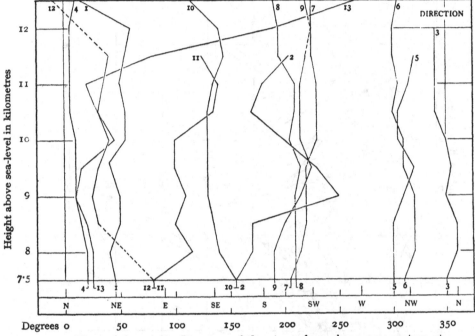

Fig. 44 *b* (VI, 10 *b*). Structure of the atmosphere above 7500 metres, at
Ditcham Park (Cave). Direction.

28 July and 29 July and 31 July, 1908, at $12\frac{1}{2}$ kilometres on 1 Oct. and at
$13\frac{1}{2}$ kilometres on 30 Sept. and 2 Oct. 1908. This falling off may be accounted
for by a change in relation of pressure and temperature between the troposphere and stratosphere and it occurs at different heights in consequence of the
variation in the height of the tropopause, to use a word which the glossary
of the Meteorological Office gives for the boundary between the two. In the
troposphere above the first kilometre level high pressure is associated with
high temperature but in the stratosphere high pressure is marked by low
temperature. A formula for the variation of wind with height which associates
the falling off of the wind with the reversal of the temperature-gradient in
relation to the pressure-gradient within the stratosphere is given later in the
next chapter (p. 199).

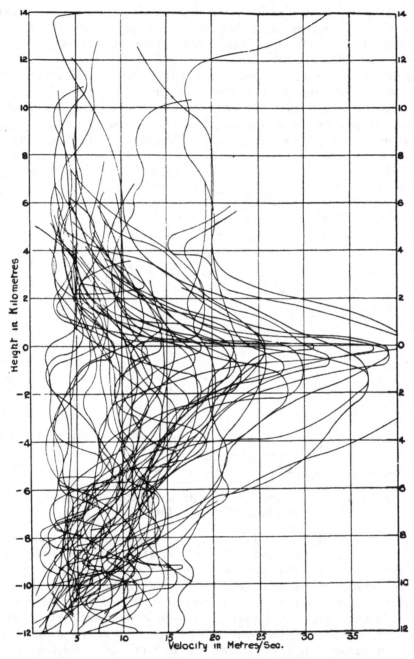

Fig. 45. Variation of velocity of wind above and below the level of the
tropopause (Dobson, *Q. J. Roy. Meteor. Soc.* vol. XLVI, 1920, p. 56).

The datum level for height marked 0 is the level of the tropopause for each of the
several soundings.

Winds in the stratosphere

From the scientific point of view the tropopause which separates the troposphere from the stratosphere is a region of peculiar interest because the thermal structure of the atmosphere certainly changes there from a region of little variation of entropy with height, and therefore nearly isentropic, to one which is nearly isothermal and consequently exhibits rapid increase of entropy with height. It is not often that the kinematic structure can be determined; it requires an occasion upon which the balloon remains visible in spite of its great height and corresponding distance, and the occasions must therefore be limited to those in which the wind-velocity in the lower regions is sufficiently small to preserve the visibility of the balloon for an unusually long time. A few results for the stratosphere are given in Cave's work on *The Structure of the Atmosphere* and a special collection is represented in a well-known diagram by G. M. B. Dobson which was reproduced in *The Air and its Ways*. It is given here as fig. 45, through the courtesy of the Royal Meteorological Society.

It must be noted that the heights of the several ascents are referred to the tropopause as a datum line and the point which is most noticeable in the diagram is the rapid falling off of the velocity in the lower regions of the stratosphere. If we remember that the variation of velocity with height beyond a certain limit entails turbulence it would seem that the rate of variation here indicated would tend to promote mixing of air in the lower part of the stratosphere and so adds a difficulty to the explanation of the isothermal character of that region.

We have now set out the problem of the variation of wind with height in those regions of the atmosphere which are at present regularly accessible to modern aircraft and in the layer of five kilometres immediately above them. Before we pass on to consider the possibilities in respect of explanation we will add two remarks. First, we have relied for our illustrations upon discussions published before the modern development of aircraft and the corresponding development of meteorological observations designed partly, though not entirely, to aid aerial navigation. We have now a great accumulation of data obtained by observations of pilot-balloons with a single theodolite and in that respect we are much better endowed than we were. It is now possible to approach the solution of the problem of the structure of the atmosphere by rigorous statistical methods, but before that can be done with any prospect of its leading to an insight into the dynamical and physical conditions we want some guidance from general meteorological principles as to the manner in which the statistics should be classified. In face of the great variety of changes which occur a direct attack upon the whole bulk of the observations by the method of statistics is not likely to lead to satisfactory results. Meanwhile the simple inspection of the data as they accumulate has added little to, and subtracted nothing from, the complexity of the problem which is put before the reader by the examples which have been selected.

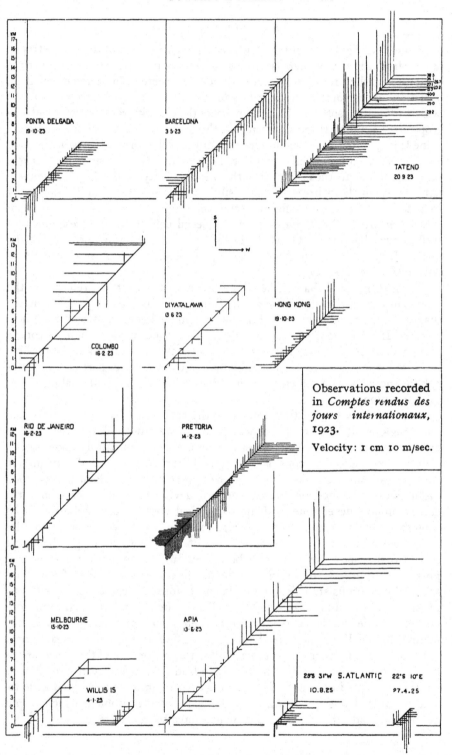

Fig. 46. Clothes-line graphs of soundings with pilot-balloons in two hemispheres (see p. 188).

Secondly we have drawn our illustrations of individual soundings from smoothed curves which have been drawn to represent the general features of the structure of the atmosphere disclosed in the soundings by giving the values at definite intervals of height, generally at each half kilometre. The interval of 1000 feet is usual in the reports from the various Services. As the original basis of the curves there are the observations of the balloon for each minute which, if included individually, would superpose upon the diagram a series of comparatively rapid variations, sometimes large sometimes small. They are well represented by the plotting of the individual points in the diagrams appended to Captain Cave's book. These variations may be real or they may be dependent upon the incidental errors of observation of the altitude or the fluctuations in the rate of ascent of the balloon. They make the results taken from "smoothed" curves, which are drawn by eye as a reasonable account of the observations, somewhat uncertain; but, once more, whatever may be the details of the fluctuations the main features remain. We may explain by minor variations of this kind some of the irregularities shown upon a synchronous map of the results of many pilot-balloons, but the fluctuations which are indicated in our diagrams are real and they call for explanation.

New data and the mode of presentation

Since the text of this chapter was first written pilot-ballooning has been adopted as a regular exercise for a normal meteorological station at which, weather permitting, observations of the winds in the upper air are made almost daily. An enormous number of results are in consequence now available for the studious meteorologist. The weekly report of the observatory at Lindenberg contained notes of upwards of 200 ascents in Germany and 900 in other parts of Europe, N. Africa and Siberia for each week from June 1926 to December 1929; the British *Daily Weather Report* contains 30 or 40 ascents at stations in the British Isles and 10 at other stations. Other sources of information on the subject are the daily weather reports of Norway, Sweden, Russia, Germany, France, Austria, Spain, Italy, N. Africa, Egypt, Brazil; the monthly publication of upper air winds by the Indian Meteorological Department and the summaries for the United States in the *Monthly Weather Review*, and the annual publications of Tokyo and Tateno, the Azores, Finland, Holland, Catalunya and Ceylon. An organised report of observations in all countries is in process of publication by the International Commission for the Upper Air.

The information there set out is obtained mainly in the immediate interest of the flying services[1] for which every part of the information is useful—the wind-speed and direction at successive heights, the greatest height reached

---

[1] The International Commission for Aerial Navigation has drawn up a special form for the summarising of observations of upper winds for air-craft. Observations in this form are published by Brazil, Czechoslovakia, Great Britain, Italy, Portugal, Poland and Yugoslavia; a summary in similar form of *Upper Winds at Cairo and Khartoûm*, by L. J. Sutton, has recently appeared.

SURFACE-WIND AT AGRA, APRIL TO JUNE, 1928

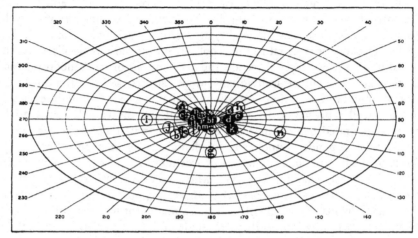

Fig. 47 a. Perspective of observations with pilot-balloons, direction and velocity at the surface.

White letters on black ground are early morning observations, 6 h or 7 h.
Black letters on white ground are early afternoon observations, 15 h to 18 h.

and the immediate cause of its limitation. For the more general interests of meteorological science, up to the present, the collection of observations has been singularly unfruitful. Information from new localities naturally brings to light new aspects of the general circulation, but the multiplication of observations in localities which have already been explored provides little that admits of compendious generalisation. Such generalisations as are obtainable by the ordinary method of mean values belong to the representation of the general circulation of the atmosphere in vol. II. Here we are concerned with the dynamics of the atmosphere as accounting for the differences in the individual ascents.

The method which we have adopted for the original version of this chapter depends upon separating the two horizontal components, W to E and S to N, of the velocity at successive levels. It has its disadvantages, the chief of which is the difficulty of combining the two components to represent the reality of the ascent. In vol. III we have suggested the clothes-line graph to remove that objection and meet the reasonable demand of Dr Hesselberg that the information given should show the stratification of the atmosphere.

The method was introduced in order to get the information about the wind in the upper air on the same sheet as the tephigram derived from a ballon-sonde. Here we give as examples (fig. 46) a dozen graphs of ascents in the two hemispheres taken from the report of 1923 of the International Commission for the Upper Air, supplemented by observations at sea through the courtesy of Commander L. G. Garbett, R.N. Heights are represented as distance along lines at 45° to the frame, and the components recorded are

WIND AT 1000 M, AGRA, APRIL TO JUNE, 1928

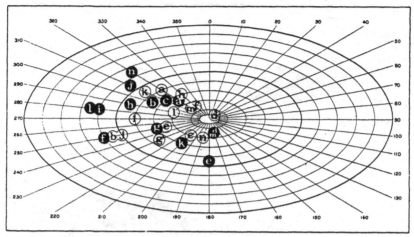

Fig. 47 *b*. Perspective of observations with pilot-balloons, direction and velocity at 1000 m.

The soundings are identified in the two diagrams by the same letter—white on black or black on white as the case may be. The outer ellipse represents a velocity of 20 m/sec.

represented on the scale of a millimetre to a metre per second by lines drawn *from* the clothes-line. It has been suggested that the clothes-line might be sloped right or left, as circumstances required.

The detailed information necessary for that purpose is not often available, here we offer for consideration a succession of diagrams that provide a representation of the stages of a number of ascents.

> ... perspectives that rightly gazed upon
> Show nothing but confusion, eyed awry
> Distinguish form.

To distinguish them from wind-roses and give the suggestion of perspective, lines of equal velocity are ellipses with the minor axis one-half of the major axis, and the angles are adjusted accordingly. The scales of the components are both linear but the scale of the W to E component is double that of the S to N component.

We use these diagrams for the representation of the results of pilot-ballooning at Agra and Mauritius[1]; and as an example of the use of the diagrams to find answers to inquiries we take the question of how far up the diurnal variation of wind extends. For the Potsdam Tower, the Eiffel Tower, and for mountains we have quoted the generalisation that, at the ground, wind has its maximum in the mid-day and its minimum in the night; the question we ask is how far the same effect is exhibited in the free air. We put on the diagram the results of successive observations on the same days at Agra.

[1] The data are taken from *Upper Air Data*, Parts 4, 5, 6, 1928, India Meteorological Department, 1929; Miscellaneous Publications of the Royal Alfred Observatory, Mauritius, No. 9.

PERSPECTIVES FOR MAURITIUS

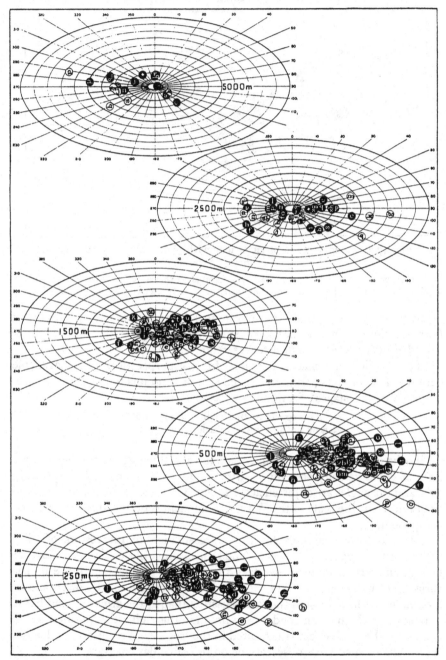

Fig. 48. Observations of direction and velocity of wind at different heights.

1927. Sept., black on white upright; Oct., sideways.
1928. March, white on black upright; April, sideways.
The same lettering is used for each level of the same sounding.

The morning and afternoon observations for the ground-level, 170 m, and for 1000 m above sea-level are plotted on the diagrams, figs. 47 *a* and *b*. The separate morning observations are each indicated by a white letter on a black ground, and the afternoon observations by a black letter on a white ground. The ellipses are drawn for intervals of 2 m/sec, and a point on the outer ellipse represents a velocity of 20 m/sec from the direction of the radius.

It will be noticed that at ground-level the black circles of early morning are clustered nearer to the centre than the white circles of the afternoons; whereas at 1000 m the early morning winds, black circles, are obviously stronger than the afternoon white circles.

For Mauritius we give a representation of the observations at five levels, 250, 500, 1500, 2500 and 5000 m in two months, September, October of 1927, white circles, and two other months, March and April, 1928, black circles, in order to ascertain whether they showed any marked seasonal change.

The picture is somewhat of a *tour de force*. In order to keep in view the separate observations which numbered about 20 for each month, letters have been set sideways for one month and upright for another. Still the general trend from south-east at 250 m to a considerable concentration with low easterly velocities at 1500 m passing over, so far as the few survivors are concerned, to westerly winds at 5000 m is obvious enough, and the individual exceptions can be traced if necessary by the code of letters, with the aid of a pocket magnifier, even on the reduced scale of the diagram.

DYNAMICS AND THE PILOT-BALLOON

One reason for the comparative sterility of the observations as a means of pursuing the study of the dynamics of the atmosphere is perhaps the following. Like many other subjects of meteorological investigation useful knowledge seems to be conditioned by the time-unit selected. The universal time-unit for pilot-ballooning is the minute. The rate of ascent for the balloons is regulated by the consideration that if the ascent is slow the balloon will be carried by the wind out of sight before a sufficient height has been reached, and if the ascent is rapid too much may happen within the minute. Again observations within a minute are not free from the local influence of the eddy-motion which is represented by gusts or squalls, and yet the minute is too long for the detailed observation of those special features of atmospheric motion.

In this case we seem to suffer from the unanimity of the observers as to the object and method of their observations. As a rule each ascent has for its object the investigation of the whole structure from what we have called the foot to the top of the trunk. Consequently the balloon must travel upward fast enough to be still visible when the trunk is reached in the stratosphere. If an observer confined his attention to the foot he might use a small lift, or no lift at all, whereas one who sought definite information about the tropopause might choose a large lift, and one who wished to investigate the limb would choose an intermediate lift.

Such an arrangement may indeed be used in practice by one or other of the various services which rely upon observations of pilot-balloons, but the deliberate selection of a special part of the atmosphere for investigation does not appear in the published results.

In the absence of any classification of that kind the reader is left to ask his own questions and find what answers he can. The questions are many; the answers are comparatively few. One of the most appealing questions is a sudden change of wind with height, a "discontinuity," either of direction or velocity. A gradual variation may be associated with changes in the distribution of temperature as explained in chap. VII; the sudden changes are on a different footing, they represent a stratification of the atmosphere.

In regard to that we should like to recall what is said in vol. III about the structure of the atmosphere in respect of entropy. We have explained that in an ordinary condition of equilibrium entropy increases gradually with height throughout the troposphere and very rapidly with height in the stratosphere. In fact the tropopause which marks the boundary between troposphere and stratosphere is indicated by a sudden increase in the rate of change of entropy with height.

And within the troposphere a leap or *saltus* in the up-grade of entropy is an indication of a surface of separation in the structure of the atmosphere as we have seen in the consideration of general meteorological theory.

Of the results of observations of pilot-balloons, one that arrests attention is J. Bjerknes's identification at considerable height of the discontinuity, along which air ascends in the front of a cyclone, as a surface with a slope of 1 in 100 or thereabout. That surface was originally indicated by Helmholtz as a surface which marked a change of entropy. In vol. III we have asked for its recognition as an isentropic surface, blurred perhaps in its upper levels by the condensation of water-vapour, etc.

Hence we may perhaps regard the prominent examples of change of wind-velocity indicated in the ascents as places of sudden increase of entropy—*saltus* in the up-grade of entropy.

It is not unusual to find an inversion of temperature-lapse—which elsewhere we have called a counterlapse of temperature—as indicating a discontinuity of dynamical structure, and here we remind the reader that the real characteristic of an inversion is the rapid increase of entropy with height a very notable *saltus* in the up-grade of entropy—it is the increase of entropy which makes the stability, increase of temperature is exaggeration of the increase of entropy and introduces no new feature in the structure.

CHAPTER VII

ATMOSPHERIC CALCULUS—TEMPERATURE AND WIND

"It may be too that from quarters of the world athwart his path two airs may stream alternately each at a fixed season." (Lucretius V, 640; tr. Cyril Bailey.)

WE have reasonable ground for supposing that the winds in the upper air are closely related to the distribution of pressure; and, in turn, the variation in the distribution of pressure at different levels is dependent upon the distribution of temperature, according to the ordinary formula for the variation of pressure with height. On the principle of the law of isobaric motion (Law IX) we can put into algebraical form the relation between the changes of wind within successive levels, as set out in the previous chapter, and the distribution of temperature in the intervening layer.

A formula for the variation of pressure-gradient with height which in differential form may be written

$$\frac{ds}{dz} = g\rho \left(\frac{q}{\theta} - \frac{s}{p} \right) \qquad \text{......(S)}$$

(where s is the horizontal pressure-gradient, q the horizontal temperature-gradient, θ and p the temperature and pressure, and g and ρ have the usual signification) was given in less conventional form in the *Journal of the Scottish Meteorological Society*[1]. The formula is deduced from the ordinary equation for the variation of pressure with height

$$\frac{dp}{dz} = - g\rho \qquad \text{......(1),}$$

combined with the characteristic equation for a permanent gas

$$p/\theta = R\rho \qquad \text{......(2),}$$

to these we may add the defining equations

$$s = dp/dx, \quad q = d\theta/dx \qquad \text{......(3).}$$

It should be noticed that s and q have negative signs when the horizontal *gradients* of pressure and temperature are taken as *positive* in the direction of *falling pressure* and *falling temperature*.

From (3) by differentiation we have

$$\frac{ds}{dz} = \frac{d^2p}{dz\,dx} = - g \frac{d\rho}{dx} \qquad \text{......(4),}$$

and from equation (2) $\qquad \dfrac{d\rho}{\rho} = \dfrac{dp}{p} - \dfrac{d\theta}{\theta},$

---

[1] Shaw, 'Upper Air Calculus and the British Soundings during the International week, 5–10 May, 1913.' *Jour. Scot. Meteor. Soc.*, vol. XVI, 1913, p. 167.

whence the change of pressure-gradient with height

$$\frac{ds}{dz} = g\rho \left(\frac{1}{\theta} \frac{d\theta}{dx} - \frac{1}{p} \frac{dp}{dx} \right)$$

$$= g\rho \left(\frac{q}{\theta} - \frac{s}{p} \right).$$

To obtain numerical values we substitute for ρ, $p/(R\theta)$; we get

$$\frac{ds}{dz} = \frac{g}{R} \cdot \frac{p}{\theta} \left(\frac{q}{\theta} - \frac{s}{p} \right).$$

Taking g as 981 cm/s² and R, for dry air, $2\cdot869 \times 10^6$ c,g,s units we obtain for the equation in c,g,s units

$$\frac{ds}{dz} = 3\cdot42 \times 10^{-4} \frac{p}{\theta} \left(\frac{q}{\theta} - \frac{s}{p} \right),$$

or if the variation be expressed in millibars per metre of height, and gradients in the variation over 100 kilometres, we get the rate of increase of pressure-gradient per metre of height in millibars per hundred kilometres

$$3\cdot42 \times 10^{-2} \frac{P}{\Theta} \left(\frac{Q}{\Theta} - \frac{S}{P} \right),$$

where Θ represents the tercentesimal temperature, P the pressure in millibars, S the horizontal pressure-gradient in millibars per hundred kilometres, and Q the horizontal temperature-gradient in tercentesimal or centigrade degrees per hundred kilometres. For our present purpose we may disregard the numerical effect of differences in R due to moisture in the atmosphere. For air saturated with moisture at 273 a the constant is $2\cdot876 \times 10^6$ instead of $2\cdot869 \times 10^6$, and at 283 a, $2\cdot884 \times 10^6$; these upper limits for the range of saturation are sufficient for the upper air and as the differences in R are quite negligible in comparison with the uncertainties in the determination of wind-velocity we may use the value of R for dry air without appreciable error.

Equation (S) was employed in the paper referred to for the purpose of explaining the dominance of the stratosphere in the distribution of pressure throughout the troposphere. The position may be set out as follows[1]. In his second report on the free atmosphere of the British Isles[2] W. H. Dines had thrown new light upon the origin of the differences of pressure at the surface by obtaining the correlation coefficient between corresponding deviations of pressure from the normal at the level of 9 kilometres and at the ground. "He had obtained results ranging from 0·67 for the last available set of a hundred soundings on the continent to 0·88 for soundings in England grouped for the winter season." Moreover, the standard deviations of pressure from the normal are of the same order of magnitude at both levels. In a more recent table standard deviations for successive levels are given by Dines as follows:

Level in kilometres	8	7	6	5	4	3	2	1	0
Standard deviation in millibars ...	11·0	11·4	11·5	11·5	10·9	10·7	10·5	10·5	10·8

[1] *Proceedings of the Royal Institution*, 1916. *Nature*, vol. XCVII, 1916, p. 191 and p. 210.
[2] *Geophysical Memoirs*, No. 2. M. O. Publication, No. 210 b, 1912.

That is to say at all levels within the troposphere pressure is subject to changes of the same order of magnitude in spite of the great difference in the normal values at the top as compared with the bottom. At the same time it was noted that the correlation coefficient between the deviations of pressure at the surface and of the mean temperature of the 9-kilometre column was small; in other words the mean temperature of a column of the atmosphere in the troposphere, between the surface and the level of 9 kilometres, has little to do with the general distribution of pressure over the country. Its effects may be regarded as occasional and local.

This aspect of the subject, which is of special interest in connexion with the explanation of the general circulation of the atmosphere, was first referred to in the official preface to W. H. Dines's report, and the equation which we are now considering affords a satisfactory explanation of the position. It is certain that the distribution of pressure at the upper levels, of which that at 9 kilometres is taken as typical, is transmitted to the surface and defines the distribution there, subject to any variations caused by the distribution of density of the air of the troposphere, between the level of 9 kilometres and the surface, in consequence of the variations of pressure and temperature. The equation (S) demonstrates that these variations are not likely to be large for the region of the British Isles because according to Dines's results Q and S are of the same sign throughout the troposphere except in the lowest layer of one kilometre and $Q/\Theta - S/P$ will depend upon the numerical difference between the two ratios. Between ground-level and 9 kilometres Θ may be said to vary from about 280 a to 220 a while P varies from 1010 mb to 300 mb; thus the denominators of the two terms vary very differently within the range of levels which has been specified; that for pressure is reduced roughly to one-third of its sea-level value, while that for temperature is only reduced to three-quarters of its sea-level value. Consequently, between the 9-kilometre level and the surface, S/P runs through a considerable range of values on account of the variation of the denominator alone whereas Q/Θ remains comparatively steady. Hence for a considerable range of values of Θ and P it is likely that somewhere between 9 kilometres and the surface the difference $Q/\Theta - S/P$ will be zero. In that case there will be positive values in the lower region and negative values in the upper region; and the effect of the whole troposphere will be small.

This accords with our knowledge of the régime of winds. If the pressure-gradient remained constant throughout the vertical height from the surface to the level of 9 kilometres the product $V\rho$ would be invariable; the velocity of the wind would increase in inverse proportion with the change of density so that the mass of air passing would be the same at all levels as Egnell[1] and Clayton were led to suggest from observations of clouds. We have seen that the actual wind shows almost every kind of variation with height but on the

[1] *Comptes Rendus*, vol. CXXXVI, 1903, p. 358, 'International Cloud Operations,' Trappes, 1896–7. The subject is discussed in vol. II, pp. 280, 416 and 420.

average perhaps less increase in the upper layers than that corresponding with a uniform gradient.

The variations of gradient between 2·5 kilometres and 7·5 kilometres may be regarded as depending mainly on the distribution of temperature because the distribution of pressure is directly related to the distribution of temperature. Hence we may regard the distribution of wind in the vertical as ordinarily controlled by the distribution of temperature, allowance being made for those cases in which the winds are affected locally by thermal convection. Such cases are not likely to be disclosed by observations with pilot-balloons because the thermal convection which produces the disturbance is likely to cause cloud and terminate the sounding.

From the equations already given combined with the equation for the geostrophic wind,

$$v\rho = s/(2\omega \sin \phi) \qquad \ldots\ldots(G),$$

a formula can be obtained for the variation of wind with height.

Thus, from equation (G), taking v the component of velocity parallel to the y-axis drawn northward,

$$\frac{1}{v}\frac{dv}{dz} = \frac{1}{s}\frac{ds}{dz} - \frac{1}{\rho}\frac{d\rho}{dz} = -\frac{g}{s}\frac{d\rho}{dx} - \frac{1}{\rho}\frac{d\rho}{dz}$$

$$= -\frac{g\rho}{s}\left(\frac{1}{p}\frac{dp}{dx} - \frac{1}{\theta}\frac{d\theta}{dx}\right) - \left(\frac{1}{p}\frac{dp}{dz} - \frac{1}{\theta}\frac{d\theta}{dz}\right)$$

$$= -\frac{g\rho}{s}\left(\frac{s}{p} - \frac{q}{\theta}\right) + \frac{g\rho}{p} + \frac{1}{\theta}\frac{d\theta}{dz}$$

$$= \frac{1}{\theta}\left(\frac{g\rho q}{s} + \frac{d\theta}{dz}\right)$$

$$\frac{s}{v}\frac{dv}{dz} = \frac{1}{\theta}\left(g\rho q + s\frac{d\theta}{dz}\right)$$

But
$$s/v = \rho \times 2\omega \sin \phi,$$

whence
$$\frac{dv}{dz} = \frac{1}{\rho\theta \times 2\omega \sin \phi}\left(-q\frac{dp}{dz} + s\frac{d\theta}{dz}\right)$$

$$= \frac{1}{\rho\theta \times 2\omega \sin \phi}\left(\frac{dp}{dx}\cdot\frac{d\theta}{dz} - \frac{d\theta}{dx}\cdot\frac{dp}{dz}\right) \qquad \ldots\ldots(5a).$$

For an elegant demonstration of the equation see *Dictionary of Applied Physics*, vol. III, s.v. 'Atmosphere, Physics of the,' by D. Brunt. The subject was included in the memoir, 'Über Temperaturschichtung in stationär bewegter und in ruhender Luft,' by Max Margules, in Hann-Band, *Meteorologische Zeitschrift*, 1906.

The equation has been put also into other equivalent forms as

$$\frac{dv}{dz} = \frac{v}{\theta}\frac{d\theta}{dz} + \frac{g}{2\omega \sin \phi}\frac{q}{\theta} \qquad \ldots\ldots(5b),$$

$$\frac{d}{dz}\left(\frac{v}{\theta}\right) = \frac{g}{2\omega \sin \phi}\frac{q}{\theta^2} \qquad \ldots\ldots(5c),$$

and
$$\frac{1}{v}\frac{dv}{dz} = \frac{1}{\theta}\left(\frac{d\theta}{dz} + \frac{gp}{R\theta}\frac{q}{s}\right) \qquad \ldots\ldots(5d).$$

We may note that equation ($5c$) implies that if q is zero, v/θ is constant and velocity at different levels is proportional to absolute temperature as set out in the article of vol. II, p. 416.

The corresponding equation for a wind along the x-axis will be

$$\frac{du}{dz} = -\frac{1}{\rho\theta \times 2\omega \sin \phi}\left(\frac{dp}{dy}\cdot\frac{d\theta}{dz} - \frac{d\theta}{dy}\cdot\frac{dp}{dz}\right),$$

because *if the pressure increases to the northward* with increasing y the corresponding change of wind will be from the *east*. We have referred the formula to an x-axis drawn eastward and a y-axis drawn northward because this resolution into components is required for computing the variation of direction with height. The equations may be taken as applicable to any direction of the wind if the y-axis be taken at right angles to the run of the wind and drawn to the left.

The several forms of equation cannot be applied generally to the numerical evaluation of special cases of the variation of wind with height in the free air because the individual values of the horizontal gradients q and s and the lapse-rates ($d\theta/dz$ and dp/dz) of temperature and pressure are not known. We could compute the horizontal gradient of pressure from the wind; and the lapse-rate of pressure can be taken from normal values without any serious error. So also can the value of ρ because that depends upon the ratio p/θ which shows as a rule very small variations from the normal for the month; they seldom exceed 5 per cent. and are generally much less: often within 1 per cent[1].

For the layers near the surface we have observations of temperature at the ground-level from which we can form an estimate of the horizontal gradient that may help us to deal with the relation of wind to the distribution of temperature at moderate heights and this part of the subject will be treated in a subsequent chapter.

And for the free air we can use the observations of variation of wind with height obtained by means of pilot-balloons to compute the values of q at successive levels with the understanding as to using normal values of θ and p/θ. Before doing so we must note an interesting application of equation ($5a$) made by W. H. Dines[2].

He has shown that on the basis of the law of isobaric motion (Law IX) there will be no variation of wind with height if the isobaric surfaces are also isothermal. The demonstration follows directly from equation ($5a$) because the conditions prescribed may be expressed in the form that the variation of pressure in any direction is proportional to the variation of temperature so that

$$\frac{dp}{dx}:\frac{dp}{dy}:\frac{dp}{dz}::\frac{d\theta}{dx}:\frac{d\theta}{dy}:\frac{d\theta}{dz},$$

and in that case the quantity within the bracket of equation ($5a$) becomes zero. It should be noted that this condition for no variation of wind with height is satisfied where the atmosphere is uniformly isothermal. Mr Dines's demonstration is as follows.

[1] See Art. 32, vol. II, p. 419.   [2] *Nature*, vol. XCIX, 1917, p. 24.

Let *ABCD* be a vertical section at right angles to the gradient-wind, *AB* and *CD* being sections of the isobaric surfaces, and *AC* and *BD* vertical straight lines (fig. 49). If v be the gradient-wind—i.e. the wind at right angles to the paper—then the tangent of the slope of *AB* is $2\omega v \sin \phi + v^2/r : g$, for $2\omega v \sin \phi + v^2/r$ is the horizontal acceleration and g the vertical. Similarly, the slope of *CD* is $2\omega V \sin \phi + V^2/r : g$. If, then, v is greater than V, *BD* must be greater than *AC*. Now the pressure-difference between *A* and *C* is equal to the pressure-difference between *B* and *D*, since *AB* and *CD* are isobaric lines; and since the corresponding elements in the two air-columns *AC*

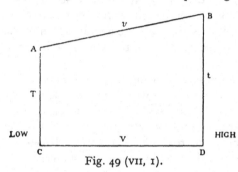

Fig. 49 (VII, 1).

and *BD* are of equal pressure, and the density in *BD* less, the temperature in *BD* must be higher than that in *AC*. That is, if v be greater than V, then t is greater than T. Or if v is the same as V, t must be equal to T, and the lines *AB*, *CD* are isothermal as well as isobaric.

An important conclusion of a very general character follows immediately from this proposition. If the isobaric surfaces are also isothermal a **vertical** cross-section of the atmosphere will show lines of equal pressure and equal temperature having the same slope in the region where there is no variation of wind with height. In any part of the section where the slope of the isothermal line is steeper than that of the isobaric line the wind will increase with height and where, on the other hand, the isobaric line is steeper than the isothermal line the velocity of the wind will diminish with height. By grouping together a large number of observations of pressure and temperature at all heights up to 20 kilometres W. H. Dines[1] constructed a diagram representing the mean distribution of temperature in relation to pressure at different levels in the upper air, from which it appears that the isothermal lines in a vertical section reach a maximum height in the highest pressure and a minimum in the lowest pressure. In these regions the isothermal lines are parallel to the isobaric lines and there is no variation of wind with height, a conclusion which is supported by observation so far as the facts go. Elsewhere the isobaric lines are distinctly more nearly horizontal than the isothermal lines. Hence it follows that in a region between high pressure and low pressure the wind in the successive layers of the troposphere should in normal circumstances show an increase of velocity with height.

We may assign a numerical estimate of the application of this proposition by using the average results for pressures, temperatures and densities at different heights, taken from Dines's diagram, as given in the *Meteorological Glossary*[2] and assuming the differences of pressure there given to be

[1] *Phil. Trans.*, vol. CCXI, A, 1911, p. 253. The diagram in another form is reproduced in *Nature*, vol. XCIX, 1917, p. 24, and *Proc. Roy. Inst.*, 10 March, 1916.

[2] M. O. Publication, No. 225 ii, s.v. Density.

distributed over a horizontal stretch of 1000 kilometres which would represent a pressure-gradient of 5 mb per hundred kilometres. Substituting the values thus obtained in formula (5 a) of this chapter we get numerical values of the increase of wind-velocity with height under the prescribed conditions as follows:

NORMAL INCREASE OF WIND-VELOCITY AT DIFFERENT HEIGHTS
UNDER THE NORMAL CONDITIONS OF TEMPERATURE FOR
A SURFACE-GRADIENT OF 5 mb PER 100 km

Height in kilometres ...	1	2	3	4	5	6	7	8
Normal increase of velocity per kilometre in metres per second	0·7	1·8	2·0	2·8	3·1	3·4	3·8	3·6

The computation of gradients from the winds

The application of the equations to the approximate evaluation of the horizontal gradient of pressure and temperature, assuming uniformity of change over a whole kilometre, is given in *Principia Atmospherica*[1] and further developed in a paper before the Royal Meteorological Society[2]. Tables for facilitating the calculations are given in the *Computer's Handbook*[3], II, § 3

We require the temperature at each kilometre and its ratio to the pressure. Tables of temperature and density can be used instead. For ΔP see p. 86.

The calculation proceeds from the formula,

change of pressure difference for 1 kilometre, $\Delta s = 34 \cdot 2 \dfrac{P}{\cdot \Theta} \left(\dfrac{\Delta \Theta}{\Theta} - \dfrac{\Delta P}{P} \right)$,

where $\Delta \Theta$ is the change of temperature per hundred kilometres and ΔP the corresponding change of pressure: and for the wind-velocity at any level

$$V = \frac{R}{2\omega \sin \phi} \frac{\Theta}{P} \Delta P.$$

Taking the components U, from W to E, and V from S to N separately, we get for the components of pressure-difference at any level

$$\Delta_N P = \frac{1}{K} \frac{P}{\Theta} U, \quad \text{and} \quad \Delta_W P = \frac{1}{K} \frac{P}{\Theta} V,$$

where K represents $R/(2\omega \sin \phi)$. For the components of temperature-difference

$$\Delta_N \Theta = \frac{\Theta}{P} \left(\frac{\Delta s_N}{34 \cdot 2} \times \Theta + \Delta_N P \right)$$

$$\Delta_W \Theta = \frac{\Theta}{P} \left(\frac{\Delta s_W}{34 \cdot 2} \times \Theta + \Delta_W P \right).$$

Θ/P and Θ are taken from tables and the calculation is applied to the change of wind-velocity in successive kilometres. It is only properly applicable when the rate of variation of wind-velocity is uniform over the range, and the precise point at which the computed value of the horizontal temperature-gradient is

---

[1] Shaw, *Proc. Roy. Soc. Edin.*, vol. XXXIV, 1914, p. 77.

[2] Shaw, 'The Interpretation of the results of Soundings with Pilot-balloons,' *Q. J. Roy. Meteor. Soc.*, vol. XL, 1914, p. 112.   [3] M. O. Publication, No. 223.

operative is rather doubtful. Still, the computations throw a considerable amount of light upon the way in which different distributions of temperature affect the structure of the atmosphere and the general conclusions to be drawn from them are not unreasonable.

Six cases are given in detail in the paper on the interpretation of the results of soundings with pilot-balloons. They deal with the soundings which were selected by Cave for illustration in his book by photographs of models. The details of the computation of an additional case representing an upper wind from north-west crossing a lower wind from south-west are given in *Principia Atmospherica*, and also in the *Computer's Handbook*.

The final result of the computation is to enable us to calculate the distance between consecutive isobars and consecutive isotherms at the several levels and to determine also the direction of the isobar and the isotherm. Hence we can draw for each level the positions of two consecutive isobars and two consecutive isotherms which give us an index of the distribution of pressure and temperature at the several levels in accord with the observed changes in the wind.

The results for four cases are given in the diagrams of figure 50.

From these diagrams we may draw certain inferences. On 1 September, 1907, when there was a surface-wind of 5·5 m/sec from 25° east of north we show that at one kilometre there was a backing to a wind of 4·5 m/sec from 290° and thereafter a rapid increase combined with backing until the wind was 16 m/sec from 270° at four kilometres; the air was always colder towards the north, but the temperature-gradient points towards the west of north except between one and two kilometres when the gradient was towards the north-east. (Fig. 43.)

On 29 April, 1908, when the wind gradually changed from a light wind from south-west at the surface to a strong wind from north-west at six kilometres, the temperature-gradient was towards the north-east except between one and three kilometres and generally the fall of temperature to the north-east was rapid. The isotherms gradually drew from being across the wind at the surface to being nearly parallel to the wind between five and six kilometres.

On 1 October, 1908, when the wind changed from SSE at the surface to SW with regularly increasing strength from eight kilometres upwards the gradient of temperature was generally towards the west with a good deal of variation in direction between SW and NW until the sixth kilometre was reached when it ranged itself according to the pressure-gradient. The two steps four kilometres to five kilometres and five kilometres to six kilometres show the isotherms across the isobars with the wind blowing directly towards the colder region. (Fig. 40.)

On 6 November, 1908, when the wind changed from south-east in the lowest layer by gradual diminution through calm, between three and four kilometres, to north-west with regular increase in the upper layers, the temperature-gradient was generally towards the north-east with a notable

Fig. 50 (VII, 2). DIAGRAMS illustrating the relation of changes in wind with height to the distribution of temperature in successive layers. Separation of consecutive isobars and isotherms computed from the observations represented in figs. 40, 42 and 43, with an example of a NW wind in the upper air crossing a SW wind at the surface.

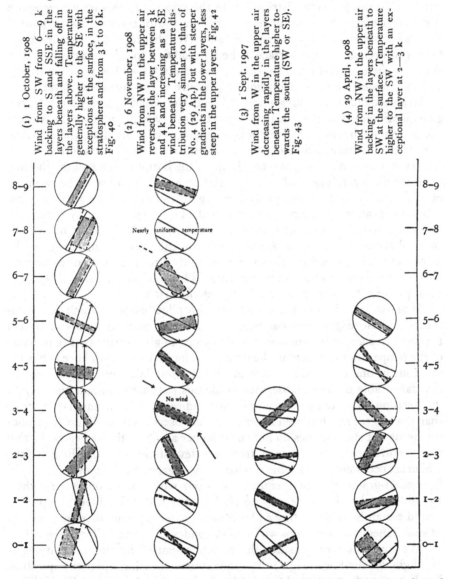

(1) 1 October, 1908
Wind from SW from 6—9 k backing to S and SSE in the layers beneath and falling off in the layers above. Temperature generally higher to the SE with exceptions at the surface, in the stratosphere and from 3 k to 6 k. Fig. 40

(2) 6 November, 1908
Wind from NW in the upper air reversed in the layer between 3 k and 4 k and increasing as a SE wind beneath. Temperature distribution very similar to that of No. 4 (29 Ap.) but with steeper gradients in the lower layers, less steep in the upper layers. Fig. 42

(3) 1 Sept. 1907
Wind from W in the upper air decreasing rapidly in the layers beneath. Temperature higher towards the south (SW or SE). Fig. 43

(4) 29 April, 1908
Wind from NW in the upper air backing in the layers beneath to SW at the surface. Temperature higher to the SW with an exceptional layer at 2—3 k

Within each circle is a plan of consecutive isobars and isotherms for the layer indicated in kilometres by the figures on either side of the diagrams. In each plan the pair of full lines represent the isobar which passes through the point of observation and the isobar for a pressure of 1 mb higher. The pair of dotted lines represent the isotherm which passes through the point of observation and the isotherm of the next higher degree on the centigrade or tercentesimal (absolute) scale. The radius of each circle represents 250 kilometres.

exception between five and six kilometres when the isotherms crossed the wind. (Fig. 42.)

The study of these diagrams suggests to us the desirability of regarding the influence of the distribution of temperature as affecting the transmission of pressure downwards from above, not as we are accustomed to think of it as building up the pressure-distribution in the upper air, because we recognise that when we have got to the top of our building, the differences to be accounted for are just as great as they were at the start. We should look upon the distribution of pressure at the surface as being modified in its transmission from above by the temperature of the air, the motion of which has to be controlled and maintained by the distribution of pressure at the successive layers.

In accordance with a suggestion of Lieut.-Col. Gold[1] we may consider the general principle that the wind at the top of any layer differs from the wind at the bottom of the layer by a vector component, depending upon the distribution of temperature within the layer, which we may call the thermal wind. Thus the top wind is the geometrical sum of the bottom wind and the thermal wind for the layer. The student may build up a working idea of the variation of wind with height by habitually forming an estimate of the thermal wind when observations of temperature are available; or, *vice versa*, by noting the difference between an upper wind and a lower wind he may obtain the thermal wind and hence a working idea of the thermal structure of the layer. But this method requires the bottom wind to be corrected in the ratio of top temperature to bottom temperature. The relation is set out in equation (5 c), p. 196; it is also given by Margules (*loc. cit.*).

If on the occasions of the soundings with pilot-balloons which are represented in these diagrams we had been fortunate enough to have observations of temperatures at a sufficient number of stations to give a trustworthy measure of the temperature-gradient in the immediate locality we might have subjected the conclusions to the direct test of observation. We have not yet reached that favourable position. Observations for this purpose require a high degree of precision. The temperature-gradients indicated seldom amount to more than one degree in a hundred kilometres and observations at different stations by means of ballon-sondes do not claim higher accuracy than a degree, so that temperature-gradients obtained from the few stations in the British Isles which are occasionally available can hardly be regarded as final evidence of the horizontal-gradient of temperature for levels within the troposphere where the isothermal surfaces are very nearly horizontal. But the direct comparison would certainly be interesting and it is much to be regretted that the occasions of simultaneous observations with ballon-sondes and pilot-balloons are so few.

We may however consider the results obtained by the calculations in the light of our knowledge from other sources bearing upon the question. We have already mentioned that for levels above four kilometres W. H. Dines has shown that there is very high correlation between the deviations from the normals of pressure and temperature obtained by the soundings with ballon-

[1] M. O. Correspondence, 14117, 1918.

sondes. That relation would be explained if the temperature-gradient at each level were always along the line of the pressure-gradient and proportional thereto, that is to say, if the isotherms were parallel to the isobars and at a proportionate distance apart. It is therefore interesting to note that in all the diagrams there is a very definite tendency for the isotherms to become parallel to the isobars especially in the upper levels. There are only two examples of a temperature-gradient nearly opposite to the pressure-gradient, those are between two and three kilometres on 29 April, 1908, and between one and two kilometres on 6 November. But there are a number of cases in which according to the computation the isotherms are at right angles to the isobars. These may be due to errors in the data upon which the computations are based, but if they are real they may be useful in explaining atmospheric processes. It was pointed out in *Principia Atmospherica*, for example, in discussing the case of 29 April, 1908, when a surface-wind from south-west was passing under an upper wind from north-west, that the south-west wind had its temperature-gradient to the north-east and the current as it went forward was continually replaced by warmer air which passed under a layer in which there was no corresponding change of temperature; that state of things must ultimately result in instability which is characteristic of the south-westerly wind of an advancing depression, and the process, which would eventuate in rainfall, is inevitable if in consequence of the distribution of temperature beneath it the gradient for north-westerly wind is transformed into a gradient for south-westerly wind near to the surface. This case happens not infrequently when a low pressure system is passing away to the north-eastward and is followed by another depression. The south-westerly wind of the coming depression appears first at the surface while the north-westerly wind remains at the higher levels.

On the other hand the cases of isotherms transverse to isobars in the upper air may be indications of local variations of temperature that are the result rather than the cause of convection, or local inversions of the lapse of temperature due to pressure changes in the stratosphere.

An example of a complete reversal of wind-velocity with height which may be accounted for by a counterlapse of temperature is given in *Principia Atmospherica* from a sounding by J. S. Dines at Pyrton Hill on 16 October, 1913. "On that day there was a sudden change of wind between 1100 and 1500 metres height from a fairly steady wind from nearly due south into one almost as steady from due north, the change being accomplished within half a kilometre. For the layer between 500 and 1100 metres the analysis in this case shows a temperature distribution in isotherms nearly north and south with the warmer air in the east and above 1500 metres an entirely different distribution with isotherms nearly east and west and cold to the northward. The intermediate layer 400 metres thick showed a very rapid increase of temperature to the west as much as 7° C per hundred kilometres.

"The complete arrest of the upper northerly current and production of a calm by the annihilation of the gradient between 1500 metres and 1100 metres is very remarkable but nevertheless a real fact. The accompanying temperature difference is probably due to a strong 'inversion' at a height of about 1500 metres at the place of observation and at about 1100 metres at a place 100 kilometres distant to the west. On that occasion it lasted for some time, as it was found an hour afterwards by a second balloon; but it must be remembered that it was a region of no velocity and therefore the warm and cold airs at those levels were not moving."

Such a distribution was by no means improbable on the day of the observation. The land area of England had been covered by cold air above which there was probably an inversion as there is above a fog. The morning observations for 7 a.m. showed a temperature of 38° at Nottingham with fog, and of 39° at Bath with blue sky, while London had a temperature of 50°, Yarmouth 50° and Pembroke 47°. The wind at Nottingham at that hour was from WSW but NE at Bath and ENE at Pembroke.

The conclusion that in the upper air the isotherms are generally parallel to the isobars and the gradients of pressure and temperature at the several levels proportional is interesting from the bearing that it has upon Egnell's or Clayton's law of the variation of wind with height. We have seen that the law requires that the pressure-gradients should be the same at all heights and from equation (S) the condition becomes

$$d\theta/\theta = dp/p.$$

We have seen that on occasions the wind shows continuous increase with height from the north, from the south and from the west, very rarely from the east. We may expect the law to be verified therefore by winds from south, west or north; if it is also verified for winds from the east it would confirm a conclusion at which W. H. Dines had arrived on other grounds that there are no preferences for direction of temperature-gradients in the upper air. In the lower layers the circumstances under which temperature increases to the north (the condition required for the maintenance of the pressure-gradient in an easterly wind) are very rare and peculiar but in the upper air the rule of proportionality of temperature-difference to pressure-difference is quite general and an easterly wind increasing with height ought not to be regarded as out of the question at those levels.

The direct relation of the gradient of pressure to the gradient of temperature which is normal in the troposphere is reversed in the stratosphere, that is to say, high pressures are cold in that region as compared with low pressures. In that case Q/Θ takes the negative sign when S/P is positive; and, in consequence, the gradient of pressure and the wind-velocity must fall off with height. We notice (fig. 45) that there is a tendency for the wind-velocity to fall off in the stratosphere. The diagram in fig. 44a, p. 182, shows six examples. From the measurements of the change of wind-velocity the gradient of temperature just above the base of the stratosphere has been computed by formula (5 b) as shown in the following table[1]:

[1] *Proc. Roy. Inst.* 10 March, 1916.

| Date 1908 | Rate of change of velocity in the stratosphere m/sec per k | Horizontal temperature-gradient | |
		Computed. Degrees per 100 k	Observed. Degrees per 100 k
27 July	—	—	2·5
28 ,,	− 13	4·0	—
29 ,,	− 11	3·3	3·3
31 ,,	− 5	1·5	—
1 Oct.	− 7	2·1	—

The days in July belonged to the international week upon which soundings were made with ballon-sondes at three stations in England, one in Scotland and one in Ireland. For 27 and 29 July enough balloons were found and returned to give the material for constructing the models which are represented in chap. IV of vol. II, and from the models the temperature-gradient can be determined with some confidence because in the stratosphere the isothermal surfaces are more nearly vertical than horizontal. The horizontal temperature-gradient as determined by measurements of the model is 3·3a per 100 kilometres, which agrees exactly with the value computed from the change in wind-velocity. The exactness of the agreement is doubtless fortuitous but it is interesting to see that results of the same order of magnitude are given for the other days for the values computed from the change of velocity and on the only other occasion available for the observed value of the horizontal gradient of temperature.

If the falling off of velocity in the stratosphere follows the distribution of temperature, as it apparently does in fact, it would also follow that the wind-velocity and with it the pressure-gradient would become zero within a few kilometres. Looking at the diagram of fig. 44a, p. 182, it would appear that zero velocity and therefore zero gradient would be reached at 13 kilometres on 28 July, 1908, at 14 kilometres on 29 July, 1908, and 6 May, 1909, at 15 kilometres on 7 May, 1909, at 16·5 kilometres on 30 Sept. 1908, and at 17·5 kilometres on 31 July, 1908. If we suppose the temperatures in the vertical to remain uniform beyond these limits, as temperature is usually nearly uniform in the vertical within the stratosphere, we must conclude that the pressure in the warm column over the "low" will diminish more slowly than that in the cold column over the "high." Hence above the level of no gradient and no wind there will be a new region in which the high is warmer than the low and the wind will be reversed in direction and gradually increase in magnitude as the heights increase until some change takes place in the distribution of temperature. Thus at great heights in the stratosphere an increasing easterly wind may be found above a westerly wind in the troposphere and so on for the other directions. Cave[1] has noted an interesting case in which a southerly wind began to show at 48,000 ft increasing to 44 miles per hour from the same direction at 58,000 ft.

It may here be recalled that as a result of the inquiry into the phenomena due to the eruption of Krakatoa in August, 1883, an easterly wind of about

[1] Cave. M. O. Correspondence, 4158, 1918.

80 miles an hour was identified in the equatorial regions[1]. Very high velocities for very high levels are sometimes indicated by the luminous trails of meteors[2].

From the examples which have been given it will be seen that the hypothesis of an atmosphere in which the wind-velocity is normally adjusted to balance the pressure-distribution enables us to explain many of the ascertained facts that have been disclosed by observations of the upper air. Among them we may recall the conditions for change of wind-velocity with height, the general arrangement of pressure-distribution according to temperature-distribution in the upper layers of the troposphere, the falling off of wind in the stratosphere and the change of wind over an "inversion." It also justifies us in regarding the stratosphere as the dominant region of the atmosphere so far as the distribution of pressure is concerned. The tendency of meteorological study in the past has been to regard the structure of the atmosphere as built upon the foundation which we see laid out at the surface. We shall probably find fewer difficulties in the path of the study if we regard the pressure-distribution at the surface as controlled by the stratosphere and only modified locally by the irregularities of temperature that are to be found in the lower layers. The proportionality of changes of pressure to changes of temperature in the section from four to eight kilometres is probably of the highest significance for the comprehension of the structure of the atmosphere.

The reader may notice that chapter IV was devoted to the explanation of the variation of wind with height as expressing the influence of turbulence without making any appeal for assistance from change in barometric gradient; whereas in this chapter, in pursuance of the idea of numerical relationship between wind and gradient in the upper air, we have offered an explanation of the change of wind with height exhibited by pilot-balloons on the basis of change of gradient due to distribution of temperature without making any appeal to turbulence for its assistance.

The reason for the difference in the mode of attack, with two problems which are apparently similar, is that in chapter IV we deal with air between the surface and 500 metres. Within those limits turbulence is certainly operative but its effect diminishes with height. In this chapter we deal with air from 500 metres upwards. So far we have found no definite reason against our assumption. The results are reasonable. Precise observation alone can show whether it is justified for the region to which it is applied.

[1] 'The Eruption of Krakatoa and Subsequent Phenomena.' *Report of the Krakatoa Committee of the Royal Society*, 1888, pp. 325–333. (See vol. II, p. 278.)

[2] See an interesting article on the travel of Meteors in *Chambers' Encyclopedia*, s.v., and a note by F. J. W. Whipple, *Meteor. Mag.*, vol. LVI, 1921, p. 292.

CHAPTER VIII

GRAPHIC ANALYSIS OF ATMOSPHERIC MOTION

In the atmosphere, pressure writes entropy upon an isothermal surface, and temperature writes entropy upon an isobaric surface; both write volume on an isentropic surface: but pressure, temperature, volume and entropy have each to write their own story on a level or isogeopotential surface.

SYNCHRONOUS CHARTS OF HORIZONTAL MOTION IN THE FREE AIR

In the preceding chapters the process of interpretation of the records of an anemogram as expressing the problem of the dynamics of the atmosphere has been as far as possible algebraical. In chap. II, which supplied us with the equations of motion, the procedure which is contemplated for the solution of the problems of the atmosphere is the integration of the equations. We may now remind the reader that an algebraical process is only applicable in so far as the atmospheric structure can be expressed by an algebraical or numerical formula. The records of pilot-balloons in chap. VI and the entropy-temperature diagrams of chap. VI of vol. III are sufficient to show that the representation of the information by arithmetical formulae is not an easy route towards the goal of our adventure.

In this chapter we propose to examine the possibilities of graphic analysis which provides a more flexible instrument, more appropriate for our extended knowledge of the atmospheric structure. At the end of the last century there would have been no hesitation about expressing the structure of the atmosphere for the purposes of arithmetical computation as isothermal or isentropic; but in modern days it would hardly seem worth while to spend time and trouble over a computation on that basis.

In 1910 V. Bjerknes and his collaborators took the lead in the application of graphical methods to the atmosphere and the hydrosphere with two volumes of Dynamic Meteorology and Hydrography[1]. The first deals with Statics and the second with Kinematics. A third on Kinetics has still to come. The process of analysing the meteorological conditions is called "diagnosis" and the tracing of the changes due to the natural transformations of energy "prognosis." The scheme employs pressure and volume, the dynamical factors of energy, or pressure and density, with velocity, as the basis of graphic representations.

Before dealing with that, and other methods of representation later on, we may note that the weather-map, which is in common use all over the world, with its isobars and its fronts, is itself a method of graphic representation. It is not a perfect analytical method because the pressure is represented as at sea-level, which may be underground, while the temperature and wind for the

---

[1] *Dynamic Meteorology and Hydrography*, Carnegie Institution of Washington, Publication No. 88. Part I, Statics, by V. Bjerknes and J. W. Sandström, 1910. Part II, Kinematics, by V. Bjerknes, Th. Hesselberg and O. Devik with separate volume of Plates, 1911.

same localities are in the air. That is why the method of forecasting by means of weather-charts is called empirical; and it must be so while the data represented are aggregated but not organised for scientific analysis. But the map shares the advantages of graphic representation and simplifies interpolation because it co-ordinates individual data without assuming an arithmetical formula. One of the general conclusions obtained from the study of maps is that for sea-level or a flat land-surface adjacent thereto there is equilibrium and no regular air-motion where pressure is uniform; but where there is a gradient of pressure, equilibrium may be exchanged for steady motion, the kinetic correlative of the static distribution of pressure.

The first step in the diagnosis is naturally to plot all the observations for the same level using a separate map for each separate level, just as we are accustomed to refer to a map of the distribution of pressure at sea-level the meteorological observations at all the available stations at the surface. The plotting was done at the Meteorological Office when the winds for five different levels, as obtained from observations of pilot-balloons at upwards of thirty stations, were set out on maps. The direction of the wind at each level was indicated on the chart for that level by an arrow drawn to fly with the wind, as usual, and the velocity by figures inserted in small circles circumscribing the points which mark the positions of the verticals at the respective stations.

The next step is to compare the figures for each station with those for the surrounding stations. The same process was regularly followed in preparing the customary map of the distribution of pressure at sea-level and the distribution of temperature at the surface. By that means the observations of pressure were subjected to a rigorous scrutiny and any outstanding exceptional reading was the subject of immediate inquiry by telegraph. The observations of temperature were similarly scrutinised and occasional errors of five degrees or ten degrees are sometimes corrected in that way. The observations of the state of the sky were practically passed without scrutiny because local cloud in a region of generally clear sky, or the reverse, is not regarded as an improbable phenomenon. And in like manner the observations of wind were not questioned unless the divergence of the reported wind from the direction or force which the gradient would lead us to expect was very marked and the situation of the station was known to be consistent with an approximate relation to the gradient. Very large deviations from the apparent gradient either in direction or force were accepted as real for stations in Iceland, the Faroë and Norway or for high level stations on the continent, and considerable deviations were allowed to pass without challenge for some of our own stations. Some stations indeed are recognised as having a peculiar local bias for which an appropriate explanation is being gradually sought.

When we come to deal with the comparison of the winds in the upper levels as plotted from observations with pilot-balloons we find sometimes considerable deviations, occasionally in direction but more frequently in the figures representing the velocity. Here a difficulty presents itself arising from the fact that the observations are made with only one theodolite. We have seen that if the balloon happens to be in an ascending air-current the horizontal velocity as computed from the observations will be too small by the amount represented by $w \cot E$, where w is the vertical component of the velocity of the air and E the angular elevation of the balloon. We must also reckon with occasional errors of reading which have the same effect upon the computed velocity as a vertical component of the motion of the air. In the circumstances there is *prima facie* no indication of the real explanation of an exceptional reading.

We are hampered to an unknown extent by the fact that in observing the flow we really want the undisturbed flow and not the actual velocity. We require in fact $u\, \partial s/\partial x$ and not Ds/Dt, and we can make no satisfactory correction of the observation on that account. Still it is worth while on occasions to make a plan of the motion and attempt the evaluation of the pressure.

In dealing with observations of pressure the isobars furnish, as a rule, a very satisfactory means of distinguishing between errors of reading and local meteorological peculiarities because the continuity of the distribution of pressure must be expressed in the isobars, but there is no such necessary continuity in the distribution of wind-velocity, although, for all ordinary circumstances, we postulate a relation between the wind and the distribution of pressure.

For the maps representing observations with pilot-balloons we have no observations of pressure in the upper air with which the wind can be compared; we have to deal with the observations of wind alone.

The reciprocity of wind and pressure in the upper air

In the preceding chapters we have relied on the map to give us the pressure-gradient and have in effect taken the isobars to represent the undisturbed air-flow somewhere above the foot of the atmospheric structure at the level of perhaps 500 metres. The examination of the relation between pressure-distribution and flow would be very much assisted if it had been possible to obtain values of pressure and flow for given values of geopotential or height. There are however not nearly enough observations of pressure to make possible the drawing of isobars. With the multiplication of pilot-balloons since 1912 there are many more observations of velocity at different levels, and it ought to be possible to use the reciprocal of the process of determining the flow from the distribution of pressure, and determine the distribution of pressure from the observations of flow.

If we could fill the map with a picture in two dimensions of a solenoidal distribution of wind-velocity it ought to afford a good representation of the distribution of pressure. By a solenoidal distribution we mean the dividing of space into a bundle of tubes along which air is flowing without turbulence. Assuming the flow to be horizontal and the structure to be two-dimensional we may attempt to deal with the observations of winds at different levels by fitting to them a system of instantaneous lines of flow for which the flow measured in terms of momentum is everywhere inversely proportional to the separation of the lines. We make in fact a map of the "stream-function" instead of a map of pressure. The difficulty in drawing a field of pressure of this kind from the charted observations of pilot-balloons is that the wind gives us only the space rate of change of pressure, and for each station we can only calculate the separation of consecutive isobars, not the actual position of either. The completed chart would in effect give us information as to the velocity at every point of the map when we have direct observations from a few. We cannot approach the solution step by step; we can only by trial submit a complete solution that fits the facts at the points of observation.

Bjerknes draws stream-lines by what is at first sight a more rigorous process. In making our own diagnosis we are not discouraged by the fact that the treatment appears less rigorous than that proposed by Bjerknes or Richardson. The course which we propose depends upon assumed relations which are not verified. Either system ultimately depends on an ideal. Bjerknes sets aside the ideal of motion which we have expressed as the law of isobaric motion with the remarks "The accordance of these curves (the isobars) with the direction of the arrows (representing the stream lines) is never complete and should be complete only in exceptional cases." "Further, the numbers representing the observed wind-intensities are never in full accordance with the formula." But he uses the acknowledged relation between wind and gradient at the surface as an auxiliary to draw lines of flow by making them cut the isobars at certain angles. And he draws the curves of equal wind-intensity so as to get certain departures from the theoretical value[1]. In view of what has been adduced in chapter III, this process seems to be based upon an unsatisfying ideal. It is not unfair to say that in the free air if we are to assign any definite value to the deviation of the wind from the isobar or any definite value to the ratio of the wind to the gradient-wind (for straight isobars), when we have not actually measured them, the only value of the deviation and the only value of the ratio that have any substantial claim are zero and unity respectively. And, moreover, in giving practical directions for constructing a picture of the field of motion in the atmospheric space near the ground, Bjerknes suggests[2] that "a point of divergence will appear where there is a descending current (centre of anticyclone) and a point of convergence where there is an ascending current (centre of cyclone)," and among the supplementary rules for obtaining the lines of flow from a limited number of observations of wind he writes "Within a barometric depression there is a probability for existence of points or lines of convergence; within areas of high pressure there is a probability for the existence of points or lines of divergence. Long ridges of high pressure will as a rule contain a line of divergence; long ridges of low pressure a line of convergence." This ideal as the basis of a representation of the atmosphere is unsatisfying for three reasons:

First, after most careful inquiry, as set out in the *Life-History of Surface Air-currents*, the central regions of anticyclones did not manifest themselves

[1] *Loc. cit.*, Part II, p. 62, § 139, 'Dynamic diagnosis of motion in the Free Space.'
[2] *Loc. cit.*, p. 48.

as regions of descending air but as masses of the atmosphere of great stability which apparently took no part in the supply of air to the surface. Secondly, ascent of air in the central region of a cyclone, as set out in the *Life-History*, is not a necessary accompaniment of the existence of the cyclone and the apparent convergence to the centre is very much modified by the motion of the centre itself. An instructive commentary on the conventional view that a cyclone is a region of upward convection whereas an anticyclone is a region of downward convection may be found in the fact that the tropopause, the lower limit of the stratosphere, is higher over an anticyclone than over a cyclone. Convection therefore reaches a higher limit and the cooling which is a natural consequence of upward convection is carried further up in an anti-cyclone than in a cyclone.

In formulating his method Bjerknes uses great precision as to the processes to which the pictures of the atmospheric fields are to be subjected, but in forming the pictures he accepts traditional ideals of meteorological situations, which doubtless have some foundation in fact but will not always bear examination in detail.

The method of reciprocal relation of pressure and wind fortunately seems most free from objection for those parts of a map where pilot-ballooning is practicable, that is to say, where there is comparative freedom from low clouds, vertical motion, and other characteristics of the neighbourhood of the singular points or lines of a field of air-motion; but let us confess that the discussion of methods is by far the easiest part of the whole programme. The practical application is beset with minor troubles and is very laborious considering that maps for many levels three times a day at least come up for consideration. We have first to form a general idea of the way in which the lines run and then decide whether exceptional values of direction and velocity shall be regarded as mistakes of reading and ignored, or as localities of ascending or descending currents which, having regard to the scale of the map, may also be ignored in drawing the lines, or thirdly, as marking some change in atmospheric structure which ought to be adequately represented and indicated with some special symbol on the finished map. Next a number of trials have to be made as the only guides are the direction and separation of the lines at certain points. Drawing lines in pencil, and rubbing them out again as may be required, is the plan adopted in drawing isobars, and is workable enough because each line can be separately and finally decided upon. But with lines to represent tubes of flow a whole line may have to be displaced with some slight modification of shape in order to accommodate the other lines in its neighbourhood. The only workable plan which has so far suggested itself is to use small squares of card of standard size which can be moved about the map and laid side by side to the proper separations and flexible wire for trial lines which can be bent to the proper shape, laid down on the map and moved bodily when necessary. When the field has been approximately mapped in this way final lines can be drawn in.

The observations of velocity in successive levels of 1000 feet come to the

14·2

Fig. 51 (IX, 2). MAPS OF THE DISTRIBUTION OF AIR-FLOW AT 20,000 ft,
15,000 ft, 10,000 ft, 5000 ft, and 1000 ft, together with a map of the isobars
at sea-level and winds at the surface at or about 13 h on 19 October, 1917.

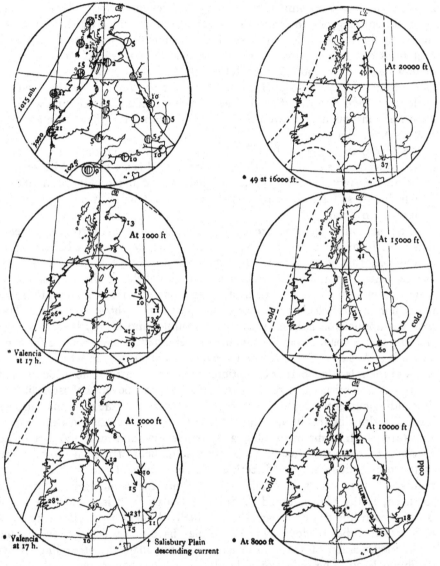

Note. The general scheme of the lines is based on the meteorological situation indicated
by the weather-maps of the day which shows that a depression was advancing from the
west. Later in the day at 17 h strong southerly winds were indicated at 1000 ft and
5000 ft at Valencia. The direction of the flow is shown by arrows and the velocity in
miles per hour is marked in figures. The lines of flow are drawn so that the velocity of flow
is inversely proportional to the separation of the lines, a separation of 60 nautical miles
indicating a velocity of 70 miles per hour. Dotted lines indicate a general idea of the
flow outside the region of direct observation.

Observations which are not in good accord with the lines are marked; so are those
which are not strictly applicable for the drawing of the lines on account of some difference
of hour or of height.

Office expressed in miles per hour with the direction given in points. The setting out of the lines therefore requires a table of distances of separation for winds of specified velocity for given values of the density which, in the absence of direct observations of pressure and temperature, must be taken from a table of normal densities at the different levels in the several months. On the average for the whole year these values expressed in grammes per cubic metre are approximately 1250 at sea-level, 1050 at 5000 ft, 900 at 10,000 ft, 775 at 15,000 ft and 650 at 20,000 ft. The table of separation which will correspond with the separation of isobars for steps of 5 mb in latitude 52° and for these densities will be as follows:

Separation in nautical miles of the lines of flow in solenoidal wind-charts

Selected values of density in g/m³

Velocity		1250	1050	900	775	650
m/sec	m.p.h.		Separation of lines of flow in nautical miles			
2·2	5	845	1006	1174	1362	1625
4·5	10	423	503	587	681	812
8·9	20	211	251	293	341	406
13·4	30	141	168	196	227	271
17·9	40	106	126	147	170	203
22·4	50	85	101	117	136	163
26·8	60	70	84	98	114	135
31·3	70	60	72	84	97	116
35·8	80	53	63	73	85	102
40·2	90	47	56	65	76	90
44·7	100	42	50	59	68	81

Afflatus, dissipati

We may give in illustration of the application of this method maps representing the conditions at the surface and the distribution of velocity at 1000, 5000, 10,000, 15,000 and 20,000 feet on 19 October, 1917, referred to in chap. VI when a very strong wind in the upper air carried a fleet of Zeppelins out of their course. In this set of maps the distribution of velocity is represented, not the distribution of momentum; the pressure-difference corresponding with the separation of the lines is 5 mb at the surface; in the upper levels it is less in the inverse ratio of the densities.

The maps of the flow of air at the different levels at about midday or 13 h of 19 October, 1917, represented in fig. 51, are appropriate to the meteorological conditions of a depression with strong southerly winds advancing in the rear of another depression which is passing eastward. They make clear the special feature of the occasion, viz., a very large increase in the northerly wind in the extreme rear of the passing depression between 5000 feet and 20,000 feet, more especially between 10,000 feet and 20,000 feet. Observations near Portsmouth extended to those heights and those near Edinburgh up to 16,000 feet. They are as follows:

Height in feet ...	5000	10,000	15,000	16,000	20,000
Wind near Edinburgh m.p.h.	NW 8	N 21	N 41	N 49	—
Wind near Portsmouth m.p.h.	—	NNW 25	N by W 60	N by W 65	N by W 87

The maps show that the strong northerly current lay over a north and south band down the middle of England at the time of the maps when the winds near the surface were the light westerly winds of the wedge of high pressure between the two "lows." At 10 o'clock of the following morning a strong northerly current of less velocity was shown in the layer from 14,000 to 20,000 feet over northern France with very light south-easterly or south-westerly winds beneath. The velocity at 20,000 feet in this strong northerly current was 47 miles per hour. Judging by the rapidity with which the lost airships travelled, the velocity in that region during the previous night or early morning was stronger than that recorded at 10 o'clock; the stronger current had probably by that time passed eastward. The rate of drift eastward may have been that shown by the westerly wind in the wedge between the departing and advancing depressions.

According to the theory which has been set out in chap. VII the rapid increase of velocity with height must be attributed to a steep horizontal gradient of temperature from west to east in the layers between 10,000 feet and 20,000 feet. According to the formula (5 a) of chap. VII the gradient of temperature required for an increase of velocity of 5 miles per hour per thousand feet at the level of 10,000 feet would be about 1·8tt or 3·3° F per hundred kilometres. Such a gradient could only arise if the temperature of the wall of air forming the western boundary of the current were nearly isothermal for a thickness of a kilometre or more because the reduction of temperature on the eastern side is limited by the lapse of temperature with height. The lapse cannot pass the adiabatic limit and is normally not far from it so that there is not much margin with which to produce an exceptionally large horizontal gradient. In other words the conditions require a wedge or tongue of air which approaches the isothermal condition or perhaps goes beyond it forming an inversion. Air of this character in the upper regions may be called "very warm"; it would lie between the lower temperature of the air of the passing low and that of the approaching low. The intermediate regions have accordingly been marked "very warm" on the maps for 10,000 feet and 15,000 feet, while "cold" has been written over the regions of the lower pressures on either side of the warm tongue in order to give an idea of the distribution of temperature necessary for the observed phenomena. The belt of very warm air with the rapid current on its right must have travelled eastward. So far as we can tell from the map for the level of 5000 feet the air underneath the warm tongue and also that to the west of it was travelling with a velocity of about 10 miles per hour from the west. At 10,000 feet in the corresponding region there is a velocity of 12 miles per hour from a point south of west; and if we can assume that the whole system, consisting of the strong current with a low on either side, was moving from west to east with a velocity of about 10 miles per hour the tongue of warm air would have travelled about 200 miles to the east between 13 h on the 19th and 10 h on the 20th. If this view is correct the "relative motion" of the strong northerly current would be the actual motion as mapped, modified by the vector subtraction of a west component of 10 miles per hour.

It so happens that we have a record of temperature in the upper air, for 10 h of the 20th, from Ipswich, about 200 miles to the east of the locality assigned as very warm at 13 h of the previous day. With all the assumptions which we have made we therefore expect to find a nearly isothermal column over Ipswich. The temperatures of this ascent are therefore of peculiar interest and we find as follows:

Height in feet	6350	8350	10,350	12,350	14,350	16,350
Temperature F	32	30	25	20	16	16

It is certainly remarkable that the air-column from fourteen to sixteen thousand feet is actually isothermal. Below those levels lapse-rates are normal except for the first step quoted.

Once more, this may be a coincidence but it is a very surprising one, and all the more surprising because the strong northerly current measured in northern France at the same hour only began at 14,000 feet instead of 10,000 feet as it did at Portsmouth on the previous day.

If these facts are more than a coincidence and are indeed what they appear to be, a justification of the assumptions which we have made, it would follow that convection must have invaded the layers over northern France and Ipswich between 10,000 feet and 14,000 feet within the 21 hours between the time of the map and the time of the observation of temperature, and have demolished the strong northerly current at the same time that it destroyed the gradient of temperature. And that is not surprising, because the formation of an isothermal layer several thousand feet thick must require very exceptional conditions. It may be reasonable to say that this tongue of "very warm" air must have been the survival of a mass of air travelling slowly over the Atlantic from west to east and free from convection long enough for isothermal conditions to be set up and gradually worn away from below to the extent of 4000 feet of its thickness while it passed across England. It had a cyclonic depression on its east side and another on its west side, but whether they invaded it laterally we cannot say.

In fact the mere existence of a strong north wind at very high levels is one of the difficult problems of the dynamics of the air. If it is true that air becomes isothermal when it is free from convection but exposed to radiation, it is clear that a knowledge of the distribution of temperature at successive levels would be of great assistance to us in the task of preparing synchronous charts of horizontal motion in the free air because they would help to guide our judgment in combining the direct observations of the wind. Observations of temperature at levels up to 20,000 feet are now possible with aeroplanes and it is time that an endeavour be made to incorporate observations of temperature with observations of wind-velocity at those stations which aim at providing means of guidance in aerial navigation. A preliminary difficulty arises from the fact that the heights at which the observations are made are not given with the accuracy that is desirable by the ordinary means of observation of height in an aeroplane, but even if that difficulty should prove insuperable in view of the slight inclination of the isobaric surfaces in ordinary conditions of weather we may still be able to identify the localities of exceptional horizontal gradients of temperature which would form the most important features of the maps. Apart from the fact that the slope of the isothermal surfaces is generally steeper than that of the isobaric surfaces we know that there are regions where the air is marked by isothermal conditions or by inversions of lapse with height and corresponding conditions are not possible with pressure. Consequently a continuous record of temperature in relation to pressure would identify these exceptional regions and help materially in setting out a map. In this connexion we may refer to the diagrams of temperature at different heights in the atmosphere on consecutive days obtained by observations with kites. They began with Teisserenc de Bort[1], who prepared one for Paris from observations made at his observatory at Trappes, and were continued at Lindenberg[2] for which a year's observations were published in separate form; and subsequently at Mount Weather near Washington[3]. They all show very striking changes of temperature setting in and lasting for some days which could easily have been identified by observations in aeroplanes made in appropriate localities. Indeed, observations by aeroplane are now used by H. G. Cannegieter[4] to provide corresponding charts of isopleths of temperature month by month over Holland.

GRAPHIC METHODS OF COMPUTATION

Let us now turn to the more general aspects of graphic analysis of atmospheric processes. The literature is not very extensive. For some recondite reason algebraical methods have greater attraction for scientific workers than graphic ones. In like manner the calculation of correlation coefficients is a more popular occupation than the exploration of the physical processes upon which they depend, though on the other hand the application of the laws of great numbers to the peculiar conditions of the atmosphere seems a lifeless occupation so long as the physical causes are undetermined and even unsuspected.

The preference for algebraical methods may perhaps be related to the difficulty of dealing on paper with phenomena in three dimensions by graphic methods, a difficulty which may be obvious in several examples in this work and which might be avoided by a table of numbers or by an appropriate arithmetical formula.

The difficulty is disposed of to a certain extent by using level surfaces for

[1] 'Sur les caractères de la température dans l'atmosphère libre au dessus de 10 kilomètres.' *Proc. verb. de la Commission pour Aérostation Scientifique*, St Petersburg, 1904, p. 110.

[2] Dr R. Assmann, *The temperature of the air above Berlin from October 1st, 1902, until December 31st, 1903*. Berlin, 1904. Otto Salle.

[3] William R. Blair. 'Free Air Data at Mount Weather.' *Bulletin of the Mount Weather Observatory*, vol. IV, 1912, pp. 176 et seq.

[4] *Hemel en Dampkring*, 'De Toestand van den Dampkring boven Soesterberg,' 1926 onwards.

graphic representation; they lend themselves perfectly for reproduction in print as in fig. 51 so long as the area to be represented does not differ appreciably from a plane, but the difficulty reappears when the level surface extends over more than a few degrees of latitude or longitude.

The limitation to level surfaces is regrettable as they are not the only possible ones nor even the best for general meteorological purposes. The alternatives are isobaric surfaces, isothermal surfaces, isosteric or isopycnic and isentropic surfaces. The great advantage of any of these, if there were any satisfactory means of representing them on paper, is that on any one of them the condition of the atmospheric structure is represented by a single family of lines. For example, on an isobaric surface an isothermal line is also isosteric and isentropic and so on, only for the isogeopotential lines is a separate family required on any one of the other surfaces.

Taking these important facts into account there is very much to be said for isobaric surfaces as surfaces on which to plot the observed state of the atmospheric structure. Such a surface cannot differ very much from a horizontal surface and there can be no complicated involutions of the surfaces in a quiescent atmosphere, whereas an isothermal surface might be extremely involved and even an isopycnic surface may be less smooth than an isobaric one.

The question then arises whether a contorted or implicated surface is a good field for the representation of atmospheric conditions. The answer depends on other considerations than the facility of representation. The popularity of the level surface and the fact that an anemometer records the horizontal wind have led to a common assumption that atmospheric motion is itself horizontal. In chap. VI of vol. III we have explained that when gain and loss of heat are excluded the motion of air is not necessarily in a horizontal surface but in an isentropic one. The use of isentropic surfaces for mapping, as Helmholtz has pointed out, keeps the important question of gain or loss of heat in full view. Supported by the action of Helmholtz and later of Margules we may regard an isentropic surface as being designed by nature to carry most effectively the representation of atmospheric structure. It is pretty clear that at high levels the isentropic surface tends to become nearly level because the entropy increases in the upper air with loss of pressure and diminishes with loss of temperature, and when we have data to examine the question effectively we may perhaps regard the contortions at the surface represented in fig. 98 of vol. III as being peculiar to the lowest levels and indicative of the state of the atmosphere there.

We may therefore ask for the consideration of graphic analysis based on entropy and temperature as well as pressure and density and the use of isobaric or isentropic surfaces instead of level surfaces for the graphic representation of the atmospheric condition over extended areas.

Meanwhile very little has been done in practice in these lines of research or in the study of maps of wind-observations at the surface as a separate part of dynamical meteorology. The two volumes on Dynamic Meteorology and

Hydrography by Bjerknes and his collaborators are in consequence very welcome. In them are described and illustrated novel methods of dealing systematically with observations of meteorological elements grouped in maps. Bjerknes's plan is very comprehensive. After defining his variables, of which those regarded as independent are the co-ordinates defining geographical position and height together with the time, and, omitting provisionally the influence of electric or magnetic fields, those required for the description of atmospheric states are five meteorological elements, namely pressure, mass, temperature, humidity and motion, with a corresponding set of five for hydrography, he states concisely the problem of meteorology and hydrography: *To investigate the five meteorological elements and the five hydrographic elements as functions of the co-ordinates and the time.* He distinguishes between the Climatological Method which consists in giving constant values to the co-ordinates and examining the effect of letting time vary (using, for example, the results of automatic recorders of the various elements set up at fixed points) and thus obtaining the normals and the periodic or secular or irregular variations, and the Dynamic Method which, using the same records, depends upon a series of synchronous representations of the field of each meteorological element. He thus presents for examination the state of the atmosphere in three dimensions for a succession of chosen epochs. ''The comparative investigation of the successive states must lead to the solution of the ultimate problem of meteorological or hydrographic science, viz., that of discovering the laws according to which an atmospheric or hydrographic state develops out of the preceding one.'' It is this method which is treated in the volumes referred to.

The method is called dynamic because, ''in virtue of the laws of hydrodynamics and thermodynamics which govern atmospheric and hydrospheric phenomena, preceding states are in relation of causality to subsequent states. Inasmuch as we know the laws of hydrodynamics and thermodynamics, we know the intrinsic laws according to which the subsequent states develop out of the preceding ones. We are therefore entitled to consider the ultimate problem of meteorological and hydrographical science, that of precalculation of future states, as one of which we already possess the *implicit* solution, and we have full reason to believe that we shall succeed in making this solution an explicit one according as we succeed in finding the methods of making full practical use of the laws of hydrodynamics and thermodynamics[1].''

These quotations illustrate the clearness with which Bjerknes treats the questions which he considers, a clearness which is equally conspicuous in his general examination of the results of the available methods of observation. The two parts already published present the treatment from the statical and kinematical standpoints respectively. They are accompanied by a series of tables for pursuing the necessary calculation. The dynamical treatment is to follow, but graphical methods for performing the mathematical operations are given in chaps. VIII and IX of Part II, Kinematics. ''These will be of the same

---

[1] *Loc. cit.*, §§ 87 to 90 (Part II, chap. I).

importance for the progress of dynamic meteorology and hydrography as the methods of graphical statics and graphical dynamics have been for the progress of technical sciences."

We have given an outline of Bjerknes's scheme for dealing with the general meteorological problem because it represents a systematic attempt to organise and combine meteorological observations in such a way as to lead directly to the inference by mathematical operations of the sequence of states of the atmosphere. The method which he puts forward is graphical; the reasoning is to be applied to a series of maps or diagrams of the distribution of elements in a set of horizontal surfaces separated by equal differences of "dynamic" level, that is to say the successive surfaces of equal geopotential, not necessarily of equal geometrical height, or some other series of surfaces defined by selected values of one of the variables. In that way a complete "diagnosis" of a succession of states of the atmosphere will be obtained which can be related the one to the other by mathematical process.

The same idea, that if we are sufficiently acquainted with the facts, and competent to deal with them on the basis of Newtonian dynamics, the causal relations of the sequence of states must be disclosed, has been pursued by L. F. Richardson to the point of dynamical operation. The matter interests us here because at the bottom of all the possibility of calculation lies the assumption that the data which form the basis of the maps or the arithmetical process are a complete representation of all the pertinent facts of the state of the atmosphere for the purpose of mathematical treatment. Bjerknes, for example, suggests that an effective organisation in regard to time intervals for pressure and temperature in the upper air would be continuous observation or observations every hour of Greenwich time at all stations at the ground, ascents at every third hour of Greenwich time from pilot-balloon stations and for every sixth hour of Greenwich time from the complete aerological stations. This would mean a considerable extension of the established meteorological practice. We may note in passing that to make the "diagnosis" complete these observations would need supplementing by corresponding observations over the sea which would require special organisation.

To anyone who has spent many years of his life in the bewildering occupation of compiling and arranging the multitude of figures and symbols which are collected for the purpose of representing the state of the atmosphere over land and sea for the hour, the day, the week, the month, the year, or a series of years, the knowledge that more than one student of the atmosphere feels that the figures are, or can be made to become, capable of arrangement in such a way as to invite a general attack upon the whole problem is very encouraging. The alternative which presents itself to those who are apprehensive that mathematical operations only develop the ideas which are intrinsically implied by the process of selecting the data is to use a different method of selecting the data, to watch for occasions when the natural phenomena arrange themselves in a manner which points to a definite classification or grouping. Reasoning may then be applied to special groups of facts that have real existence for a sufficiently long period to furnish a definite mental picture even if the grouping be not apparent on other occasions and therefore not strictly speaking general. In other words, we select examples for which a special train of reasoning may be improvised rather than prescribe beforehand the course of reasoning to be applied on all occasions with the condition that the data shall be so organised and selected as to make the prescribed course of reasoning applicable.

For no element is the difference of standpoint more easily illustrated than the motion of the free atmosphere at different levels, the representation of which belongs to the volume on Kinematics. The measure of the motion is a vector quantity, its direction and speed must be defined. Bjerknes represents a field of atmospheric motion by means of instantaneous lines of flow, or "isogonal" lines instead, and lines of equal speed. In this way the field of motion of the air on any occasion is completely mapped. The mapping of a large area for air-motion will generally disclose a series of lines or points of convergence or divergence of the instantaneous lines of flow with which must be associated instantaneous upward or downward motion.

Taken in successive layers the whole of space would be filled with a solenoid of tubes of flow of which the lines in the successive planes represent the sections; and, according to the work of Sandström, a particular pattern of lines of flow in any plane can be associated with atmospheric motion in three dimensions of recognised character such as wave-motion, or the convergence of a cyclone or the divergence of an anticyclone.

For those who are unfamiliar with this mode of procedure the patterns formed by the lines of flow for selected types of motion are sometimes surprising and they are useful to us because they remind us that the information contained in a chart of instantaneous motion, of which a synchronous weather-map is an example, may not at first sight disclose all the information implicitly contained in it. We have taken the liberty of reproducing in fig. 52 four diagrams A, B, C, D, representing wave-motion travelling across a current of air. As represented in the first pair of diagrams A, B, the train of waves is travelling in the direction of the arrow at the top directly across the flow of the air-current. Diagram A comprises two figures representing the wave-motion in horizontal and vertical section and diagram B the lines of flow in horizontal and vertical section for the wave-motion combined with the motion of translation of the current which is represented by the small thick arrow near the middle of the diagram. Of the other pair of diagrams D represents the horizontal lines of flow for the same wave-motion combined with a translation, also represented by an arrow, oblique to the direction of advance of the wave, while C gives the components of motion in the direction of advance of the wave. The transverse lines in the resultant diagrams show the lines of convergence and divergence where there is no horizontal motion in the waves. There is however vertical motion in the wave and the horizontal motion of translation. The crests and hollows of the waves are regions of maximum horizontal velocity in opposite directions.

These diagrams are reproduced because the study of wave-motion in the atmosphere is one of great interest and many meteorologists are unaware that it would be indicated by a set of observations with pilot-balloons which, when plotted on a map, conformed to one of the varieties of the patterns represented; and they make clear that the systematic study of atmospheric

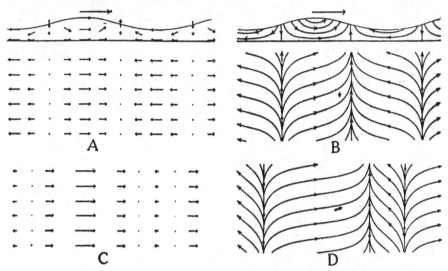

Fig. 52 (IX, 1). Wave-motion crossing a current of air. (From Bjerknes, *Dynamic Meteorology and Hydrography*, Part II, chap. v.)

data as proposed by Bjerknes is a practical part of meteorological science. The peculiarity of wave-motion is that, provided conditions for the transmission of waves exist, it may be found very far away from the locality where it was produced and may therefore be observed quite independently of any local exciting cause. It has already been pointed out in chap. v with reference to the photographs of clouds attributed to eddy-motion that the forms of the lenticular clouds suggest a combination of wave-motion and translation such as that represented in fig. 52. Many other typical forms of motion are represented for which reference should be made to the original work.

The construction of diagrams of this kind for air at any level is an agreeable occupation when the direction and velocity of motion at successive levels are known. Adequate information for a single country would require a large number of simultaneous observations of pilot-balloons, for the northern hemisphere almost an infinity of observations. And yet already there are a vast number of observations for many countries, and we await with great expectation, and perhaps with some impatience, the exhibition of a trial synchronous map of the circulation of air, at the level of say 4 kilometres, by some friend of science who has access to the observations, and the leisure to co-ordinate them. Bjerknes and his colleagues have done it already for the surface of a large part of the earth.

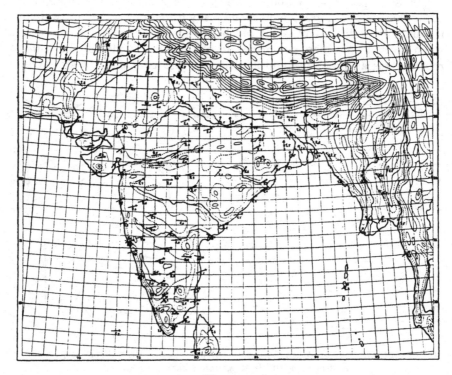

Fig. 53 *a*.

Kinematics of the surface air

In illustration of Bjerknes's method we give a reproduction of his analysis of the south-west monsoon of India and the inferences drawn from it in regard to vertical motion.

Fig. 53 *a*. Discontinuous representation of the air-motion over India in July from Sir John Eliot's Atlas (*Dynamic Meteorology*, by Bjerknes and collaborators).

The arrows represent the average wind-directions for the month and the figures the velocities in metres per second. Moderately idealised contour lines are shown in the background.

Fig. 53 *b*. Lines of flow (thick lines) and curves of equal wind-intensity in metres per second (fine lines) derived from the data shown in fig. 53 *a*.

Fig. 53 *c*. Forced vertical velocity at the ground. Areas of ascending (positive) currents are shaded, those of descending (negative) currents are unshaded. The figures give the velocity in centimetres per second.

Fig. 53 *c* is derived from fig. 53 *b* by graphical differentiation and graphical algebra on the hypothesis that the vertical velocity is the observed velocity of the wind multiplied by the contour-gradient.

The reader may be interested to compare these pictures with the diagrams in Dr S. K. Banerji's paper on 'The effect of the Indian mountain ranges on the configuration of the isobars,' *Indian Journal of Physics*, vol. IV, part VI, Calcutta, 1930.

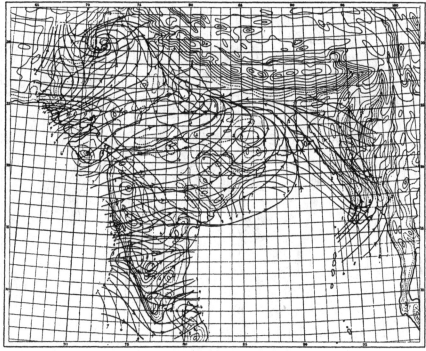

Fig. 53 *b*.

If the motion were confined to two dimensions so that each plane could be regarded as independent of the other planes we could map the field of each level as a solenoidal field in which the separation of the lines of flow is inversely proportional to the flow measured in terms of momentum. If we map the field on this hypothesis, we get of course into difficulties wherever the air-currents have a vertical component. Using a familiar analogy those readers who are acquainted with the practice of a physical laboratory will recollect that the same kind of difficulty arises in the use of lines of force to represent the intensity as well as the direction of the horizontal magnetic force due to a bar magnet in the earth's field.

The other elements are scalar and a single set of lines is sufficient to represent the field. Thus barometric pressure is scalar but the barometric gradient is a vector and in meteorological practice we are accustomed to associate closely the vector wind with the vector gradient, so that having a map of the distribution of pressure we should not hesitate to draw a map of the winds which from experience we should expect to be quite as effective in representing the winds, or even more effective than one which was based upon a series of actual observations of winds at a limited number of stations say 100 kilometres apart. Our map of the winds so drawn according to the isobars would certainly be in difficulties at any point where there was vertical motion, but whether it be that the vertical component of motion is generally so small that

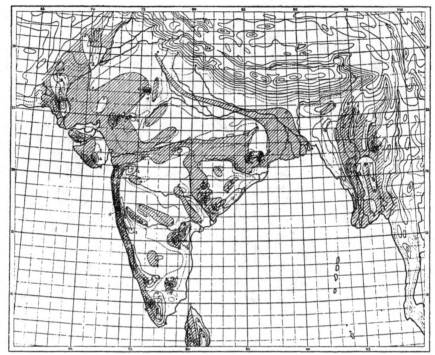

Fig. 53 c.

its effect does not seriously interfere with the local wind or that it is so local or transient that its effect is not noticed in our maps, experience has not taught us to distrust the gradient as an indication of the wind. By assuming the relation between the horizontal wind and the horizontal distribution of pressure in the free air as a law of atmospheric motion, we are really assuming that those portions of the field where vertical motion or other complications invalidate the relation are to be left out of the survey or postponed for future consideration while the parts of the field where the law is practically applicable are being dealt with. But if the distribution of pressure enables us to draw a map of the winds then equally a map of pilot-balloon observations, in so far as they give us a correct representation of the horizontal winds, ought to enable us to form a working idea of the distribution of pressure.

A very effective exposition of the process of dealing with daily weather on this basis which was subsequently developed into the Norwegian method of forecasting is given in a paper on 'The structure of the atmosphere when rain is falling[1].' It traces the flow of air over the surface of southern Norway and the Baltic and thereby indicates the localities of convection and their associated weather. It forms the starting-point of the physical interpretation of weather introduced by the Norwegian school of weather-study as set out by Gold in vol. II, p. 383.

[1] V. Bjerknes, *Q. J. Roy. Meteor. Soc.*, vol. XLVI, 1920, pp. 119–38.

Atmospheric equilibrium and motion

From the point of view of statics the structure of the atmosphere, as regarded by Bjerknes, is expressed by isobaric and isosteric or isopycnic surfaces and pictured by horizontal sections or maps and vertical sections or profiles of the structure. We propose to couple with that the method of expression by means of isentropic and isothermal surfaces, and representation by horizontal sections and profiles of those surfaces.

Either of these conveys information about energy because the changes in the condition of the air marked out by a cycle on the profile or the plan would require or afford an amount of energy which could be calculated from the diagram.

We note the following inferences.

At any level the atmosphere is in equilibrium if pressure-gradient is zero; at successive levels if the temperature-gradient at each level is also zero. Hence the condition of equilibrium is that the isobaric and isothermal surfaces should be level and consequently the volume surfaces also level. The condition for stability is that what we may call "potential volume" should increase with height. Potential volume, on the analogy of potential temperature, here means the volume as reduced to a standard pressure.

If the pressure surfaces cut the volume surfaces the defect from equilibrium is expressed as wind. In a profile section the wind is westerly for a northerly gradient of pressure, easterly for a southerly gradient whatever the temperature may be.

If pressure is level and temperature is level, entropy is also level, and in consequence a condition of concurrence of surfaces of entropy and temperature in a level surface represents equilibrium.

If entropy surfaces cross temperature surfaces statical equilibrium is lost and the structure requires air-motion for its maintenance. If the isentropic surfaces are sloped, motion along them upward or downward has to be allowed.

If the kinematics were properly adjusted the conditions might be regarded as steady and the system would be permanent if it travelled without change of form; it might be "reduced to rest" and permanence of shape by adding to the kinematics the reverse of the velocity of travel. Presumably no natural system in atmospheric motion would satisfy the conditions of permanence but the modes of approximation might be studied with advantage if the data were available.

Unfortunately in making an attempt of this kind we have to avoid the surface layer for which the data are comparatively numerous, because our analysis of the foot of the structure in chaps. iii–v shows it to be subject to so many disturbances both thermal and dynamical that its motion is irregular beyond hope of expression, and we must seek a more amenable part of the structure for our essay. The part which we should choose, if choice were allowed, is that part of the limb between 4 km and 8 km or 4000 and 8000 dynamic metres about which we have already learned from W. H. Dines, on the analysis which he made, first that the standard deviation of pressure is of the same order of magnitude as at the surface, a result which accounts for the Egnell-Clayton law of increase of velocity with height, and secondly that changes of pressure are proportional to changes of temperature. It will be remembered (p. 204) that the Egnell-Clayton law asserts that mass-transport at each level is the same, or the velocity is inversely proportional to the density, a condition which, by the geostrophic equation, would give the same gradient at each level.

Hence it would follow that isobars would be also isotherms and therefore isentropes too if the chronological changes recorded by Dines expressed sufficiently the geographical changes in all directions round the place of observation. That is not necessarily the case, the rates of geographical change are $\partial/\partial x$, $\partial/\partial y$, $\partial/\partial z$, but the rate of change recorded at the station is $\partial/\partial t$ experienced by the arriving air. Perhaps we may regard Dines's conclusions as correct for $\partial/\partial x$, the change in the W to E direction, because atmospheric systems generally travel from west to east in north temperate latitudes without much change of shape within twenty-four hours, but we must await further information about $\partial/\partial y$, the S to N variation.

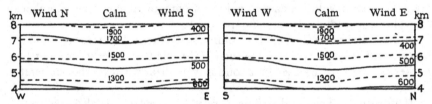

Fig. 54. Sectional profiles W to E and S to N of a generalised cyclonic depression at the levels of 4 km to 8 km referred to pressure and volume of unit mass.

Full lines, isobaric surfaces in mb; broken lines, isosteric surfaces in cc/g.

In the meantime we may illustrate the graphical analysis by a hypothetical profile of a section from 4 km to 8 km across a travelling depression with some closed isobars. We give first (fig. 54) profiles in two directions at right angles, W to E and S to N, with pressure and specific volume as the elements of reference. In order to compare the profiles with an ordinary map the reader must refer the changes to a horizontal section. Elevation of an isobaric surface must be interpreted as increase of pressure on a map and elevations of a volume line as decrease of volume. Both sections show a loss of pressure as the centre is approached but little change in specific volume because in the upper air the effect of reduced pressure would be compensated partly at least by reduced temperature. The west to east section is shown symmetrical with regard to the centre, the south to north section also shows a recovery of pressure which however only holds for a little distance northward of the centre.

If we attempt a similar plan with the representation of the structure by entropy and temperature we may obtain according to vol. III, fig. 98, the following (fig. 55). In this case again the structure of the cyclone has been taken to be symmetrical with regard to the centre in a west to east direction, and unsymmetrical in a south to north direction.

The entropy-temperature diagrams are much more showy than the pressure-volume diagrams with the same vertical scale. They show a diaper of crossing lines which mean defect of equilibrium compensated by wind. The same information is embodied in the pressure-volume diagrams but is not so easily seen.

It must be confessed that the diagrams as presented are freehand drawings rather than precise plots of observations. That is because no suitable observations exist. An exact check of the accuracy of the representation when it is possible will be a gratifying addition to our knowledge.

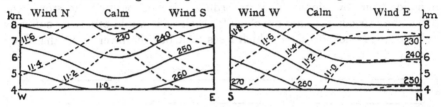

Fig. 55. Sectional profiles W to E and S to N of a generalised cyclonic depression at the levels of 4 km to 8 km referred to temperature and entropy.

Full lines, isothermal surfaces tt; broken lines, isentropic surfaces c, g, s × 10⁶.

Slope-effect on the tephigram

We may illustrate graphic computation further by the consideration of the dynamics and thermodynamics of air which is descending a slope in consequence of the effect of radiation. In the absence of actual data we may quote a number of auxiliary assumptions, which however do not affect the principle of computation.

We first limit the computation to two dimensions by imagining air to descend a slope through about 1 kilometre from 900 mb pressure to 1000 mb. We suppose the descending air to leave a patch one square dekametre in area at the level of 900 mb and the downward flow to be continued until, apart from surface-eddies due to turbulence, it is steady. Suppose the slope to be losing heat by radiation at the rate of ten kilowatts per square dekametre (vol. III, p. 160). Suppose the temperature of the air to be recorded during its descent. It will be warmed dynamically by the increase of pressure and cooled thermally by loss of heat to the slope which is colder than itself. If the flow were prevented, the surface would cool by radiation without any limit except that imposed by conduction from below ground, which we will neglect. If the flow were infinitely quick, the ground-temperature would approximate to that of air warmed isentropically by the descent.

Between those two extremes we may suppose the temperature of the descending air and of the surface of the slope to be uniform throughout the journey; what might be gained from the compression is lost by contact with the slope and transferred by the slope to space by radiation.

We have to give some account of the air which forms the environment of the slope; we may suppose that to be approximately isentropic in consequence of previous dynamical churning which has ceased with the daylight.

The conditions supposed are represented in standard form by fig. 56. A is the starting point of the descending air, AB the condition of the environment, AC the assumed condition of the descending air. If the conditions are not as assumed but are known from observation the lines AB, AC may be adjusted to give an accurate representation.

Taking them as they are represented, AC represents the change of entropy of the air in its descent, and the area of the rectangle AOoC extending to zero temperature represents the heat lost by the descending air. The heat so lost does not raise the temperature of the slope which takes it but is lost by the radiation from the slope. That accordingly is one part of the story; the descending air compensates the slope for the loss it would otherwise suffer by radiation; it is not the slope which loses heat but the layer of air that flows down within touch of it.

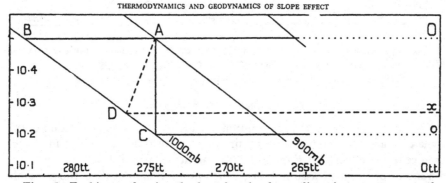

THERMODYNAMICS AND GEODYNAMICS OF SLOPE EFFECT

Fig. 56. Tephigram showing the heat lost by descending air to compensate the loss by radiation from the slope and the geopotential available to supply kinetic energy to the descending air.

Besides losing heat, the air in descending gets rid of the energy corresponding with the change in its geopotential represented by something like 1000 geodynamic metres. This is spent in producing the kinetic energy of a wind at the bottom or in helping in the compensation of heat frittered away by turbulence.

The energy corresponding with the change of geopotential is known because it is represented on the diagram by the same area as that which represents the heat lost, viz. AOoC, but the amount of energy thereby derived is quite independent of that expressed by the loss of heat.

The equality arises from the fact that the temperature does not change during the descent. If AC were not isothermal but were drawn to some other point as D, the heat lost would be represented by AOxD and the energy derived from geopotential by AOoCD.

Hence by graphic analysis we can form an estimate of the course of events in the air which descends the slope, including the contribution of the descending air towards the loss by radiation, an item of considerable importance in fruit-farming, and also the expenditure of energy in producing the wind at the bottom which in some circumstances may take the form of a blizzard.

The tracing of the expenditure of this energy is a matter of some difficulty. If we could regard it as belonging entirely to the air which has descended from top to bottom the calculation would be simple. The energy per gramme would be equal to the loss of geopotential and with steady motion, each portion flowing independently of its predecessor or follower, if friction and turbulence could be avoided, the gramme at the bottom would have kinetic energy equivalent thereto, or equivalent to the heat lost and represented by AOoC. But turbulence spreads energy over a large mass by the redistribution of momentum. The computation seems to be as follows:

Each unit mass that goes down gives out heat corresponding with the change of entropy which can be estimated from the pressure and temperature at top and bottom, $tt\ (E - E_0)$; it will be of the order $275 \times \cdot 3 \times 10^6$ per gramme, and the same amount, the equivalent of a velocity of 130 m/sec, is expressed as kinetic energy or its equivalent derived from the geopotential.

The volume over which the equivalent of the kinetic energy is distributed depends upon the turbulent motion produced by the descending air so that the full velocity 130 m/sec will not be reached. The distribution takes place by redistribution of momentum and if the thickness of the layer at the top be 1 metre and at the bottom x metres with a velocity v, in place of 1 metre with velocity 130, we have by the equality of momentum $xv = 130$ or $v = 130/x$.

Hence a 10-metre layer at bottom would have a velocity of 13 m/sec, a 20-metre layer a velocity of 7 m/sec. The energy at the bottom will be $1/x$ that of the work expended, i.e. $8 \cdot 25 \times 10^7/x$ ergs per gramme.

The amount of air involved can be calculated in another way. The amount of heat lost by 1 g is $82 \cdot 5 \times 10^6$ ergs: this passes to supply the loss by radiation. The radiation from a strip of surface 10 m wide extending over a vertical fall of 1000 m is 1000 kw or 10^{13} ergs per second, hence $10^{13}/(8 \cdot 25 \times 10^7)$ g/sec must pass down the slope, which is equivalent to $1000/8 \cdot 25$ cubic metres per second.

The flow of a layer 1 metre thick and 10 m wide would be $100/8 \cdot 25$ or 12 m/sec.

The ideas which are here developed for the loss of heat from air moving downward as a katabatic wind over a slope can be applied with some modification to the reverse action, namely the supply of heat to air which develops an anabatic wind.

Gliding in the lower atmosphere

The upward motion of air along a slope is of fundamental importance for the art of gliding in an aeroplane without engine-power. It supplies the momentum necessary to keep the glider in the air, which could otherwise only be obtained by speed artificially developed by the engine-power of an ordinary aeroplane.

The vertical component of motion which is essential for the practice of gliding may be developed in three different ways. The first is the dynamical effect of the impact of air upon a slope which diverts the flow upwards and produces a layer of isentropic air with a bounding isentropic surface along the slope and leading upward (fig. 57). A special form of the same effect is to be found in the cliff-eddy or other eddy of regular position and large diameter due to large obstacles.

The second form is the thermal effect upon the air of solar radiation on a slope—the converse of the effect of loss of heat by radiation which we have discussed at length in this chapter. For the converse circumstances we have an increase of entropy in the rising air due to heat absorbed by the air from the surface of the slope, the influence extending to some distance from the slope by the effect of turbulence. The condition may be represented by the diagram (fig. 58).

The lines of the two diagrams should be crossed by isotherms: those for either diagram would be not very dissimilar from the isentropes of the other.

The third form is the rising of air by convection to form clouds of the cumulus type if carried far enough.

Fig. 57.   Fig. 58.

Fig. 57. The effect of an obstacle on the isentropic surfaces of an air current.

Fig. 58. The effect of solarisation of a slope on the isentropic surfaces of the air.

The art of gliding has been developed to a remarkable extent in Germany in connexion with the Forschungs-Institut der Rhön-Rossitten Gesellschaft, which publishes an annual report and memoirs. It is in connexion with the Institut für Flügmeteorologie an der technischen Hochschule, Darmstadt, of which Dr Walter Georgii is Director. The name of the Society is derived from the Rhön mountains on the west of Thuringen, lat. 50½, and Rossitten on the spit of sand in the Baltic in front of Königsberg.

By a careful study of the conditions in the neighbourhood of the site, light aeroplanes, like sea-birds, can be made to glide upwards without any power, and have reached a height of 10,000 ft, remaining in the air for hours.

The practice of gliding implies a very intimate knowledge of the structure of the air-currents in circumstances indicated by wind, sun or shadow.

The dynamical effects of the motion of air near slopes are sometimes illustrated in a striking manner. An example of a corkscrew whirl of three turns with vertical range of 2000 ft described by floating paper in the Yosemite Valley is figured by B. M. Varney in the *Monthly Weather Review* for June, 1920.

CHAPTER IX

CURVED ISOBARS

It is pressure that turns the corners of the air's path. Air flow from W to E or E to W may be the expression of the earth's rotation; flow from N to S or S to N is the kinetic index of the distribution of pressure.

HITHERTO no account has been taken of the second or cyclostrophic term in the equation of p. 87, which gives the relation between wind and pressure for the motion of the air under balanced forces. It represents that part of the gradient which is balanced by the deviation of the path of the air from a great circle; its numerical importance is inversely proportional to the radius of the small circle which osculates the path of the air and is therefore directly proportional to the curvature of the path in the horizontal plane. In ignoring the effect of curvature we have supposed that the numerical value of the term is, in general, sufficiently small for that course to be followed in view of the uncertainties of the measurements of the wind in the upper air. It has been convenient for us to ignore it because the determination of the curvature of the path is not possible when we have only a map for a single epoch, and what we have written hitherto has dealt with the features of the single map. The fact that the velocity of the air at the particular epoch can be fitted into a scheme of velocities arranged as lines of flow in the form of circles or spirals is not evidence that the circle or the spiral itself represents the path of the air. It can only do so if the features of the map remain stationary. We cannot assume that they do so without referring either to previous maps or subsequent ones, and whenever we make the reference we find that the condition is not exactly fulfilled, generally speaking not even approximately so. There is nearly always a considerable change in the distribution of pressure and wind except in those regions which are represented by the isobars of a large anticyclone.

We now pass on to consider the effect of the curvature of the path upon the relation of wind to the distribution of pressure and our first step shall be to consider what the curvature of the path means for us. The maps which we use to represent the state of the air at any epoch are apt to mislead us unless we are careful. The lines of flow, which in the upper air we have considered in the previous chapters to be in agreement with the isobars and at the surface to cross the isobars at certain finite angles depending upon various conditions of turbulence, are represented by curves with a very great variety of curvature, but the lines representing the paths of air whenever they have been constructed are not at all likely to be mistaken either for lines of flow or for isobars unless we happen to be dealing with a part of the map where the isobars are straight and parallel, and the lines of flow either lie along the isobars or cross them at a uniform angle. On the other hand there is nothing in the appearance of a line of flow or an isobar that disqualifies it as a path. It becomes, in fact, a path if the situation is permanent, but that means a set of conditions for which, so far as we know, there is no rigorous example in our maps. A near approach is represented in the series of maps for the end of July and beginning of August 1917, which are referred to in the next chapter

Regarded from the mathematical point of view to which reference was made in chap. VIII, the curvature of the lines of flow and the associated curvature of the isobars are part of the interplay of the inertia of the air and the forces which operate upon it, but in pursuance of our plan of selecting conditions which are amenable to treatment as special cases, we may note certain types or groupings of isobars which exhibit the characteristic of stability in the sense that they may last for days together without much change, and move bodily across the map. Pre-eminent among them, so far as our Islands are concerned, we have the group of isobars running from west to east across the Atlantic giving us a westerly type of weather which was strikingly exemplified in the stormy winter of 1898–99, and represented by a band of westerly current across the Atlantic, the high pressure on the south and the low pressure on the north. On the side of the low pressure there were considerable fluctuations which appeared as successive cyclonic depressions with great local intensity. With this type we include the south-westerly type of weather in which the band of isobars runs from south-west to north-east.

In contrast with these we have the easterly or north-easterly types which are represented by bands of isobars running from east to west or, more often, from north-east to south-west with a high pressure to the north or north-west. These also may last for weeks together and in that way be classed among the stable types[1]. A fine specimen was experienced in England in February 1929.

In a sense it may be said that the westerly type is oceanic and the easterly type continental because the band of west to east isobars seldom penetrates far over the continent, it generally turns northward on reaching the land, and the east to west band turns partly northward alongside the flow from the ocean and partly southward to feed the north-east trade wind.

Sometimes in our Islands we find ourselves in the region between a specimen of either of these types represented by a large quasi-permanent anticyclone to the north of us and another to the south of us. In that case we find the region between the two covered by a succession of rainy depressions. A good example will be found in the succession of maps in the month of July 1918, particularly from the 15th to the 27th day of the month. This state of things might prove to be typical of the conditions for high latitudes for the band of west to east isobars if our maps extended far enough to the northward to show fully the northern anticyclonic region.

In any case the three sets of conditions mentioned, if we supplement them by the addition of the type of a quasi-permanent anticyclone directly over us, give a general idea of the classification of our weather as represented by locally persistent isobars and the lines of flow which accompany them.

---

[1] The travel of weather changes along the northern or southern side of a persistent band of high pressure is illustrated in *Weather of the British Coasts* (M.O. Publication, No. 230), chap. XI, § 9, 1918.

We must add the type represented by a band of isobars running from north to south which sometimes represents a quasi-permanent condition although northerly winds are often only the transient accompaniment of the last stages of a passing cyclonic depression.

This will appear to be a digression but it has a bearing upon the relation of curvature to the permanence of type of air-motion in the atmosphere. There is the curvature of the earth's surface on the one hand and the curvature of the path of the air as it passes over the surface on the other hand; and both of them have to be considered in making a selection of possible permanent conditions underlying local and temporary fluctuations of weather. Perhaps the chart of mean isobars of the northern hemisphere for the month should be regarded as the mean value from which the local conditions deviate from time to time, but the mean is made up from so many different types of map that it is better to classify the conspicuous types.

In dealing with the geostrophic wind we have taken the air as moving along a great circle; but isobars drawn in a band for considerable distances along great circles converge[1], and a broad band arranged as the small circles of plane sections parallel to and on either side of a great circle would have different values for the latitude in different parts. The simplest form of distribution which we can regard as free from difficulties of that kind and the most stable form of atmospheric motion imaginable is that which is represented by a band of isobars from west to east or from east to west. A distribution of that kind represents part of a cap or series of rings of air rotating round the polar axis. The actual average distribution of isobars figured for four levels in chap. VI, vol. II, roughly represents such a cap for the northern hemisphere, and in the southern hemisphere the distribution of pressure over the surface of the great Southern Ocean corresponds even more nearly with that ideal[2].

Within such a cap, revolving like a solid about the polar axis, there are no elements of instability. Any disturbances that arise must come from causes outside. Let us therefore consider a cap rotating as a solid round the pole as an ideal of stable atmospheric motion and the type to which the actual motion tends to revert when freed from the causes of disturbance to which it is subjected. The direction of motion may be either cyclonic or anticyclonic. The distribution of pressure for such a rotating cap would be as follows:

$$p_\phi = -\int_0^\phi V\rho \left(\omega + \frac{1}{2}\frac{V}{\epsilon}\right) \sin 2\phi\, d\phi \text{ for a cyclone and}$$

$$p_\phi = \int_0^\phi V\rho \left(\omega - \frac{1}{2}\frac{V}{\epsilon}\right) \sin 2\phi\, d\phi \text{ for an anticyclone,}$$

where V is the velocity of the wind at the equator.

[1] A suggestion as to the effect of the convergence in the northward component of air-motion is given in *Principia Atmospherica*, loc. cit., p. 88.

[2] *National Antarctic Expedition*, 1901–1904. *Meteorology*, Part II, Royal Society, 1913.

Making an approximation, for the density, at 1250 g/m³ the distribution would be made up as follows:

TABLE I. VELOCITIES AND CORRESPONDING VALUES OF THE PRESSURE AND ITS GRADIENT FOR DIFFERENT LATITUDES IN A HEMISPHERICAL CAP OF AIR REVOLVING LIKE A SOLID ABOUT THE POLAR AXIS.

| | | Pressure | | Gradient | | | |
| | | | | Geo-strophic component mb/deg | Cyclo-strophic component mb/deg | Total for cyclonic system mb/deg | Total for anticyclonic system mb/deg |
Latitude °	Velocity m/sec	Cyclone mb	Anticyclone mb				
0	15	1054	967	0	0	0	0
10	14·8	1051	970	0·52	·01	0·53	0·51
20	14·1	1044	977	0·98	·015	1·00	0·97
30	13·0	1032	989	1·32	·02	1·34	1·30
40	11·5	1018	1003	1·50	·02	1·52	1·48
50	9·7	1002	1017	1·50	·02	1·52	1·48
60	7·5	988	1031	1·32	·02	1·34	1·30
70	5·2	976	1043	0·98	·015	1·00	0·97
80	2·6	968	1050	0·52	·01	0·53	0·51
90	0	966	1053	0	0	0	0

Such a cap is obviously not fully represented by the distribution of winds over the earth's surface. There are westerly winds in middle latitudes and in the southern hemisphere they extend round the earth, but not in the northern hemisphere; and in neither do they extend to latitudes below 35°. Sometimes it would appear as though the winds in the southern hemisphere might constitute a cap, rotating from west to east, over a lower layer rotating from east to west or over the Antarctic continent which occupies a great part of the room that would belong to the lower rotating anticyclonic cap. And in the northern hemisphere in the region of the British Isles there is sometimes, as already mentioned, a great anticyclone to the north of our Islands covering Greenland, Iceland and the north of Scandinavia suggesting a portion of an anticyclonic cap, while the pressure to the south of us is arranged in west-east lines and corresponds with that of a portion of a cyclonic cap in middle latitudes south of the anticyclonic cap. Between the two is a belt bounded by isobars of the same designation between which local cyclones are shown[1]. The conditions in the northern hemisphere are singularly unpropitious for a regular anticyclonic circulation of the lower layers of the air round the pole because the land mass of Greenland ten thousand feet high stretching from beyond the eightieth to the sixtieth parallel effectually blocks the way. The easterly wind of the lower atmosphere cannot go over it and must be diverted round it.

It is generally allowed that the maintenance unchanged of the length of the period of revolution of the earth is conclusive evidence against an average rotation of the whole atmosphere in one direction. Hence the existence, in one

[1] Shaw, 'On the General Circulation of the Atmosphere in Middle and Higher Latitudes,' *Proc. Roy. Soc.*, vol. LXXIV, 1904, p. 20, and *National Antarctic Expedition*, 1901–1904, *Meteorology*, Part I, Royal Society, 1908, p. xiii.

part, of a belt with one kind of rotation implies the existence of its opposite equivalent somewhere else.

We have given attention to the idea of a cap, or a belt of air forming part of a cap round the globe, because it may have some bearing upon the situations which we have to consider although no complete cap or belt is shown. There is nothing at all in the way of mutual influence to bind one part of a cap to another part on the other side of the world; each part must be separately maintained by an appropriate environment. Just as a small sector of a rainbow will exhibit all the essential properties of the complete arc so in atmospheric motion a sector of a belt may exhibit all the essential properties of a complete belt. Wherever a group of isobars is formed which, so far as they extend, would correspond with a portion of such a cap, the air within the portion will have the same kind of stability, and be subject to the same conditions as the cap would satisfy, although it is maintained in its successive positions by a different environment. It follows that the bands of air-flow which maintain themselves sometimes for days, sometimes even for weeks, may have the kind of stability that belongs to a rotating cap and be quite as important elements of the atmospheric circulation as anticyclones or cyclonic depressions and have the same kind of permanence if their environment permits.

For these bands of limited extent, which may be treated as sectors of belts, the effect of curvature is small. According to the formula of p. 233, the ratio of the cyclostrophic component to the geostrophic component is one-half of the ratio of the equatorial velocity of the cap to the equatorial velocity of the earth's surface. In the case represented in Table I it amounts to 1·6 per cent.

Spin in small circles

The other type of atmospheric motion which claims attention is rotation round a centre in a circle of much smaller radius than 40°, indeed it may be taken as being from 10° down to 1°, or even less in the case of tornadoes and water-spouts. There is evident stability in motion of this character because beginning with examples of whirls lasting for some seconds there is apparently an uninterrupted sequence by way of revolving sandstorms or dust-devils, tornadoes or whirlwinds, to tropical revolving storms and large cyclonic areas with radii of 10° or more. The only limit of the series is a revolving air-cap covering a hemisphere or a large part of it. And just as a belt of west wind or a belt of east wind may lie over these Islands for weeks so the other type of quasi-permanent atmospheric motion, which has always been thought of as a column of air in continuous revolution, may preserve its identity for days or weeks. Through the kindness of Professor McAdie, of Blue Hill Observatory, Harvard University, we are enabled to give two notable examples (p. 281). The first is that of a tropical revolving storm which started on a westerly track towards the Philippine Islands (where visitations of that kind are known as "Baguios"), turned round towards the north and north-east, crossed the Pacific Ocean and after some vagaries on the North American continent continued its journey eastward and crossed the Atlantic in the usual track of cyclonic depressions over that ocean. The whole journey lasted from 20 November, 1895 to 22 January, 1896. The second is a cyclonic depression of October 1913, in the outer region of which the tornado was formed which caused so much destruction in South Wales on the twenty-seventh of that month[1]. The track of the main depression shows an anomalous path from Canada to the north of the British Isles. The tracks of the centres of these depressions are shown upon the map which forms fig. 74 in chap. xi.

To these notable examples has been added the long track of a cyclonic depression which was figured in the Meteorological Office chart of the North Atlantic and Mediterranean for August 1904[2]. The cyclone was first noted on 3 August, 1899 in that part of the North Atlantic Ocean where West Indian hurricanes often take their rise. It moved westward to the West Indies, skirted the coast of Florida and turned eastward over the Gulf Stream. After some hesitation about latitude 40° W it made for the mouth of the English Channel and, missing that, crossed to the Mediterranean where it lost itself on 9 September after a life of thirty-eight days.

Maps of the courses of depressions over the world are given in vol. ii and for the northern hemisphere the subject is treated more fully by Mitchell[3]. We have included in fig. 74 one that circumnavigated the world.

Thus out of the kaleidoscopic features of the circulation of air in temperate latitudes two definite states sort themselves each having its own stability. The first represents air moving like a portion of a belt round an axis through the earth's centre. It is dependent upon the earth's spin and the geostrophic component of the gradient is the important feature; the curvature of the isobars is of small importance. The second represents air rotating round a point not very far away: it is dependent upon the local spin, and the curvature of the isobars with the corresponding cyclostrophic component of the gradient is the dominant consideration.

In reality of course both components are operative in all cases except at the equator where the geostrophic component is zero. Only the geostrophic component depends upon the latitude. The numerical importance of the cyclostrophic component increases rapidly with the velocity of the spin and is paramount for rotational systems of small radius. The following table shows the velocities in different latitudes for which the two components are equal when the radius is 100 kilometres.

Latitude in degrees	90	80	70	60	50	40	30	20	10	0
Velocity m/sec	14·6	14·3	13·7	12·6	11·2	9·4	7·3	5·0	2·5	0

Just as in the case of the belts of air revolving about a diameter of the earth so, in the case of a mass of air in rotation with a small angular radius, there is no influence of one part of the whirl upon another part in another sector which holds them together; each sector, or truncated part of it, must satisfy independently the conditions which are applicable to any and every

---

[1] *Geophysical Memoirs*, No. 11. M.O. Publication, No. 220a. [2] M.O. Publication, No. 149.
[3] 'Cyclones and anticyclones of the Northern hemisphere, January to April inclusive,' 1925. *Monthly Weather Review, Washington*, vol. LVIII, 1930, pp. 1–22.

part of the whole whirl; and this it can do, without the other sectors, provided that its environment is suitable for its persistence. Experience of maps would seem to indicate that the complete circular form is not necessary for the persistence of a part. We may make use of this principle later on but it is convenient for the present to deal with those cases in which the circular form is complete. We shall therefore proceed with our study of curved isobars by considering the case of a short circular column or disc of air which is in rotation about a centre.

Rotation with translation

We shall suppose that the disc of air rotates like a solid, in which case the angular velocity of each portion of the disc will be the same. Subsequently we can consider whether that assumption is reasonable from the meteorological point of view. We may take the common angular velocity to be ζ. The linear velocity v at any point distant r from the centre will be $r\zeta$. Its direction will be at right angles to r. If the motion belongs to a cyclone, in the northern hemisphere it will be counter-clockwise. For the sake of brevity we will identify such a distribution of winds as a normal cyclone. Solid spin would be a better name because there is no evidence either for or against its representing the actual conditions of motion of a cyclone in the upper air. At the surface there is always some incurvature of the wind with respect to the circular path round the centre.

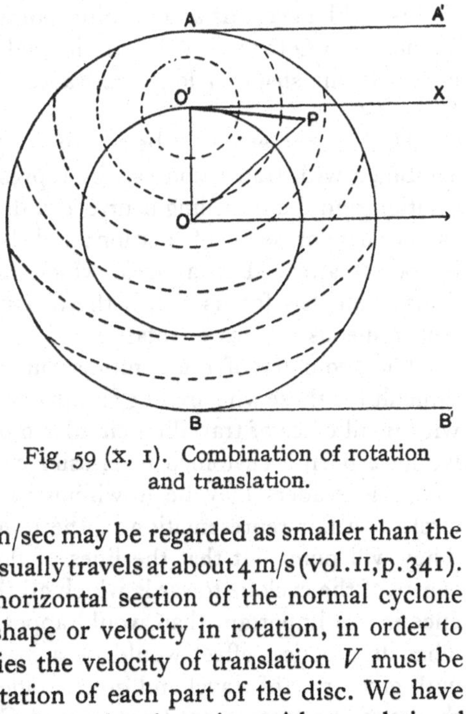

The characteristic feature of our cyclonic depressions is that they travel across the map. The velocity of travel is very varied but when the depression is represented by circular isobars it generally has a rapid speed of travel. A velocity of 20 m/sec for the centre of a cyclonic depression is large but not

Fig. 59 (x, 1). Combination of rotation and translation.

unknown, a velocity of less than 10 m/sec may be regarded as smaller than the average. A tropical revolving storm usually travels at about 4 m/s (vol. II, p. 341).

If the rotating disc forming a horizontal section of the normal cyclone travels bodily without altering its shape or velocity in rotation, in order to obtain the scheme of actual velocities the velocity of translation V must be superposed upon the velocity in rotation of each part of the disc. We have thus to deal with the combination of rotational motion with translational motion. That is the subject of certain well-known propositions of which we require the following.

1. A disc which is rotating like a solid round a point with angular velocity ζ and is travelling in its own plane with a velocity of translation V will have

a point O' *instantaneously at rest*, where OO' is equal to V/ζ and O' is on the radius of the disc drawn transverse to the line of travel and to the left; the disc will be *instantaneously in rotation round the point O'* with the angular velocity ζ.

This proposition depends upon the geometrical addition:

the vector $O'O$ + the vector OP = the vector $O'P$.

If the vector in each case is the velocity proportional to and at right angles to the lines named the conclusion follows directly from the parallelogram of velocities.

2. The centre of instantaneous rotation O' will "travel" along a line parallel to the path of the centre of the disc with the same velocity of travel as the centre of the disc. The word travel is conventional in this statement. Nothing really travels along the line $O'X$. New systems of rotation are developed with successive points on the line as centres.

3. The actual **paths** of the particles of the disc will be the series of curves traced out by the several **points of** the disc when the circle with radius $O'O$ rolls along the line $O'X$.

These curves are well known in geometry. The path of O will be a straight line, the paths of the points in the circumference of the circle with $O'O$ as radius will be cycloids and of other points trochoids. Those points which are farther from O than O' will form looped curves: those within that limit form curves in the shape of long sea-waves. The curves are represented later as figure 65.

All this geometry may be summed up by saying that rotation like a solid combined with translation can be represented by consecutive instantaneous rotation round a travelling centre at a distance V/ζ on the left of the path of the permanent centre of rotation; and the actual paths are the curves formed by points attached to a circle which rolls along the line of instantaneous centres and has for its radius the distance between the permanent and the instantaneous centres of rotation.

The geometry of the combination of rotary motion with translation is difficult for those who are not familiar with it, but obviously it has to be dealt with in all cases of travelling circular motion with which for sixty years now we have been accustomed to classify cyclonic depressions.

Some readers may be unwilling to regard instantaneous rotation in a circle as a fair representation of the motion of air in a cyclonic depression. They will point out that the lines of flow of air are spirals meeting in the centre of the cyclone towards which all the air directs its motion. The difference is less important than would appear at first sight. In order to find out what its practical effect would be a machine for drawing the actual paths, with given rate of travel and given angle of incurvature, was constructed by the Scientific Instrument Company of Cambridge[1]. The effect of the incurvature was not that any paths led directly to the centre but that the loops on the curves had a curious tilt to the left. The subject was treated analytically by the late Professor W. H. H. Hudson[2].

[1] *Life-history of Surface Air-currents*, p. 100.   [2] 'Anemoids,' *B.A. Report*, 1906, p. 483.

It will be noticed that the circles showing the instantaneous rotation round the point O' are incomplete because the real rotating disc is not sufficiently extensive. That is of no consequence. We are quite accustomed to it in the case of a cart-wheel for which the centre of instantaneous rotation is the point which touches the ground and, in consequence, is instantaneously at rest. No inconvenience arises: the arcs of circles representing the instantaneous motion of a cart-wheel are merely aids to comprehending the situation; so are the circles round the centre in fig. 59: they have no permanent existence, yet they represent the lines of flow of the actual winds of a normal cyclone.

On the other hand, we may note that the actual disc in its rotation will use the whole space between the lines AA', BB'; so that, in order to represent a complete normal cyclone, there must be something besides the complete circles of instantaneous rotation, namely the parts of the instantaneous

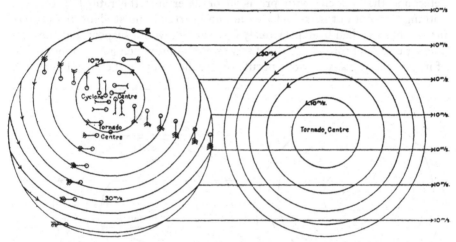

Fig. 60 (x, 2). Component and resultant fields of velocity in a normal cyclone. The arrows show surface-winds computed in the manner indicated on p. 243.

circles required to fill the circular boundary of the disc. We can represent the component and resultant motions by fig. 60, the first diagram of which represents the instantaneous resultant motion and the second the separate instantaneous motion of a rotating disc and of a uniformly flowing stream which carries the disc along with it.

It should be noticed that in the resultant distribution there is a discontinuity of velocity at the junction of the disc and the stream. In the diagram the finite angle seems to be between the flow outside and the flow inside the disc, making it appear that fluid crosses the boundary of the disc; that, of course, is not the case. The component in the direction of flow is common to both and the relative motion is the rotation of the boundary of the circle. The discontinuity would in actual cases cause eddies and the nature of the field of flow in the immediate neighbourhood of the rotating disc requires some adjustment on that account which is not yet made out.

Thus, in a normal cyclone there are two centres, one the centre of instantaneous rotation, the point instantaneously at rest which we call the *kinematic centre*, and the other the proper centre of the rotating disc which we call the *tornado centre*. They lie in a line at right angles to the path of the cyclone and are separated from each other by a distance V/ζ, where V is the velocity of travel and ζ is the angular velocity of rotation of the disc.

On a map representing a normal cyclone the observed winds would give the resultant velocities for those points at which the stations happened to be placed. A complete system of observations would enable us to construct the diagram of instantaneous lines of flow represented in fig. 60. It must be noted that this diagram gives no means of identifying the tornado centre. It is, in fact, the point where the direction and velocity of the wind are the same as the direction and velocity of the travel of the whole disc, but unless we can determine that velocity with precision or determine the ratio V/ζ there is nothing to point out its actual position and certainly no student of weather-maps, unless he had been previously forewarned, would suspect that the permanent centre of rotation of the disc was at a point which gave no indication of its presence by any obvious peculiarity of the winds in the neighbourhood.

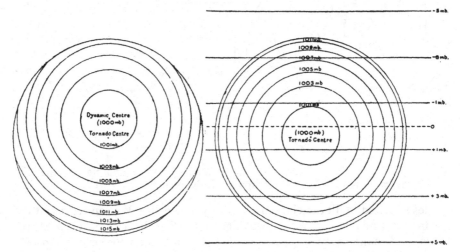

Fig. 61 (x, 3). Component and resultant fields of pressure for a normal cyclone.

The next question for consideration is the fitting of a system of isobars to the lines of flow of the normal cyclone. The answer is that if we consider the travel of the cyclone to take place along a horizontal plane and neglect the curvature of the earth's surface, the variation of the geostrophic relation with the latitude due to the variation in sin ϕ, and also the small variations of density that occur during the travel of the air in the horizontal layer, the appropriate system of isobars will consist of a field of circular isobars, corresponding with the rotation of the disc, embedded in a field of straight isobars corresponding with the velocity of translation, but that the centre of the circular isobars will coincide neither with the centre of the rotating disc, the tornado centre, nor with the centre of instantaneous rotation, the kinematic centre. It will be at a point on the line joining those two centres at a distance from the kinematic centre equal to $V/(2\omega \sin \phi + \zeta)$.

This will constitute a third centre for the normal cyclone. It will be at the centre of isobars as drawn on the map and therefore quite easily identified. We call it the *dynamic centre*. Since cyclones were mapped it has always been regarded as the centre of the cyclone, but clearly it is not unique in that. Our ideas about the relation of the features of a normal cyclone to its centre will not be complete unless we recognise that the permanent rotation is centred at another point, the tornado centre, and the instantaneous winds are centred about a third point, the kinematic centre.

The demonstration of the position of the dynamic centre, the centre of isobars, with reference to the tornado centre, the centre of permanent rotation of the disc, follows simply from the combination of the field of pressure representing the velocity of travel according to the geostrophic law and the field of pressure representing the rotation of the disc with its cyclostrophic and geostrophic components.

Taking the centre of the rotating disc as origin with the axis of x to the east and that of y to the north, since for uniform eastward motion the pressure *diminishes* uniformly to the north at the rate $2\omega\rho V \sin \phi$ the geostrophic field of pressure for a velocity of translation V to the east will be

$$p' - p_0' = - 2\rho\omega Vy \sin \phi \qquad \qquad \text{......(1)},$$

where p' is the pressure at any point, p_0' the pressure at any point on the axis of x. The equation to the circular field which would balance the rotation of the disc round a stationary centre at the origin is

$$p - p_0 = \tfrac{1}{2}\rho\zeta \, (2\omega \sin \phi + \zeta) \, (x^2 + y^2) \qquad \qquad \text{......(2)}.$$

Equation (2) is the direct integration of the gradient equation

$$dp/dr = \rho \, (2\omega v \sin \phi + v^2 \cot \theta/\epsilon).$$

Neglecting the curvature of the earth the second term becomes v^2/r, and v is equal to $r\zeta$; hence

$$r^{-1} dp/dr = \rho\zeta \, (2\omega \sin \phi + \zeta),$$

whence, since ρ and ϕ are taken as constant

$$p - p_0 = \tfrac{1}{2}\rho\zeta \, (2\omega \sin \phi + \zeta) \, r^2,$$

and $\qquad \qquad \qquad \qquad r^2 = x^2 + y^2.$

Combining the two fields by adding equations (1) and (2) and writing P for the resultant pressure, we get for the equation of the resultant field

$$P - P_0 = \tfrac{1}{2}\rho\zeta \, (2\omega \sin \phi + \zeta) \, (x^2 + y^2) - 2\rho\omega Vy \sin \phi \quad \text{......(3)}.$$

This equation represents a circular field of pressure round the centre

$$x = 0, \; y = \frac{V 2\omega \sin \phi}{\zeta \, (2\omega \sin \phi + \zeta)},$$

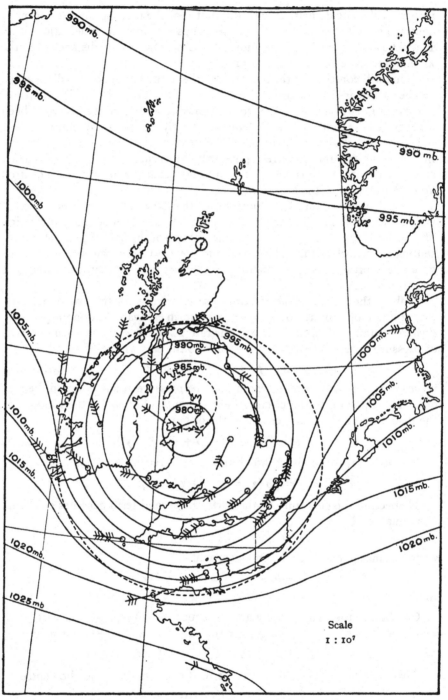

Fig. 62 (x, 6). Solid spin of 38 m/sec at 380 kilometres, in a westerly current of 20 m/sec.
    The larger dotted circle shows the boundary of the revolving fluid; the smaller, the
position of the kinematic centre of instantaneous rotation.
    A composite of a solid spin with simple vortical environment is given in chap. x.

and the pressure P_0 is at the centre instead of the origin. The distance of the dynamic centre from the tornado centre which was chosen as the origin is

$$\frac{V}{\zeta} \times \frac{2\omega \sin \phi}{2\omega \sin \phi + \zeta},$$

and the distance of the kinematic centre from the tornado centre is V/ζ. Hence the distance of the dynamic centre from the kinematic centre is the difference, that is $V/(2\omega \sin \phi + \zeta)$.

It follows that a system of circular isobars embedded in a field of straight isobars corresponding with the velocity of travel, having its centre at a properly selected point, will provide the field of pressure necessary for the disc to go on rotating. The component and resultant fields of pressure are represented in fig. 61.

That the centre of isobars is not coincident with either of the centres of rotation is a peculiarity of atmospheric motion arising from the fact that in consequence of the rotation of the earth a field of pressure is required to balance what appears in our reasoning as rectilinear motion in a horizontal plane rotating uniformly with an angular velocity $\omega \sin \phi$.

These conclusions enable us to approach the true position with regard to the relation of winds to curved isobars which we were unable to deal with previously owing to our ignorance of the curvature of the path. For a travelling system of isobars in the form of circles the winds computed according to the gradient equation of page 87 would be a system of winds arranged instantaneously in circles round the kinematic centre distant $V/(2\omega \sin \phi + \zeta)$ from, and on the left-hand side of, the path of the centre of isobars. It would indicate the existence of a tornado centre of permanent rotation at a distance

$$\frac{V}{\zeta} \times \frac{2\omega \sin \phi}{2\omega \sin \phi + \zeta}$$

on the right of the path of the centre of isobars.

The next question is what resemblance the results of this calculation bear to reality as represented on charts of pressure and wind. We cannot make a direct comparison without a further application of theory because what we have dealt with relates to a normal cyclone in the free atmosphere and our weather-maps represent the pressure and wind at the surface. The representation of the pressure will serve for the surface as well as for the free atmosphere, but the winds at the surface will be affected by the eddy-motion due to the friction of the ground. We can make an allowance for the effect, according to Taylor's theory as in fig. 60, by assuming a deviation of the wind from the direction of the isobars of 20° and a corresponding reduction of the velocity to two-thirds of that corresponding with the gradient in the free air. With this further application our theory is complete, except that we have not dealt with the discontinuities in velocity and pressure-gradient at the boundary of the rotating disc. In order to test the conclusions a theoretical map has been constructed showing a normal cyclone of 10 m/sec at a distance of 100 kilometres

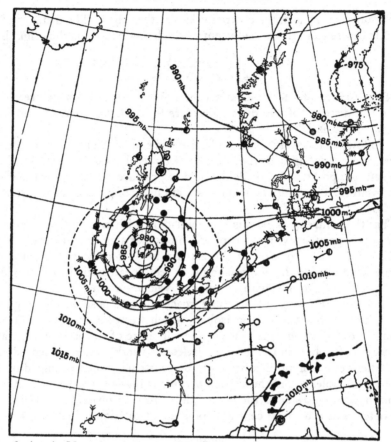

Fig. 63 (x, 4). Map of cyclone of 10 September 1903, 18 h. (Scale 1 : 2 × 10⁷

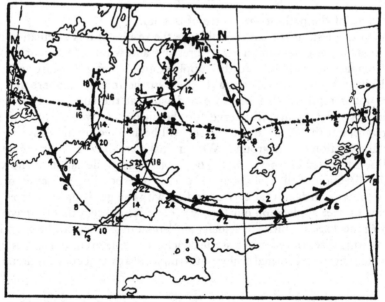

Fig. 64 (x, 5). Trajectories, or actual paths of air. (Scale 1 : 10⁷.)

which travels in a broad westerly current at the rate of 20 m/sec. The isobars in which the normal cyclone is formed are drawn curved but that should make little difference. The region of discontinuity has been treated by bending the isobars in their course on either side of the boundary of the revolving disc. The map is reproduced in fig. 62.

For comparison with this theoretical map we reproduce in fig. 63 the weather-map for 18 h of 10 September, 1903, representing an actual cyclonic storm of that date.

Let us note the comparison. The general similarity of the two maps is obvious, perhaps it has been made improperly so by the smoothing of the discontinuity and by drawing an arbitrary isobar outside the region of revolution, but it is confirmed by the agreement between the actual velocity of travel of the storm which is given in the *Life-history* as 16 m/sec and the geostrophic velocity deduced from the isobars of its path which is 17 m/sec. The scale of winds which defines the angular velocity in the theoretical cyclone, namely 10 m/sec at a radius of 100 kilometres, is perhaps too large, as force 11 is shown in the outer rings of the theoretical cyclone and is an unusual figure for a surface-wind. Little change, however, in the general character would be introduced by adjusting the ratio for a more suitable value of the wind-velocity in the outer rings.

There are two features of agreement which are very striking, one is the existence in the real case of close isobars in the form of incomplete circular arcs very like those which complete the mapping of the revolving disc in the theoretical map. They will be recognised as characteristic of small cyclones which travel rapidly. The other is the peculiarity of the wind indicated as passing outward from the innermost isobar of the real cyclone. A wind with similar disobedience to the run of the isobars on the northern side of the centre is shown in many other actual maps of rapid travelling storms. There are several included in the carefully drawn maps of the *Life-history*. We have been accustomed to regard them merely as unimportant irregularities to be expected from the light winds which are found near the centre of a cyclone, but if the theoretical map is correct so also are these hitherto irregular winds. They mark the rotation of the winds round a centre not coincident with the centre of isobars but on the left of its path, and that conclusion is confirmed by our previous inference that light winds are not excluded from the influence of the distribution of pressure but show closer agreement with it than stronger winds. Another interesting feature of similarity is the incurvature shown by the winds in the rear of the storm as compared with the stricter agreement with the run of the isobars in the front. These considerations lead us to accept the conclusion to be drawn from the conditions of the normal cyclone, namely that the wind *calculated from the gradient by the full formula, using the curvature of the isobars, gives the true wind in the free air not at the point at which the gradient is taken but at a point distant from it along a line at right angles to the path and on the left of it by the amount* $V/(2\omega \sin \phi + \zeta)$. For the particular cyclone represented in fig. 63, the distance is about 50 kilometres. It increases with the velocity of translation V, and when ζ is equal to $\omega \sin \phi$ and the air in consequence forms part of a cap rotating as a solid round the pole it is equal to $V/(3\omega \sin \phi)$.

And more striking still is perhaps the agreement to be found between the actual paths of the air in the cyclone of 10–11 September, 1903, constructed many years ago without reference to any theory whatever from successive hourly maps for the *Life-history*, for which the original diagrams are reproduced in fig. 64, and the calculated paths of the air in the normal cyclone as shown in fig. 65.

Corresponding agreement is to be found in the case of the circular storm of 24–25 March, 1902, which is also figured in the *Life-history*, but in that

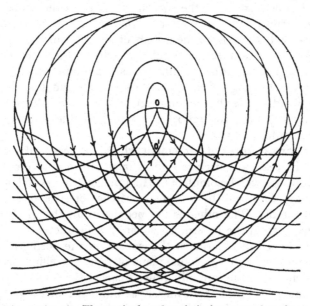

Fig. 65 (x, 7). Theoretical paths of air for normal cyclone.

case the isobars which governed the rate of travel are found in the rear of the storm, not in the front. It must be remarked here that what has been said about the travel of circular storms cannot be applied to the cyclonic depressions which often succeed one another in a belt between two isobars one on the north and the other on the south, as in the case of the series of depressions of July, 1918, and in that of the slowly moving depression of 11–13 November, 1901, also figured in the *Life-history*. The environment of the cyclone is quite different in these cases and the control of the travel must be different. Cordeiro[1] has pointed out that the gyroscopic effect of a revolving column on a rotating earth necessitates a motion of the column in order to keep its axis in the progressive vertical, so that if a cyclone is to persist it must travel

[1] F. J. B. Cordeiro, *The Atmosphere*, New York, Spon and Chamberlain. London, E. and F. N. Spon, Ltd., 1910.

even if it is formed in a body of calm air, but this subject belongs more properly to the next chapter. In this chapter we have assumed the motion to be over a plane rotating surface and the examples are taken from cyclones with rapid motion of translation.

There is another aspect of the combination of a circular field of pressure with a linear field to produce the distribution of pressure which has been shown to be necessary for the travel of a normal cyclone which is interesting from the point of view of travelling groups of curved isobars. For the reasons which we give below it appears that if circumstances were so arranged that a linear field of force with gradient s suddenly came into operation upon a stationary normal cyclone it would forthwith produce the field of force appropriate for a normal cyclone travelling with the velocity V, where V is equal to $s/(\zeta\rho)$, along a line at right angles to the superposed gradient and *from left to right* of a person facing the high pressure. And, since the travelling

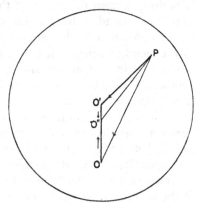

normal cyclone has a distribution of velocity round the instantaneous centre indistinguishable from that of the normal cyclone round its tornado centre, it follows that the sudden superposition of a gradient would *ipso facto* transform the stationary cyclone into a cyclone travelling with the assigned velocity.

In *Geophysical Memoirs*, No. 12, this conclusion was reached from the consideration of a proposition set out by Gold[1] to the effect that in a travelling cyclone which consists of rings of fluid in instantaneous rotation about a moving axis the relation between X, the radius

Fig. 66 (x, 8). Three centres of normal cyclone.

of curvature of the path, and r, the radius of the instantaneous circle, is given by the equation $\qquad X = r/(1 + V \sin \alpha/v),$

where α is the angle which the radius makes with the line of path of the kinematic centre. But the conclusion follows very easily from the properties of the three centres of the normal cyclone, fig. 66.

Let O' be the centre of the stationary normal cyclone, O the tornado centre for a velocity of travel V, so that $OO' = V/\zeta$. Let P be any point of the rotating air. The acceleration at P will be $\zeta(2\omega \sin \phi + \zeta) OP$. Suppose a field of uniform gradient s (equal to $V\rho\zeta$) to be superposed upon the system in the direction $O'O$; the acceleration of P corresponding therewith is $V\zeta$. It will be represented by a line $O'O''$ where O'' is defined by the condition

$$O'O'' : OP :: V\zeta : \zeta(2\omega \sin \phi + \zeta) OP.$$

Hence $\qquad\qquad O'O'' = V/(2\omega \sin \phi + \zeta).$

[1] *Barometric Gradient and Wind Force*, M.O. Publication, No. 190, p. 43.

Thus O'' is also the dynamic centre of the normal cyclone with tornado centre O, travelling with the velocity V, and the resultant acceleration $\zeta (2\omega \sin \phi + \zeta) O''P$ is in every respect the same as the combination of an acceleration $\zeta (2\omega \sin \phi + \zeta) OP$ towards the tornado centre O with an acceleration $V/(2\omega \sin \phi)$ proportional to OO'' across the line of travel from right to left. That is to say, the resulting gradient or distribution of pressure is the same in direction and magnitude as that required for a travelling normal cyclone centred at O. The difference between the curvature of the path and the curvature of the circle of instantaneous motion is expressed by a linear field of pressure which tends to push the rotating air towards O. The effect of the push is not to displace the system in the direction of the push but to make the instantaneous centre of rotation travel from left to right across the line of the push carrying, of course, the rotating system with it.

At the moment of the superposition of the new transverse gradient there is nothing (except the extent of the area affected) to distinguish the motion of rotation round O' from an identical motion of rotation round O combined with a translation V: they are simply two aspects of the same field of motion. Consequently when the new field is superposed all the forces will be accommodated if instead of continuing the original rotation round the point O', another aspect of the same motion is continued, namely, rotation round O and travel with the velocity V.

If instead of a finite gradient being suddenly superposed the superposition is gradual as, for example, by the passing overhead of a system of isobars belonging to the region of the stratosphere, the development of the corresponding travel of the normal cyclone would be similarly gradual. If we pursued the matter further we should have to recognise that only a part of the stationary cyclone forms complete circles round the point O; the new permanent rotation would be lop-sided. Also the stationary cyclone has to be imagined in calm air, and in the new conditions it is through the surrounding air that it would have to make its way and we can make no estimate of the reactions that would ensue. The new gradient, if it extended beyond the area of the cyclone, would not help matters as regards the environment; for the balancing motion corresponding therewith is in the opposite direction from that in which the cyclone has to move.

What modification in the subsequent motion these considerations would introduce we are, therefore, at present unable to say, but without that further development the result obtained is of considerable interest to us in our pursuit of an answer to the question of the relation of the wind to the distribution of pressure. That the wind should always be regarded as balancing the gradient is a hard saying for many meteorologists. It has even been said that the assumption simply ignores the causes through whose operation the changes which we wish to study are brought about. We may attempt to devise circumstances under which finite changes of pressure would conceivably come into operation so quickly that the theoretical adjustment could not have been approached. The reader must be content to judge whether the conclusions which have been drawn from the assumption throw sufficient light on some of the hidden atmospheric processes to make it worth the risk. From the results of our work it seems possible that as a general rule in the free atmosphere in ordinary circumstances the disturbances of the balance are not large enough to interfere with the conclusions to be drawn from it, but there may be in special localities singular points or lines, points or lines of convergence or divergence, and therefore of convection, to which we cannot apply the assumption of a balance between the field of pressure and the field of velocity.

The last proposition enables us to get some insight into the effect of a difference between the actual field of pressure and the field required for the balance, which may be generalised by saying that with curved isobars the uncompensated part of the field will express itself in the travel of the group of isobars. The only case that we have dealt with algebraically is that in which the uncompensated field is uniform, but we may suppose that in general the effects will be similar when the condition is not satisfied; and here we may with advantage revert to the principle that each sector which preserves its identity is subject to the same conditions as a complete rotating system. If we find on the map a persistent group of isobars consisting of segments of circles and if we can by any means identify the instantaneous centre of its winds, the difference between the actual field and the field appropriate to permanent rotation round the instantaneous centre will be expressed by a proper motion of the group of isobars. The identification of the kinematic centre must depend upon further study of the field of motion in actual cases.

Our consideration has been limited by the restriction to motion over a plane surface, we have not yet been able to extend it to a spherical surface. It would be interesting to know what the effect of a specified uncompensated field would be upon the motion of a rotating cap of air such as those which we considered in the earlier pages of this chapter, but we are not aware of any investigation which answers this question.

A further matter of interest is the travel of anticyclones regarded from the standpoint here adopted for that of cyclones, but that must be left until we find an opportunity for more detailed study of pressure-distribution and wind-velocity in the free air as correlative indices of the state of the atmosphere, the one static and the other kinetic. It is briefly initiated on p. 302.

CHAPTER X

REVOLVING FLUID IN THE ATMOSPHERE

Nature is often hidden, sometimes overcome, never extinguished. *Francis Bacon.*

Conventional cyclones and anticyclones

THE literature of dynamical meteorology is largely concerned with the idea of revolving fluid as represented perhaps primarily in tropical revolving storms and in other whirls on a smaller scale, and then by a process of analogy in the phenomena of the cyclones and anticyclones which had taken an established place as the primary features of the weather-maps of middle latitudes. In introducing the name *anticyclone* for the regions of closed isobars surrounding a centre of high pressure Sir Francis Galton[1] expounded the idea that as an area of low pressure or cyclone was a locus of light ascending currents and "therefore" an indraught of surface winds with counter-clockwise motion, similarly an area of high pressure was a locus of dense descending air and of dispersion of cold dry air with motion becoming eventually clockwise.

Based upon the representation of the process which these words convey there gradually grew the conception on the one hand of the central area of a cyclone on the map as a centre of advective motion, a focus of attraction for the surrounding air and the general idea of a cyclone as a region of ascending warm air producing rain or snow; round the central region the air moves inward with a counter-clockwise motion in spiral curves. On the other hand the conception of the central area of an anticyclone is of a centre of directive motion, a region of repulsion; the general area of an anticyclone as a region of descending cold air which moves with a clockwise motion spirally outwards. The conventional representation of cyclones and anticyclones included the instantaneous lines of flow as a series of double spiral or reversed S-shaped curves leading from centres of high pressure to centres of low pressure. We have explained elsewhere the objections which may be urged against these conceptions of the physical nature of cyclones and anticyclones. In the *Life-history of Surface Air-currents* the paths of air over the surface were shown to be quite different in actual shape from the instantaneous spirals and to include motion from low pressure to high pressure, despite the instantaneous incurvature towards the "low," in consequence of the travel of the isobars. And in previous volumes of this work we have given reasons drawn from the conditions of thermal convection which controvert the idea of a descending column of air, assumed to be cold, in the central region of an anticyclone in direct relation with an ascending column of air, assumed to be warm, in the central region of a cyclone.

[1] *Proc. Roy. Soc.*, vol. XII, 1862–3, p. 385. See also vol. I, p. 312.

(250)

Here we wish to remind the reader that the idea has been found singularly sterile as a means of developing our knowledge of the physical processes which are expressed by the weather which we experience. The grouping of the phenomena with reference to the centre of isobars of travelling cyclones has proved ineffective for this purpose. There was no symmetry with respect to the centre for any of the meteorological elements with the exception of pressure and, to a certain extent, of the winds. The difficulty lay in the fact that the feature of the phenomena for which it was necessary to find an explanation was the travel of the groups of isobars across the map. That essential part of the phenomena was ignored, being regarded as a matter that could be dealt with separately without disturbing the elements of the cyclone visible on the map. When allowance was made for the motion of the group of isobars which form a travelling cyclone the residual velocities were not those of air in rotation round the centre of isobars. We had no use for the details of the properties of fluid in permanent rotation because we could not find examples on our maps to which they could be applied.

But further inquiry showed that examples might be found in localities where they had not previously been looked for[1]. It was first realised that the isobars corresponding with a column of fluid in permanent rotation and travelling bodily across the map would not necessarily be indicated by concentric circles but might be shown by local deviations of isobars from their regular run with reference to the centre of a large cyclonic depression, such as we are accustomed to call "a small secondary." Two examples were adduced. One which was sufficiently indicated in the isobars of the maps for 24 March, 1895, travelled from Cork Harbour to the mouth of the Humber at an average speed of 55 miles per hour, and then went on to the west coast of Denmark with an average speed of 82 miles per hour. Its diameter was probably about 150 miles at the beginning of the journey and 300 miles at the end. The velocities were in rough agreement with the geostrophic winds of the isobars in which the local circulation was formed. A notable peculiarity of this case was that no rain fell in the travelling cyclone. The other case was that of the tornado which visited South Wales on 27 October, 1913. It was less than 10 miles in diameter and showed no disturbance of the isobars as drawn on the maps for the day. But it travelled along the line of the isobars with a velocity about three-quarters of the computed geostrophic wind. It was accompanied by very heavy rain in various localities on its route.

These examples were sufficient to show that the properties of revolving fluid are pertinent to the phenomena of travelling depressions provided that the proper centre of permanent rotation can be identified. The indication was confirmed by the occurrence of tornadoes in the southern portion of the large cyclonic depressions of the United States; the trough-line of a large depression was indicated as a probable locality for their formation. At the time velocity tangential to the isobars and uniform over the area of the section of the travelling column was assumed as being a sufficient generalisation of the

[1] Shaw, 'Revolving Fluid in the Atmosphere,' *Proc. Roy. Soc.*, A, vol. XCIV, 1917, p. 33.

irregular velocities observed at the surface in different parts of a travelling cyclone. Subsequently it proved to be desirable to examine the phenomena of the travelling cyclone in relation to the properties of the normal cyclone as defined in the previous chapter and the propriety of including in a similar category other cyclonic depressions with properly chosen centres became evident. It may be noted that in doing so we approach the question of the cyclone from a standpoint which is different from that which Galton indicated. We regard the incurvature of the surface-winds not as the primary step from which all the rest of the phenomena are derived but merely as incidental to the retardation of the lowest layers of the revolving air by the friction of the ground. We make no hypothesis as to how the air at some unknown height above the ground comes to be in rotation. The question to which we address ourselves is what happens to the revolving air during its life-history as a definite travelling mass.

We shall evidently be on safe ground in applying the properties of revolving fluid provided we think of the properties as related to the centre of the rotating mass and not to the centre of instantaneous rotation.

Dynamics of revolving fluid

"Such simple conclusions from the dynamics of revolving fluid as are within our reach" have been set out by Lord Rayleigh[1] for the reason that "so much of meteorology depends ultimately upon their study"; we cannot, therefore, do better than appropriate his conclusions. They deal with rotation about a fixed axis and make no allowance for the rotation of the earth. This limitation must be borne in mind in considering the distribution of pressure appropriate to the field of motion which is indicated.

The reasoning is based upon the fundamental equations of hydrodynamics, adapted to cylindrical coordinates r, θ, z with velocities u, v, w reckoned respectively in the direction of r, θ, z, increasing. For the present purposes assuming symmetry with regard to the vertical axis the equations become

$$\frac{\partial u}{\partial t} + u\frac{\partial u}{\partial r} - \frac{v^2}{r} + w\frac{\partial u}{\partial z} = -\frac{\partial P}{\partial r} \qquad \ldots\ldots(1),$$

$$\frac{\partial v}{\partial t} + u\frac{\partial v}{\partial r} + \frac{uv}{r} + w\frac{\partial v}{\partial z} = 0 \qquad \ldots\ldots(2),$$

$$\frac{\partial w}{\partial t} + u\frac{\partial w}{\partial r} + w\frac{\partial w}{\partial z} = -\frac{\partial P}{\partial z} \qquad \ldots\ldots(3),$$

where P (assumed to be single-valued and also independent of θ) is equal to

$$\int dp/\rho - V \qquad \ldots\ldots(4).$$

V is the potential of the extraneous forces of which only the force of gravity need be considered for present purposes.

We may take in order the properties which Lord Rayleigh deduces from these equations.

---

[1] 'On the Dynamics of Revolving Fluid,' *Proc. Roy. Soc.*, A, vol. XCIII, 1917, p. 148.

1. *Persistence of " circulation " and conservation of angular momentum*
Equation (2) may be written

$$\left(\frac{\partial}{\partial t} + u\frac{\partial}{\partial r} + w\frac{\partial}{\partial z}\right)(rv) = 0 \qquad \ldots\ldots(5).$$

This signifies that rv may be considered to move with the fluid. If r_0, v_0 be the initial values of r, v for any particle of the fluid, the value of v at any future time when the particle is at a distance r from the axis is given by

$$rv = r_0 v_0.$$

2. *Motion in vertical planes through the axis of z*, as for example the ascent of air near the axis and its descent in the outer region or *vice versa*.

The motion is the same as that which might occur if $v = 0$ with the addition of the centrifugal acceleration v^2/r along r.

3. *Distribution of pressure when there is no radial or vertical motion.* In this case $u = 0$ and $w = 0$, and it follows from (3) that P is independent of z and therefore a function of r and t only. From (1) it follows that v is also a function of r only and $P = \int v^2 dr/r$. Accordingly by (4)

$$\int dp/\rho = V + \int v^2 r^{-1} dr \qquad \ldots\ldots(6).$$

V is the potential of the impressed forces; and, if gravity (acting downwards) is the only extraneous force, V is equal to $C - gz$ whence

$$\int dp/\rho = C - gz + \int v^2 dr/r.$$

For the solution of this equation we require to know the conditions of the motion in respect of the relation of p and ρ. For an incompressible fluid ρ would be constant. For isothermal variations Boyle's law ($p = a^2\rho$) would hold and Lord Rayleigh deals with no other alternative. In the free atmosphere the changes must be regarded as approximately adiabatic and the relation of p and ρ will depend upon whether the air is saturated or not. Assuming that the changes do not pass saturation we may write

$$p = (\kappa\rho)^\gamma,$$

where κ depends upon the entropy of the sample of air. Hence

$$\frac{\kappa\gamma}{\gamma - 1}p^{(\gamma-1)/\gamma} = \rho - gz + \int v^2 dr/r \qquad \ldots\ldots(7).$$

At a constant level dp/dr is always positive. Pressure and consequently density diminish as the axis is approached. But the rarefaction near the axis does not cause the fluid there to ascend. The denser fluid outside is prevented from approaching the centre by the centrifugal force. This conclusion would be modified for the bottom layer of the atmosphere where there is loss of circulation in consequence of the eddy-motion of the air over the ground.

4. *Stability of the motion.* The equilibrium represented by equation (7) will hold good whatever may be the relation between v and r, but the motion might be unstable. The instability in the atmosphere with which we are most familiar is that of layers of air of which the heavier is above the lighter or to speak more technically the higher has less entropy or potential temperature

than those beneath. If the atmosphere is isentropic the equilibrium is neutral. So a series of rings of fluid may be in equilibrium but unstable because the outer rings are not revolving fast enough. The rotational equilibrium is neutral if vr is uniform over the area at any level because when motion takes place in one level vr remains constant and equal to k for a ring consisting always of the same matter and the centrifugal force acting upon a given portion of fluid is k^2/r^3. It is stable for fluid revolving one way between coaxial cylindrical walls only under the condition that the circulation $2\pi k$ increases with r. The conclusion is confirmed by the consideration of the change of kinetic energy on the interchange of two rings.

The result is applied to two cases of fluid moving between an inner cylinder and an outer cylinder. If the inner one rotates while the outer is fixed the equilibrium requires that the circulation should diminish outwards and therefore the motion of an inviscid fluid in that case would be unstable. On the other hand, if the outer cylinder rotates while the inner is fixed the motion satisfies the condition that the circulation increases outwards and so the motion of an inviscid fluid would be stable.

From these conclusions we infer that, regarding the motion of the margin of a column of revolving fluid forming a cyclone as the rotation of the inner cylinder, the motion caused in the air which surrounds it will be unstable, and presumably its energy will gradually be dissipated, whereas if we may regard the air-currents round the margin of an anticyclone as a rotating outer wall the motion caused in the air within will be stable.

Considerations of the stability of the motion of revolving fluid must be regarded as of vital importance in the study of the dynamics of cyclones and anticyclones in so far as they are examples of revolving fluid. The mere fact of the obvious persistence of the motion of rotation of cyclones is in itself remarkable considering that it is maintained in an environment that has no intrinsic cohesion. An outer cylinder can only be represented by the pressure of the surrounding air and an inner cylinder by the pressure of a column of revolving air. In watching weather-maps we sometimes see "secondaries" absorbed into one primary and sometimes, as in one of the cases in the *Life-history*, what was originally a secondary may absorb its own primary. These are apparently cases in which the stability of the motion round the ultimately successful centre was overpowering. What we would like to know is whether the energy of temporary disturbance of the motion due to local convection at some point away from the centre can be absorbed in the motion round the original centre and the intensity of the system grow in that way by successive slight additions in different parts of its area in the same way as a boy's whip-top can acquire speed of rotation from impulses that would meet with no such response if it were not for the stability of the motion already possessed by the top. But the air has no rigidity nor has its motion the stability which rigidity gives, and in spite of that it can acquire stability. Apparently sometimes one centre, sometimes another is favoured and the conditions of preference are unknown to us.

5. *The distribution of velocity and the variation of pressure in the outer region when there is convergence towards the axis*. This is approached by assuming that u is a function of r and t only and that $w = 0$ or at most a finite constant. Pressure may be supposed to be kept constant at the axis or preferably at an inner cylindrical boundary by the removal of fluid from within a certain radius. This is the idea of the central portion of the air of a cyclone being removed by upward convection which is in the minds of many meteorologists as the fundamental conception of a cyclone but which can be only vaguely supported by the actual phenomena of weather which accompany an ordinary cyclone.

On the hypothesis that $w = 0$ or constant, the motion is two-dimensional and it may be conveniently expressed by means of the vorticity ζ, which moves with the fluid, and the stream-function ψ connected with ζ by the equation

$$\frac{1}{r}\frac{\partial}{\partial r}\left(r\frac{\partial \psi}{\partial r}\right) + \frac{1}{r^2}\frac{\partial^2 \psi}{\partial \theta^2} = 2\zeta \qquad \dots\dots(8)$$

The appropriate solution is

$$\psi = 2\int r^{-1}\,dr\int \zeta r\,dr + A\log r + B\theta \qquad \dots\dots(9),$$

where A and B are arbitrary constants of integration. Accordingly

$$u = -\frac{\partial \psi}{r\partial \theta} = -\frac{B}{r}, \qquad v = \frac{\partial \psi}{\partial r} = \frac{2}{r}\int \zeta r\,dr + \frac{A}{r} \qquad \dots\dots(10).$$

In general A and B are functions of the time and ζ is a function of the time as well as of r.

If ζ is initially and therefore permanently uniform throughout the fluid

$$v = \zeta r + Ar^{-1} \qquad \dots\dots(11),$$

and Lord Rayleigh remarks that this equation is still applicable under appropriate boundary conditions even when the fluid is viscous. In the case of the normal cyclone $v = \zeta r$ and therefore $A = 0$. But if the central portion of the cyclone be removed and the outer boundary closes in from R_0 initially to R at time t, since vr remains unchanged for each ring of fluid we get

$$v/\zeta = r + (R_0{}^2 - R^2)\,r^{-1} \qquad \dots\dots(12).$$

And thus convergence towards the axis in a normal cyclone causes the fluid to acquire in addition the motion of a simple vortex of intensity increasing as R diminishes.

Brunt[1] has shown the effect of the earth's rotation in the determination of the value of A. It increases the strength of the superposed vortex. In equation 12 for example $v = \zeta r + (\zeta + \omega \sin\phi)\,(R_0{}^2 - R^2)/r$.

If at any stage the convergence ceases (6) gives $dp/dr = \rho v^2/r$ and neglecting the variations of density

$$p/\rho = \zeta^2\left\{\tfrac{1}{2}r^2 + 2\,(R_0{}^2 - R^2)\log r - \tfrac{1}{2}\,(R_0{}^2 - R^2)^2\,r^{-2}\right\} + \text{const.} \quad (13).$$

[1] *Proc. Roy. Soc.*, A, vol. xcix, 1921, p. 397.

Since v^2 as a function of r continually increases as R diminishes the same is true for the difference of pressures at two given values of r, say r_1 and r_2, where r_2 is greater than r_1. Hence if by removal of air or by any other process the pressure is maintained constant at r_1, it must continually increase at r_2, or in meteorological language convergence to a central region at which the pressure remains constant will require an increase of the gradient of the cyclone by increase of the pressure in the outer rings.

Lord Rayleigh concludes by explaining that when the u, w motion is slow relatively to the v motion we may formulate a general idea of the solution of the problem. When revolving fluid is drawn off from a point near the axis of rotation there is a tendency for the surfaces of constant circulation to retain their form and position, the more pronounced the greater the speed of rotation. The escaping fluid is therefore drawn off along the axis and not symmetrically from all directions as when there is no rotation. Thus we may conclude that convection near the axis of a column of rotating fluid in the atmosphere will increase the gradient of pressure in the column provided that the air which is removed is disposed of without altering the other conditions of the environment.

Experiments on the development of spin

The purpose of Lord Rayleigh's exposition was to give an analytical representation of experimental results which J. Aitken[1] had previously put forward as contributions to the study of cyclones and anticyclones. We therefore briefly recapitulate the experiments described. What concerns us chiefly in this part of the subject is the apparatus with which the experiments are conducted because in such cases the reader requires to think for himself whether the conditions and arrangements prescribed for the experiments have their counterpart in the atmosphere.

In Part I, in order to show that some initial motion of the fluid is necessary for a vortex to be formed, a vessel of water is used with a plug at the bottom that can be operated from the outside, and subsequently provision is made for the position of the opening with reference to the circumference to be changed in order to show that if there is a difference of current on two sides of the axis the motion of the vortex is with the stronger current. The same apparatus is used to show that the actual velocity of motion in the vortex is increased in a notable degree as the centre is approached, and this conclusion is further illustrated by a rotating system consisting of two balls which can be made to approach the axis of rotation while the whole system is spinning, when it is seen that the actual energy of motion of the balls

[1] 'Notes on the Dynamics of Cyclones and Anticyclones,' by John Aitken, F.R.S., Parts I and II, *Trans. Roy. Soc. Edin.*, vol. XL, 1901, p. 131; Part III, *Proc. Roy. Soc. Edin.*, vol. XXXVI, 1916, p. 174. 'Revolving Fluid in the Atmosphere,' *Proc. Roy. Soc.*, vol. XCIV, 1918, p. 250. The study of vortices in water and its application in the atmosphere has been pursued experimentally by S. Fujiwhara, *Q. J. Roy. Meteor. Soc.*, vol. XLVII, 1921, p. 287, and vol. XLIX, 1923, pp. 75 and 105.

increases as they get nearer together in accordance with the law that vr is constant.

For making experiments upon cyclonic movements in air a metal tube 15 cm in diameter and 2 m high was used. At the lower end of the tube was a circular disc 75 cm in diameter supported on three legs 15 cm high, thus leaving a space of 15 cm between the disc and the table on which it rests. To produce an up-draught jets of gas were fitted inside the tube near the lower end. To study the circulation between the disc and the table a number of light vanes, or fumes of hydrochloric acid and ammonia, were used. With this apparatus it is shown that the air moves radially towards the chimney if there is no initial movement in the air before the up-draught is started. An initial movement equally strong at all points is not of much use in generating cyclonic movement; but if the current on one side is cut off by a screen a violent cyclonic motion results in which the fumes are carried to the chimney in graceful ascending spirals. It is the lowest stratum of air that is drawn to the very centre. A further effect of the tangential motion is that the lower end of the cyclone bends away from under the centre of the apparatus, moving in the direction of the tangential current. Similar experiments can be made simply with a good fire, a free going chimney and a wet towel with a suitable arrangement of the draughts of the room. In this case when the wet towel is held vertically in front of the fire the steam is formed into a horizontal column of revolving air leading from the towel to the chimney.

In Part II the apparatus used is a large sheet of metal 75 cm square forming a platform which can be heated by gas burners underneath, or otherwise, and from which wreaths of steam rise irregularly, when the hot surface is covered by wet cloth or paper, unless there is a definite current of air which passes over one part of the platform and misses the other. In that case the rising steam is gathered up into small cyclones which may reach a height of a metre or more above the platform and which travel across the platform with the characteristic features of the eddies that are sometimes seen in the open air.

In Part III the formation of a horizontal whirl between a wet towel and the chimney over a good fire is depicted; and further a useful modification of the arrangement for forming a vortex in water, by replacing the plug in the bottom by a siphon drawing water from the top, is described and figured and attention is called to the narrowness of the vortex produced in that way. This feature is dealt with in Lord Rayleigh's paper and it will be useful to bear it in mind when we come to consider the transmission of a circular field of pressure from the upper air to the surface.

Many experiments with eddy-motion in closed vessels are described by C. L. Weyher[1], and other authors have given descriptions of experimental illustrations. But the question whether they are really representative of cyclonic motion in the atmosphere depends upon the precise analogy of the conditions as we have explained in chap. I. In the free atmosphere we miss the rigid boundaries and discontinuities which so often form an essential part of

[1] C. L. Weyher, *Sur les Tourbillons*, Gauthier-Villars, 1889.

the experimental apparatus. On the other hand we have the ubiquitous effect of the rotation of the earth. There is a quasi-rigidity attaching to the motion of flexible chains under tension, but in developing its own species of rigidity with only a single coefficient of elasticity the atmosphere has to rely entirely upon the distribution of pressure for preventing general disruption. Hence we are thrown back upon the relation of pressure to wind as after all the chief consideration of the dynamics of the atmosphere.

In his paper on revolving fluid before the Royal Society, Aitken describes a new experiment of great interest designed to illustrate the effect of rain in causing a downward motion of air at moderate heights. A large circular flat-bottomed vessel was filled with water in which a little fine sawdust was mixed to show its movements. The water was set in circular motion with a steady flow: sand was then dropped into the water to imitate falling rain; the current was followed round with the hand so as always to drop the sand in the same part of the rotating water. "When this was done a quite well-formed eddy or cyclone was observed which travelled round in the vessel at the distance from the centre at which the sand was dropped in. It was in miniature like a secondary cyclone moving in the current of the large cyclone revolving in the vessel. Only a small quantity of sand is required to produce the result, 1 or 2 g. being sufficient." It is usual to regard the convection which precedes the rain as the cause of the secondary and the idea that the secondary may be produced by the dynamical effect of the falling rain is certainly novel and adds another to the marvellous varieties of eddy-motion.

In considering the meteorological application of experiments the fitting of the several models to the actual phenomena of the atmosphere requires the utmost caution. One has only to consider the difference between the ordinary phenomena of convection as illustrated in the laboratory and the corresponding phenomena on the large scale in the atmosphere where the conditions of the environment are of at least as much importance as the local conditions of the air which rises or falls, in order to realise that the scale of the free atmosphere introduces conditions peculiar to itself.

With regard to the application of Lord Rayleigh's equations and Aitken's experimental illustrations to the phenomena of cyclonic depressions or anticyclones as we find them represented by meteorological observations we ought to devote our attention especially to the details of the motion of the air and of the distribution of pressure and density within the cyclone or anticyclone because those are the elements which enter into the equations and are represented in the experiments. This object is by no means easy of attainment with the material at our disposal because there is not enough information to enable us to complete the picture with the accuracy that a rigorous comparison requires. We have already cited two instances in which the idea of rotating columns of air appears justified but they are on a comparatively small scale and the details of distribution within the areas identified are too meagre for a proper comparison. In the cases of the fast travelling storms of 10–11 September, 1903, and 24–25 March, 1902, we have sufficient detail to show that the phenomena are fairly well represented by a rotating column of fluid with an appropriate centre and we shall presently give some noteworthy evidence as to the distribution of velocity with reference to the centre of the September storm at one point of its course. In considering cyclones of larger diameter we are faced with the difficulty that convection which may be represented by rainfall alters, for the time being at least, the regularity of the distribution of pressure and wind. We have no satisfactory expression for the effect of local convection outside the central region. We may surmise that it will cause a local circulation and if sufficiently prolonged may give rise to local rotation of sufficient extent and intensity to become recognised as a secondary depression and ultimately, if conditions are favourable, may absorb the circulation of the original system.

We have, however, some guidance from Lord Rayleigh's equations as to the effect of convection near the core of the revolving column and it will be useful to add some further considerations as to the process, regarding separately in accordance with proposition (5) that part of it which is operative during the convection and that part of it which follows when the convection has ceased and the rotation continues.

While the convection is operative within the core of a revolving column in the upper air it has been shown that air will be drawn from the core of the column below, not from the cylindrical walls which surround the rising air; the question of the statical equilibrium of the core of the column beneath the original locus of convection does not enter immediately into the solution.

Convection will take place spontaneously when the successive layers of air are so arranged that the entropy or potential temperature of the lower strata is greater than that of the higher and will go on until that state of affairs is ended. How the ascended air distributes itself we cannot say because we have no adequate means of forming an opinion. The shape of an anvil cloud forming the top of a cumulo-nimbus cloud may give us an indication on a comparatively small scale. We have assumed that the motion of clouds in the upper levels gives us an indication of the distribution of pressure at those levels. We cannot therefore regard it as independent of that distribution and we cannot coordinate their motion with the ascent of air from down below until we know what the distribution of pressure in the upper air is.

Our knowledge of the phenomena at the top of a revolving column of air when convection is active at the core is very defective; we can only say vaguely that the convection will go on until the conditions for thermodynamic stability, controlled by entropy, are satisfied. Ascending air cannot pass a layer of counterlapse of temperature nor penetrate far into an isothermal region, a region of no lapse. When the thermodynamic conditions are satisfied the rotation of the air prevents the flow of air from the sides to the core and the thermal conditions prevent the air filling up the low pressure from the top unless the column travels from under the air which has risen to other air of lower entropy. There will be a certain amount of flow inwards at the bottom but that will be restricted by the isentropic conditions of the ascent and some considerable time will be required to disintegrate the rotation. If, therefore, the column be finished off at the top in such a way as to prevent the low pressure filling up there the revolving column might travel for a considerable period without material change. Protection of that kind might be afforded if the column were surmounted by a cap which travelled along with the column and in which the rotation gradually diminished with height.

We have no satisfactory information as to the level at which the development of a column of rotating fluid begins nor how far up it extends. Apart from slope-effect we should attribute effective convection exclusively to water-condensation which may begin at any level in the air leaving dry air to form pools in convective equilibrium. But radiation upon a slope may carry dry air upwards, so likewise the movement of air in a sloped isentropic surface. Where condensation would begin to reinforce convection is uncertain in view of the influence of hygroscopic nuclei.

A circular distribution of pressure once set up will be transmitted to the layers beneath. Those layers at the first setting up of the circular isobars within them will not have the rotation which keeps the fluid from moving towards the axis to fill up the low pressure; consequently the air will move inward towards the core and the part immediately below the core of the whirl will pass upward into the core; the air moving inwards in the layer beneath will gradually develop rotation which will balance the distribution of pressure, and so the core will gradually be drawn out of the column beneath and rotation set up provided that the convection up above is strong enough to carry with it the air supplied from the core of the column beneath. Thus the sudden creation of a circular field of pressure due to the convection will set up a sort of trunk of "suction" along the core which will extend further downwards as the rotation gradually develops and ultimately reach the ground. If the original instability is very marked it seems possible that the suction at the ground, when the core reaches it, might be very strong and the inrush and uprush of air near it very powerful. This process on a large scale might account for the carrying up of dust, sand, small fish and other objects into the core and so upwards into the air[1].

And here it is important to notice in continuation of the proposition, explained in chapter IX, as to the superposition of a uniform linear field of pressure upon a circular field, that the transmission to the surface of a circular field of pressure suddenly created by convection or otherwise in the upper air would form a circular field at the surface within a field of straight isobars, but the centre would not be vertically underneath the centre of the circular field in the upper air. There would be displacement of the centre of circular isobars through the distance $V/(2\omega \sin \phi + \zeta)$ where ζ is the angular velocity of the original rotation and V is equal to $s/(\rho\zeta)$, s being the gradient of the superposed field, in this case the gradient of the field due to the layer between the original level of the whirl and the ground. The added gradient of this layer will be represented by the change in the flow of momentum in the layer.

---

[1] Cf. *Nature*, vol. CII, p. 46. *Q. J. Roy. Meteor. Soc.*, vol. XLIV, 1918, p. 270.

Hence we may conclude that the core of a column descending to the surface will not be along a vertical line but to a point on the surface displaced from the vertical across the wind of the lower layers. The air will be drawn out along the line of the core whether it be vertical or sloped at the angle defined by the superposed gradient. In this we may find an explanation of the resemblance of water-spouts to elephants' trunks, and we may also conclude in general that the core of a column of revolving fluid will not be vertical and that the position of the core at any level will depend upon the gradient, at that level, of the isobars within which the rotation takes place. It has long been surmised that the axis of a cyclone is inclined to the vertical and the considerations here set out add definiteness to the meaning of that idea. Opportunities for definite evidence upon the subject are rare but we may cite an instance of a pilot-balloon ascent close to the core of a cyclone at 17 h 15 m on 28 March 1918.

The point of observation, near Leith, was about 100 kilometres due north of the centre of a well-marked circular depression which was complete up to 800 kilometres in diameter.

The observations were:

Height	1	2	3	4	5	6	7	8	9	10	km
Wind-velocity	4	3	3	3	4	6	9	12	9	8	m/sec
Wind-direction	E	SE	S	SE	S	SW	S	S	S	S	

Thus the results give a south wind above the easterly wind on the surface and point therefore to the displacement of the core towards the north-west in the upper air.

REVOLVING FLUID IN ACTUAL CYCLONES

From these preliminary considerations we may pass to Lord Rayleigh's equation (12) as representing the distribution of velocity in a cyclone originally normal in which convection at the core, and consequently convergence, is operative, and equation (13) as representing the distribution of pressure at the stage when the convergence ceases. Since the distribution of pressure in a cyclone of considerable dimensions will alter very slowly, we can regard the pressure equation as holding good for the specified distribution of velocity although the convergence may not have ceased. And the two equations may be regarded as general equations for the velocity and pressure in a cyclone originally normal but affected by convection at the core or any other process which is equivalent thereto and causes the convergence towards the centre of the circles of fluid in any layer.

For the purpose of rigorous comparison we ought to make allowance in the equations for the rotation of the earth. For this purpose as a first approximation, neglecting the earth's curvature, we should increase the radial acceleration v^2/r in equation (1) by $2\omega v \sin \phi$, and subsequent equations depending upon it would be modified in consequence. And if the cyclone is travelling

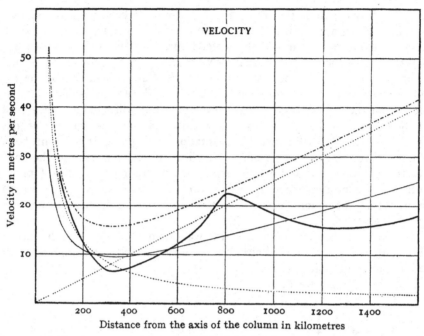

VELOCITY

Velocity in metres per second

Distance from the axis of the column in kilometres

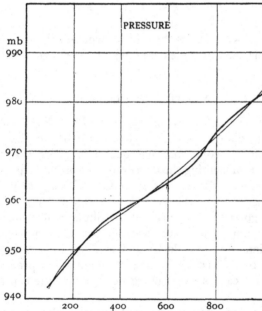

PRESSURE

mb

The dotted lines .... represent the components of velocity depending upon ζr and $\zeta A r^{-1}$ respectively, and the chain line ▬·▬·▬·▬ the combination of the two. The thin full lines represent the computed velocity (reduced in the ratio of 6:10 to allow for friction at the surface) and the computed pressure, respectively. The thick lines represent the velocity and pressure as recorded on the map on 20 February, 1907, in the SW quadrant of the storm. The hump at 800 km in the curve of observed velocity with no counterpart in the computed curve is of special interest. If the appearance of revolving fluid be due to local convectional impulse the hump would be the surviving effect of a previous impulse in the same part of the fluid.

Fig. 67 (XI, I). Graphs of the distribution of velocity and pressure in the base of a column of revolving fluid, in lat. 60°, according to the formulae $v = \zeta\,(r + A r^{-1})$ and $dp/\rho dr = v^2 r^{-1} + 2\omega v \sin\phi$, where $\zeta = 2\cdot5 \times 10^{-5}$ radians per second and $\zeta A = 2\cdot5 \times 10^{6}$ c, g, s units, compared with the distribution of velocity and pressure in the SW quadrant of the storm of February 20, 1907, at 8 h.

as our cyclones travel we should allow for the effect of the motion of translation. That may perhaps be sufficiently provided for by understanding that the velocity should represent the velocity relative to the tornado centre and not the resultant velocity of the air and bearing in mind the displacement of the centre of circular isobars with reference to the centres of permanent and instantaneous rotation. We cannot assume, even for the purposes of rough approximation, that the effect of the geostrophic term which ought to appear in equation (1) may be ignored in our consideration of the use of the two equations to represent the phenomena of the cyclones of our experience.

We can give a general idea of the distribution of velocity and pressure in a cyclone that has experienced convergence towards the centre by presenting graphs on an arbitrary scale of the several terms of the right-hand side of equation (12) and their resultant and the graph for the resultant distribution of pressure as given by equation (13), modified by the addition of $2\omega v \sin \phi$ to the pressure equation from which equation (13) is derived by direct integration. These graphs are all included in fig. 67.

To provide material for comparison with actual cyclonic conditions we can appeal in the first instance to maps because they present the actual position at a specific epoch. We select the most conspicuous example known to us of a well-formed cyclone of the largest scale which is that of 20 February, 1907, memorable for the gale which wrecked the s.s. *Berlin* off the Hook of Holland. We can regard it as a huge cap of air instantaneously in rotation about a centre very close to the west coast of Norway in latitude 60° N in the North Atlantic not far from the centre of isobars. Taking the figures for pressure and velocity from the observations at exposed stations in the south-western sector of the cyclone from the map for 8 h of the day of the storm[1] we get the curves represented by the thick lines of fig. 67. The resultant graphs obtained from the equations are represented by continuous thin lines: the auxiliary graphs by dotted lines. The similarity of the actual to the theoretical curves is sufficiently well marked to justify the comparison. It must be remembered that the winds are surface-winds and the velocities of the theoretical wind of the free atmosphere have been reduced by subtracting a third for the comparison. We ought also to note that there are in the map some local peculiarities which are practically of the highest importance because among them was the line-squall near the trough-line which caused the wreck referred to. Such local disturbances are not represented in our diagrams, because the south-western sector was generally free from them.

Another mode of comparison with the theoretical results may be based on the autographic records of wind and pressure obtained at the various observatories within the area covered by a depression. This method has the great advantage attaching to continuous records, but it is not quite so appropriate as a perfectly complete map would be because the record which we obtain depends upon the travel of the disturbance over the station and changes in the distribution of velocity and pressure certainly may take place while the

[1] *Weather of the British Coasts*, chap. XI, § 7; *Forecasting Weather*, 2nd edition, 1923, p. 67.

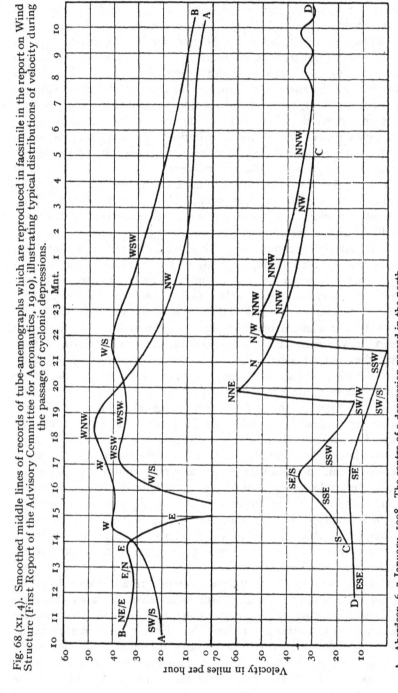

Fig. 68 (xi, 4). Smoothed middle lines of records of tube-anemographs which are reproduced in facsimile in the report on Wind Structure (First Report of the Advisory Committee for Aeronautics, 1910), illustrating typical distributions of velocity during the passage of cyclonic depressions.

A. Aberdeen, 6–7 January, 1908. The centre of a depression passed in the north.
B. Scilly, 9–10 January, 1901. The centre of a depression passed directly over the anemometer from south to north.
C. Holyhead, 10–11 September, 1903. The depression represented in fig. 63 of chap. ix passed from west to east with the centre of winds close to the anemometer.
D. Holyhead, 3 August, 1900. Representing a case of discontinuity of horizontal motion.

record is in progress and probably more or less material changes occur in every case.

Four anemograph records

Taking first the information provided for us by the records of anemographs let us consider the four curves of fig. 68 which give smoothed curves of variation of wind during the passage of depressions on four occasions for which facsimiles of the original records appear in the First Report of the Advisory Committee for Aeronautics[1]. The curves have been obtained by drawing lines along the middle of the ribbon of the original traces.

The first, A, represents the changes in the velocity of the wind as a depression passed Aberdeen on the north. It is characteristic of records of such occurrences. The variation of direction was from SW by S through W to NW, so that the actual centre missed the Observatory; and it should be noted that the velocity increases to a maximum when the wind is from WNW and the centre is nearest. From that point it falls off again. The shape of the curve suggests therefore the simple vortex with velocity proportional to $1/r$ rather than the normal cyclone with velocity proportional to r.

The next example, B, is from Scilly for 9–10 January, 1901, which shows a sudden lull and recovery in the wind. The shape of the curve suggests a normal cyclone with great angular velocity comprised within a space equivalent to an hour-and-a-half on each side of the centre which passed from south to north: beyond those limits the velocity is nearly uniform for a long time before and after the passage of the centre but in opposite directions on either side of it.

The third example, C, shows remarkable changes in wind at Holyhead during the passage of the storm of 10–11 September, 1903, which has already been represented in fig. 63 of chap. IX and which will be within the recollection of all those who were present at the meeting of the British Association at Southport in that year. The sudden change of direction from SW by W to NNE, N and NW shows that there was a centre to the north of the anemometer which passed from west to east but there is no proper symmetry with regard to the centre. At first we supposed that an explanation of this want of symmetry might be found by assuming that there were two whirls both of the order A/r superposed and not quite concentric, and that the smaller one escaped notice in the map of the depression in chap. IX.

And this may remind us that there are endless possibilities for one column of revolving fluid within another because, as we have seen in chap. IX, a revolving column can be carried along with the flow along the isobars. Hence if the primary analysis of the distribution of winds over the northern hemisphere be a rotation from west to east round the pole covering in the more northern regions a rotation from east to west represented by a great anticyclone, then within either the westerly or the easterly current or between them there may be examples of revolving fluid of which the whirl of 20 February, 1907, is a most notable specimen, and within the area of a cyclonic

[1] *Reports and Memoranda*, No. 9, Details of Wind Structure. Figures 11a, 16, 9d and 17.

depression of that type there may be travelling columns of revolving fluid which are identified on the map as small secondaries. Even in those we may have still smaller whirls which at most are represented by embroidery of the barogram.

But the real explanation in this case proved to be quite different. The orientation of the winds on the two sides of the centre of the depression is nearly opposite and it was realised that, as the record of wind is obtained from an anemometer only 32 feet from the ground, the record will be subject to the peculiarities of the exposure; and that, according to the figures given in Table I of chap. III, and the curve of relation of surface-wind to the gradient in fig. 19, the effect of the exposure at Holyhead is very different for different orientations. It seemed desirable therefore to "correct" the readings taken from the record in order to obtain the "free" wind for the several orientations by the application, as a multiplier, of the reciprocal of the corresponding ratio of the surface-wind to the geostrophic wind. The record, C, was therefore treated in this manner and a graph of the free wind of the depression thus obtained using the figures of Table I which were derived from an analysis of eight years of observation. The "corrected" graph is represented in fig. 70 and it will be apparent that the original want of symmetry has disappeared and we obtain a very striking curve representing the distribution of velocity in the depression of 10 September, 1903, which consists of an inner portion or normal cyclone with a large angular velocity, like that in the central portion of the graph B, and an outer portion in which the variation of velocity with distance from the centre is more nearly according to the law Ar^{-1}. The depression was travelling at the time at about 44 miles per hour or 20 metres per second; and the diameter of the central portion which took three hours to pass the instrument may be estimated at about 250 kilometres. On reference to the map on p. 244 we find that the centre of isobars was a considerable distance south of Holyhead and we therefore have further direct evidence of the separation of the dynamic centre from the instantaneous centre in the case chosen as an example in chap. IX. The diameter of the complete revolving system is about 1000 kilometres. The nearly uniform velocity on either side of it in the record represents the velocity of translation corresponding with the isobars of the larger system in which the rotating column moved. They were from west at first but subsequently from north-west. Record C has therefore furnished us with another type of depression different in its distribution of velocity from that of 20 February, 1907.

The result is interesting in another way. The velocity at its lowest does not reach the zero line as it would do at the actual centre of instantaneous rotation of a travelling column. Making a correction for the deviation by friction, the minimum velocity is from 260° (W by S) and previously from 221° (SW). Hence we may conclude that the instantaneous centre at the time of the minimum velocity was a little to the north of the anemometer at Holyhead though the centre of the circle of maximum velocity was just to the south of that station.

It is only fair to say that in Table I of p. 100 there is another line of values for the ratio of the surface-wind to the geostrophic wind at Holyhead which gives a much higher factor of correction for winds in south-easterly quarters. The result of using these corrections instead of those which were employed for fig. 70 is that the SSW wind when corrected reaches the very high figure of 130 miles an hour while the NNE wind on the other side only reaches 100 miles an hour. The want of symmetry instead of disappearing is reversed. The discrepancy is to be regretted but, as we have already said, its existence is not surprising because the determination of the ratio of the surface-wind to the geostrophic wind depends upon a very rough-and-ready mode of procedure. Certainly when the figures were taken out it was never supposed that they would some day be used to correct individual readings of an anemometer of the first class; and yet, the conclusion that there was a symmetrical circulation round an instantaneous centre fits in so well with all the rest of the known circumstances of the occasion that it claims recognition in spite of the crudeness of the observations by which it has been reached.

Let us therefore pursue the study of this occasion a little further. We have already seen in chap. IX that a normal cyclone with vorticity ζ travelling with velocity V is represented by instantaneous rotation round a kinematic centre distant V/ζ from the tornado centre. With a rotating system in which the velocity depends upon the distance from the centre but is not directly proportional to it the result is not so simple, but we can obtain a solution graphically. If we consider a "simple vortex" in which the velocity is proportional to $1/r$, a ring with radius r will have a vorticity proportional to $1/r^2$, and when the system travels with velocity V the particles in that particular ring will have the velocity which corresponds with rotation round a centre at a distance proportional to Vr^2. By finding the centres for a series of rings we can obtain enough indications of the velocity at different points of the resultant field to enable us to draw lines of flow for the combination of the motion of translation and the motion of rotation in the "simple vortex."

Suppose, therefore, that we regard the conditions of the travelling cyclone of 10–11 September, 1903, as represented by a normal cyclone with a radius of 125 kilometres and marginal velocity of 43 metres per second occupying the central portion and surrounded by a simple vortex in which vr is constant and equal to the value indicated in the marginal ring of the normal cyclone, where v is 43 metres per second and r 125 kilometres. The product in c, g, s units is $5\cdot4 \times 10^{10}$. Combining the motion of this rotating system with the translation of 20 metres per second we get the results which are represented by the lines of flow charted in fig. 69 on the scale of the larger map of the *Daily Weather Report*. There is purposely a little want of symmetry in the lines, introduced by the change in the direction of the stream in which the rotating system was drifting. The centres which have been used for constructing the elements out of which the lines of flow are made are indicated on the chart; the figures against the centres give the radii of the original rings.

Fig. 69 (XI, 5). Chart for 18 h (on the scale of the 7 h chart of the *Daily Weather Report*, 1 : 2 × 10⁷) of the lines of flow of air in a stream running at 20 metres per second, carrying with it a normal cyclone which has a diameter of 250 kilometres and vorticity ζ equal to $3 \cdot 4 \times 10^{-4}$ radians per second, and is surrounded by a "simple vortex" in the labile condition as regards stability having the product vr constant and equal to $5 \cdot 4 \times 10^{10}$ c, g, s units. (43 metres per second at 125 kilometres.)

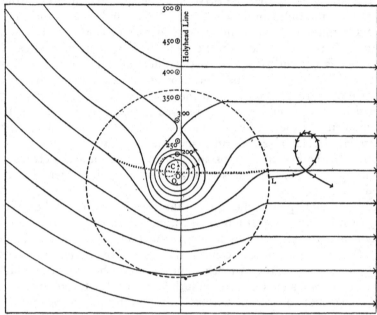

Note. In fig. 69 the circle marked by a chain line has been taken to mark a column of revolving fluid which had a velocity in rotation of ten metres per second. O is the tornado centre. O′ the instantaneous centre of the normal cyclone of 125 kilometres radius. C the centre of a small circle representing the actual circulation within the normal cyclone as recorded by the anemometer at Holyhead on 10 September. The instantaneous centre of rotation for the rings of the simple vortex of different diameters is indicated by points against which the corresponding radius is marked in kilometres. The section of the vortex which passed over the anemometer is indicated by a line of crosses.

Then taking a line across the chart indicating as nearly as can be ascertained the section of the system which passed over the anemometer at Holyhead we obtain the velocity for successive stages as the velocity in rotation round the centre proper for the moment with the proper vorticity and radius. By these means a graph in a thin line has been added to the diagram of fig. 70 representing the velocity at different stages of the passage of the hypothetical system over the anemometer. We have already mentioned that the thick line represents the record of the Holyhead anemometer corrected for the friction at the surface. The reader must remember that the anemometer curve is read from left to right so that the later stages of the proceedings are on the right; but the chart represents what passes over the anemometer from left to right and hence the earlier stages are represented on the right and the two representations must be compared by reading in opposite ways.

As an analysis of meteorological conditions the agreement disclosed by the comparison is remarkable. No serious misrepresentation would result if we regarded the cyclonic depression of 10–11 September as consisting of a central core of normal cyclone surrounded by a simple vortex, with vr equal to $5\cdot4 \times 10^{10}$ c, g, s units.

It will be remembered that for stability vr ought to increase with distance and the actual example shows vr to be apparently just beyond the limit of stability. This is a very interesting circumstance in view of the fact that some day we may learn what the consequence of instability must be in such a case. The normal cyclone is a very stable form of motion and the wavy form of the

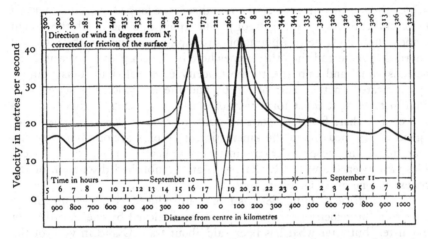

Fig. 70 (XI, 6). Graph (in thin line) of the computed record of the wind during the passage of the hypothetical cyclonic system represented in fig. 69, together with a graph (in thick line) of the record of the wind at Holyhead from 10 September, 5 h, to 11 September, 9 h, 1903, corrected for the influence of the friction of the surface by the data of Table I, chap. III, to give the velocity of the wind in the "free air."

The time-scale is adjusted so that (if the graph were reversed) the velocity at any point of it would correspond with that of a point in fig. 69 at the same distance from the centre on the line of crosses which marks the section of the system represented on the anemogram for Holyhead.

On revising the calculation of fig. 70 it has been noted that in drawing the theoretical curve represented by the thin line it was assumed that the velocity of rotation in the dotted circle of fig. 69 was 1 m/sec instead of 10 m/sec. The result of the error is that the curve as drawn descends more steeply from the maximum than was intended. Comparing the two curves, computed and observed, we note that the velocity in the curve of observation falls off more rapidly with increase of distance from the axis than is regular in a stable vortex— that is not surprising—also that the velocity of the external medium is 20 m/sec in theory but only 16 m/sec as indicated on the anemometer. The 20 arises from a special estimate of the velocity of travel and may be too large.

A striking feature of the diagram is the occurrence of two pairs of humps in the observation curve, similar to the more conspicuous hump of fig. 67—one pair at from 500 to 600 kilometres from the centre and the other at about 900 kilometres. These again are suggestive of the residuary effects of convective impulses previous to the one which caused the double peak of maximum velocity. On examination of the *Tägliche Synoptische Karten* of the Atlantic for the period, in order to ascertain whether the three successive impulses could be discerned, it appeared that there were three successive cyclonic depressions forming what is called a "family," but clearly those were developed, on separate lines, from sinuosities south-west of Greenland in the isobars of the original large depression—a most interesting feature of the charts. They could not be interpreted as successive impulses within a coherent air-mass. The successive impulses recorded on the anemometer must rather be compared to successive puffs of steam from a steam-engine with the same atmospheric environment.

graph of velocity is very suggestive of successive rings of stable motion with intermediate rings of less stability. It would of course be absurd to lay great stress upon the details of the shape of a graph arrived at by the method described, but the shape and its symmetry with regard to the centre are too attractive to be passed without remark.

Perhaps we may regard the fast travelling cyclones, at least, as resembling more or less clearly the structure indicated by a simple vortex surrounding a normal cyclone, and gradually degenerating for lack of stability. That hypothesis explains another feature of the cyclone of 10–11 September which has for a long time seemed rather mysterious; that is the looped trajectory marked L on the chart of trajectories, fig. 64 of p. 244 It appears that, with the system described, there must be two points of zero motion, one within the region of the normal cyclone and the other somewhere in the simple vortex where the velocity is equal and opposite to the velocity of translation. This point is clearly indicated on the chart of fig. 69 and the air which has a westward motion just below it must form part of a trajectory such as that which is drawn on the chart from the starting point marked L. Looped trajectories are generally formed round the kinematic centre of a cyclone but they can also be formed in the peculiar circumstances indicated in fig. 69. A few examples have been found by R. Corless in an unpublished discussion of the complex cyclone of 23 October, 1909.

The fourth trace, D of fig. 68, was merely introduced as an example of the discontinuity of velocity in the atmosphere as disclosed by the record of an anemometer, but after what has been said about trace C and in view of the general similitude of the two curves it seems possible that the discontinuity was more apparent than real. It should be noted that the time-scale for the trace differs from that of the other traces by twelve hours. The sudden increase of velocity occurred at 10 h on 3 August, 1900. A better example of discontinuity in velocity is shown in the record at Pendennis Castle for 11 August, 1903, which is reproduced in the *Life-history*. It would clearly be desirable to examine the effect of treating the other records of fig. 68 in the same way as those for 10 September. A cursory inspection shows that the irregularity of the two maxima of curve A for Aberdeen would also disappear.

The distribution of pressure

To the distribution of velocity represented in fig. 69 may be fitted a distribution of pressure because, as we have seen in chap. IX, the field of pressure appropriate for a travelling rotating system is obtained by the combination of the field appropriate to the rotation with that appropriate to the translation. In the case of the normal cyclone it has been shown that the combination is a system of circular isobars identical with that appropriate to the rotational component but centred at a new point. The same mode of procedure enables us to deal with the central region of the scheme of fig. 69, within 125 kilometres of the centre. But in the region of the outer "vortex," where each ring of air has a different angular velocity, a whole series of new centres will be required, not merely a single one; and the equation to the family of curves representing the isobars becomes complicated. The resultant distribution of pressure can, however, be obtained graphically. The component field of pressure in the region of the vortex appropriate to the rotation can be obtained by the integration of the gradient equation of p. 87 in the form

$$dp/dr = \rho v \,(2\omega \sin \phi + v/r)$$

by substituting A/r for the velocity in rotation, v, where A, in c, g, s units, is $5 \cdot 4 \times 10^{10}$. The other component field, appropriate to the translation, is given by the geostrophic equation (G)

$$dp'/dy = -\,2\omega\rho V \sin \phi,$$

where for V is substituted the velocity of translation, which in c, g, s units is 2×10^3. Plotting these two component fields upon squared paper and combining them we obtain a chart of isobars representing the resultant distribution of pressure which is quite similar in its general features to the chart of velocity as represented in fig. 69. But there are differences in detail which are specially noteworthy. There are two points of zero gradient where the gradient for the motion of rotation is equal and opposite to that for the motion of translation, just as there are two points of no velocity; but the points of zero gradient do not coincide with the points of no velocity. In the "normal cyclone" the point of zero gradient will be nearer to the centre than the point of no velocity; and in the "vortex" it will be further away. The separation of the two points of zero gradient in the resultant field of pressure will therefore be greater than the separation of the two points of no velocity, and the isobars must be adjusted to represent the situation. This state of things arises from the fact that the equation for the gradient of pressure in the field for rotation includes a term v^2/r which does not appear in the geostrophic equation (G) for the field of translation. In other words, motion along a curved path balances a greater gradient than motion with equal velocity along a straight path.

The positions of the points of zero gradient will be on the y-axis as selected for the representation of the translation, and can be computed by writing y for r in the gradient equation and equating the right-hand sides of the two equations. It is evident that the two gradients are equal and opposite when the velocity of rotation v is less than the velocity of translation V. Thus, where the gradients are equal and opposite there will be a residual velocity in the direction of translation and, on the other hand, where the velocities are equal and opposite there will be a residual gradient of pressure in favour of the rotational component. Since the smaller velocity is to be found nearer the centre in the normal cyclone and farther from the centre in the simple vortex the points of zero gradient will be nearer to the centre in the one case, and farther from the centre in the other case, than the two points of zero velocity respectively. The isobars which represent the distribution must be elongated to provide for the increased separation.

The equation for the determination of the point of zero gradient in the region of the vortex is a cubic which has a real root giving a distance of approximately 370 kilometres from the centre of the revolving fluid. This solution agrees well, as of course it should do, with a chart constructed to represent the distribution of pressure. The displacement of the centre of isobars for the interior normal cyclone, for which ζ is equal to $3\cdot4 \times 10^{-4}$ radians per second, in a current with a flow of 2×10^3 cm/sec, as computed by the formula of p. 241, is 15 kilometres. The calculated distance between the two points of zero gradient is therefore 355 kilometres. The separation of the two points of zero velocity shown in fig. 69 is 250 kilometres. Hence we see that the point of zero gradient in the region of the vortex may be a considerable distance from the point of zero velocity.

The conclusion is very well illustrated by the actual map of the distribution of pressure on the occasion of the cyclone of 10–11 September, 1903, which is given in fig. 63 of p. 244. We have there the two points of zero gradient indicated, one very definitely by the small circle near to Holyhead and the other very vaguely by the bending of the isobars in the neighbourhood of Aberdeen. Making an estimate of the position of the second, the distance apart measures about 500 kilometres which is again considerably greater than that obtained from calculation, but in this connexion we ought to bear in mind the recrudescence of velocity in the actual cyclone of 10–11 September, 1903, shown in the record of velocity represented in fig. 70, at 600 kilometres from the centre on the eastern side and at 500 kilometres on the western side. We have already explained that too much stress must not be laid upon these irregularities in the curve; yet it is curious, if nothing more, that a ring of enhanced velocity at 500 kilometres distance, with the pressure-gradient which would accompany it, would be extraordinarily useful in bringing the theoretical map into accord with the observed phenomena.

Indeed, the two external rings of the depression of 1903 indicated by the humps in the corrected anemometer record referred to in the note on fig. 70 taken in conjunction with the more pronounced hump in the velocity curve for 20 Feb. 1907 (p. 262) which was derived from wind-observations at exposed stations furnish the most striking evidence that we have at our disposal for the reality of revolving fluid as the characteristic of the structure of some of the cyclonic depressions of our area, and also of the probable influence of convection, presumably of saturated air, in developing the energy of the local circulation. We have compared the successive impulses which produced the outer rings and the final core of the depression of 1903 to successive puffs from the exhaust of a steam-engine, but the scales of the two phenomena are very different. The puff of a locomotive is a matter affecting at most a few kilogrammes of air; the amount of air removed to form the depression of 10–11 September would be of the order of half a billion tons and that removal was accomplished during the transformation in the course of one or two days of a slight sinuosity of the isobars south-west of Greenland into a well-developed depression over the north-west of England. Apparently the transformation was achieved by three successive impulses of convection as recorded by the three recuperative rings of velocity in the trace of the anemometer.

Comparison of theory and observation·

And now, having obtained, in a very unexpected manner, further insight into what the structure of an actual travelling cyclone really is we may profitably turn back to the representation of the theoretical normal cyclone as set out in fig. 62 of p. 242. In describing the map we explained that the theoretical cyclone therein represented ended in a ring of maximum velocity which had to be adjusted to its environment and we had no information as to what the nature of the adjustment was. We left it as a discontinuity of velocity which had to be accommodated and now we see that the accommodation is arrived at by including in the column of revolving fluid an outer region which is approximately represented by the law of the simple vortex with vr constant. It follows that the area of the column of revolving fluid extends far beyond the boundary which was drawn in figs. 62 and 63 to mark the limit of the normal cyclone and includes on the northern side regions of zero velocity and of zero gradient of pressure which in themselves are not at all suggestive of continuous motion in rotation. In order to include the whole rotating mass the boundary circle in the map of fig. 63, chap. IX, ought to have extended northward beyond the Scottish mainland and in that case would have taken in also the curved isobar which crosses central France. It would follow that the whole mass of the revolving column occupied the outer margin of the general system of isobars running from west to east across the map. When we think of the travel of a cyclone we must include those exterior regions which hitherto have seemed to be only part of the environment.

On the other hand, as a suggestion for the boundary of the normal cyclone which formed the central portion of the complete revolving system in the particular case of 10–11 September, 1903, the dotted circle of fig. 63, chap. IX, is much too large. Its radius is 380 kilometres, whereas, according to the record of wind at Holyhead, the normal cyclone extended only to 125 kilometres from the centre. The limiting circle, as drawn in fig. 63, was taken from the theoretical map of fig. 62 (chap. IX), and we may now see the explanation of some of the differences between the two maps. There is a gradually increasing compression of the consecutive isobars in the outer region on the southern side of the theoretical cyclone which is not borne out in the comparison with the actual map. The changes in wind-velocity with distance from the centre stand likewise in need of adjustment. When the theoretical map was drawn it was thought (p. 245) that too large a figure had been taken for the vorticity, but now it appears that for this particular occasion the opposite was the case; the vorticity for the central region was in reality three-and-a-half times the assumed figure; but the radius over which it remained constant was only one-third of that represented. The ring of maximum velocity was much closer to the centre; the immediate environment of the normal cyclone formed part of the revolving system and was a much more extensive and important part of it than was then supposed.

We have noted some curiosities in the relation of wind to the distribution of pressure along the axis of y, which in this case is drawn northward from the centre of the revolving fluid; there will be peculiarities of another kind in other parts of the area, which we have not yet traced and which show that even in the free air, and when the balance of pressure and wind is completely adjusted, the relation of the wind to the distribution of pressure for curved isobars is by no means a direct connexion between the lines of run and separation of the isobars and the direction and velocity of the wind in the same locality. Many more examples must be analysed before the complicated relationships of pressure to wind in the case of revolving fluid can be regarded as established, but if our diagnosis of the conditions of the cyclone of 10–11 September, 1903, be correct we may derive some satisfaction for the discomfort which it caused to the International Meteorological Committee, at the Meeting of the British Association at Southport in that year, from the consideration that it was the first to yield a complete analysis of the true internal structure of a travelling cyclonic depression.

The evidence of the barogram

Turning to the records of pressure for evidence of the existence of revolving fluid, barograms with very steep fall and recovery of pressure such as those reproduced in chap. VIII of vol. II have been held to support the idea as representing the structure of tropical revolving storms and tornadoes.

Records of pressure of travelling depressions of more northern latitudes are familiar to meteorologists but the analysis by inspection of their evidence in illustration of revolving fluid in the atmosphere is not an easy matter. We may first call attention to a peculiarity that is noticeable in many of them in connexion with the passage of the trough and may be attributed to the effect of the juxtaposition of bodies of air of different temperatures in the lower layers which destroys the symmetry of the distribution of pressure with regard to the centre. The motion of these heterogeneous masses of air is the chief line of approach to forecasting on the Norwegian system. Its relation to the distribution of pressure above them need not come very definitely into consideration of forecasts but it is interesting from the point of view of the dynamics of the atmosphere. The inherent difficulty about surface-temperature in respect of the dynamics of the atmosphere depends upon its unilateral behaviour. In the free air all changes of temperature can find their adjustment in a mutually reciprocal manner. Cold air may go downward, warm air upward, but at the surface the air which is cooled must remain and in consequence the air of the lowest layers beneath a revolving column may show the combination of portions of markedly different temperature. Where these different bodies of air are in juxtaposition line-squalls occur and the symmetry of the curve representing the variation of the barometer is destroyed. In illustration of this interference with the symmetry characteristic of revolving fluid we may show the differences noticeable in six barograms representing the same cyclonic depression (fig. 71). On this account, ordinary barographic records do not generally help us in our study of the relation of the phenomena of revolving fluid to those of cyclonic depressions in our latitudes.

With regard to these curves we must explain that a mass of fluid in solid rotation passing at uniform speed over a barograph would leave as its record a trace in the shape of a parabola. The trace of a simple vortex surrounding the solid core would be a complicated curve of hyperbolic type. Thus the outer portions of the records displayed fit well enough with an outer simple vortex; but just about where we should expect the parabola to begin there is a notable revulsion followed in at least three of the diagrams by a second and more successful effort to impress the barograph with a parabolic minimum. It is natural to attribute this sudden interruption of the fall to some irruption of cold air.

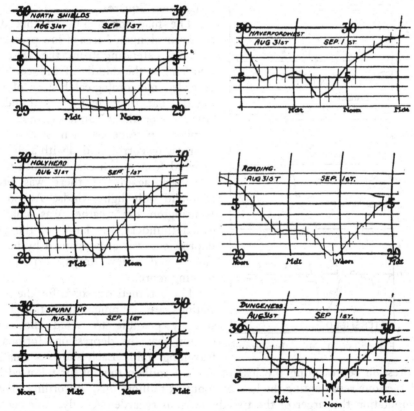

Fig. 71 (XI, 2). Graphs of pressure showing interference of the surface layers with the symmetry of revolving fluid, 1908.

Occasionally however at different stations we have obtained curves representing variations of pressure which may be attributed to revolving fluid. As an example we may cite the simultaneous records, represented reduced to half scale in fig. 72, from four micro-barographs which magnify the pressure-variation twenty fold. They are compiled from the records of 12 July, 1908, and are reproduced in the Report of the British Association for that year[1]. The most conspicuous of these small depressions shows a well-marked minimum about 16 h 20 m at Petersfield; at Reading (Leighton Park) in a more flattened curve about 17 h; at South Kensington a sharp minimum at the same time and at Cambridge less pronounced but equally sharp at 18 h 10 m. The times are not easily read because the ordinates of the diagrams are curved and the timing of the clocks is not above suspicion. The photographic records at Kew show successive depressions of $1\frac{1}{2}$ mb and 3 mb. The observations are supported by a record at Camden Square[2].

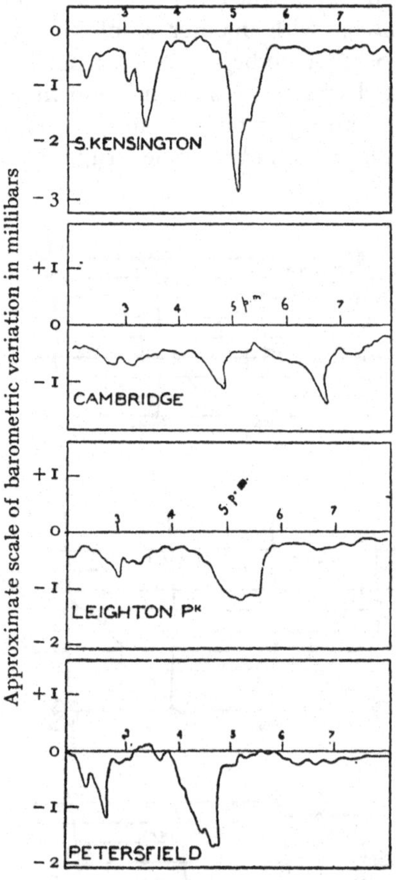

The map for 6 p.m. of that day shows an elongated area of low pressure, 29·6 in, extending from the west of Scotland to the middle of France. The recorded meteorological history of the day is that there had been thunderstorms over the Channel islands. At Kew the passage of each of the two depressions corresponded with a change of wind from north-east to south-west and back again but there was little rain and only light winds. At South Kensington the weather was what is called thundery, the larger of the two depressions was accompanied by heavy clouds and a few drops of rain but nothing more.

The barometric records for the four stations are so much alike that it is scarcely possible to avoid the conclusion that the depressions formed a travelling train. Each shows a smaller depression representing the loss of about 8 million tons of air followed by a larger one representing the loss of about 25 million tons. According to our computation the larger of the two depressions regarded as a dynamic centre might indicate a tornado centre one kilometre to the east with a velocity of spin of 15 m/sec at a radius of 10 km. On that hypothesis the train travelled from Petersfield over South Kensington to Cambridge in about two hours at about 50 miles an hour, but was so long that its head was about seven miles beyond Cambridge station when its tail was leaving Petersfield.

Fig. 72 (XI, 3). Micro-barograms.

[1] British Association Report, 1908, p. 608.
[2] Symons's Meteorological Magazine, vol. XLIII, 1908, p. 135.

At Epsom, on the line, there was a sudden drop of the barometer, about a tenth of an inch (3 mb) at 5 p.m., and likewise at New Malden; and to the east of that belt there was heavy rain.

The air-current of 50 miles per hour that carried the train from Petersfield to Cambridge was not indicated at the surface at any of the stations, nor was there any notable wind while the two depressions were passing. The current must have been in the upper air carrying the local spins in the same way that a travelling cloud carries water-spouts.

We have taken the circular form as typical of the stable condition of rotation but apparently there may be other forms too. The isobars of permanent cyclones are of very diversified shapes and are subject to continual change. The elliptic shape is, however, often persistent and possibly an elliptic form of line of flow is a stable form. A noteworthy example occurred on the map for 7 h on 29 August, 1917, in which the major axes of successive isobars were just twice the minor axes and the distribution of wind showed proportionality to the distance of the isobar from the centre.

The reader will gather from the fragmentary nature of the discussion of the application of equations of revolving fluid to the phenomena of cyclones that the subject is as yet almost unexplored. So far as we have gone, it would appear that the distribution of velocity in the ordinary cyclones of our maps suggests the "simple vortex" with velocity A/r for the outer margin of a cyclone with velocity ζr and the examples of the fast travelling storms of the *Life-history* were not exceptional in that respect. The cyclones of our maps show local distortion of the isobars in the form of secondaries or line-squalls and we have yet to learn how these disturbances become incorporated in the larger general circulation and what is their ultimate effect upon that circulation. A cyclone like that of 20 February, 1907, certainly represents the instantaneous motion of revolving rings about 2600 miles in diameter, or 40° of latitude and a mass of several billion tons; and such a mass must have a good deal of stability. At the surface the variations of temperature are very considerable and cause local phenomena of various kinds, but up above we may suppose the distributions of temperature to become as symmetrical as those of pressure or wind.

It has long been supposed that the variations of temperature at the surface are themselves the cause of the original circulation of the cyclone, but it is much more easy to explain convection along the core as the effect of an existing circulation above than *vice versa*, and there are so many examples of convection attended even by copious rainfall which produces no visible circulation that it is difficult to regard convection from the surface as a sufficient cause of our numerous depressions. A useful example may be given from the remarkably wet period of 27 July to 5 August, 1917, of the influence produced by rainfall[1]. At the beginning of the period very heavy rainfall occurred in a region of strong winds near Glasgow without any apparent effect on the circulation. But after four or five days of rain about the Straits of Dover, in

[1] For another example see *Daily Weather Reports* for 3–10 January, 1919.

a persistently quiet environment, a moderate cyclonic depression did appear near that area and it disappeared again after a few days' existence. As the equivalent of the energy set free by the condensation of water-vapour it was very inadequate and we cannot suppose that there was any proportionality between them. Some other conditions than simple convection from the surface are necessary for the development of our cyclonic storms and when they are developed other conditions than convection at the warmest spot provide for their maintenance.

Thus it would appear that the trace of an anemometer record after correction for surface disturbance in certain notable cases leads us to the identification of a mass of revolving fluid in the atmosphere carried along in a current belonging to a more general circulation. The identification is however only an occasional pleasure.

Horizontal motion and vertical disturbance

Looking back upon the chapter now completed and its predecessor the reader can hardly fail to note that all the influences which are cited as affecting the distribution of velocity in the atmosphere are related to motion in the vertical, and all the movements of the air cited from anemograms or introduced by symbols into the algebraical equations are horizontal.

In fact the revolving fluid of these chapters seem to have been regarded as a column, reaching up to the stratosphere, of which any horizontal section is a layer of fluid to which the reasoning of horizontal motion can be applied. Probably there are examples to be found which would justify the assumption because the distributions of pressure may on occasions extend from the ground to the stratosphere, but the process of formation of a column of revolving fluid of that vertical extent has never yet been deciphered by observation.

The distribution of pressure at sea-level is the integral of all the distributions above it, but to assume that the motion in each successive layer is horizontal while vertical motion is cited as its ultimate cause is hardly reasonable. Upward motion is implied both by penetrative convection which may take place in spiral curves round the solid core and by motion along those isentropic surfaces which are inclined to the horizontal. Both require satisfactory arrangements for the delivery of the air when it has reached the limit of its convection. The arrangement sketched in the model of fig. 220 of vol. II may represent such an arrangement which is the more probable when we consider that in the frontispiece to *The Life-history of Surface Air-currents* the surface air which forms the leaving portions of the loops of the trajectories as represented theoretically in fig. 65 is of different character from that of the entering portions; it has in particular a lower temperature.

In that case the horizontal winds on the map for sea-level are not a complete representation of the behaviour of surface-air. The pressure lines are the resultant effect of the distribution at all levels, but at some level in the upper air there may be a horizontal section of limited thickness which corresponds approximately with the behaviour of a horizontal disc of revolving fluid.

The fact that it has been necessary to select the cases warns us that we cannot analyse all records into recognisable flow and spin, and indeed our analysis does not extend beyond the diagnosis of travelling spin in a single layer of the atmosphere at about the level of 500 metres. In chapters III to V we have made some approximation to the filling of the space between the plane of the spin and the ground, we have still left for future exploration the relation of a travelling spin to the free atmosphere above it.

The work which this chapter represents was prompted by Lord Rayleigh's opinion of the importance for the science of meteorology of the theory of revolving fluid. In view of the striking effects of tropical hurricanes and sub-tropical tornadoes the opinion is natural enough; but so far as the meteorology of middle latitudes is concerned it may be said about revolving fluid that the inverse proportion of its importance to its linear dimension, the difficulty of detecting it in the flow that carries it, the disturbance of the lowest layers by turbulence, and the modern assurance that when it seems to be recognisable it is dying, make one hesitate to acknowledge the importance that it used to have. It is not easy to point to any part of modern practice that turns upon the theory of revolving fluid and it is not unfair to say that at the present day the kinematics of the surface-air is of more importance than the dynamics of the air above—trajectories are more regarded than lines of flow.

The outstanding question of the cyclonic depression, whether it be represented by revolving fluid or not, is the relation of convection to its development, the part of the subject which is noticeable in our experiments but not in our theory. Whatever the theory may be the disposal of hundreds of thousands of millions of tons of air removed from a limited area of the surface to make a depression is a practical reality and a serious matter for meteorologists. To this question we must give some attention before taking leave of our reader.

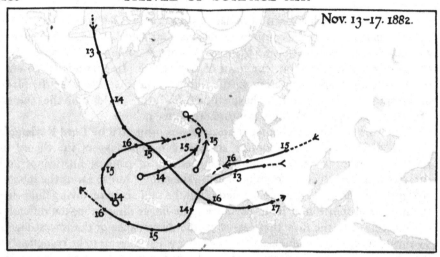

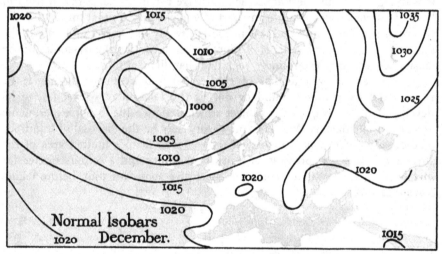

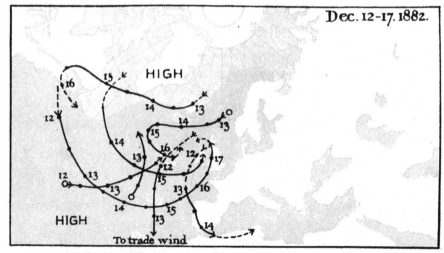

Fig. 73. Trajectories of air, November and December, 1882.
From *The Life-history of Surface Air-currents* (H.M.S.O.).

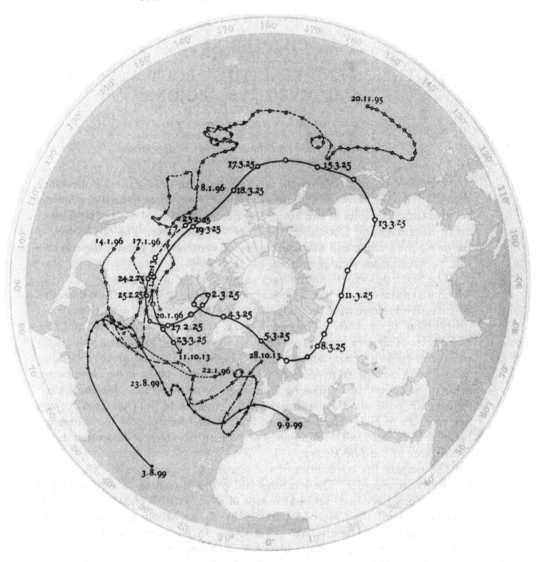

Fig. 74. Paths of cyclonic depressions with long life-histories.

1. From the Western Pacific on 20.11.95 to middle North Atlantic on 22.1.96, with duplication in the Pacific after 8.1.96 and redistribution on 14.1.96 at the Rocky Mountain barrier (McAdie).

A similar example is given by H. Harries, *Q. J. Roy. Meteor. Soc.* vol. XII, 1886, p. 10.

2. From the Great Lakes 1.10.13 by an irregular path to the Faroë on 28.10.13 (McAdie).

3. From equatorial Atlantic, long. 35° W, on 3.8.99 to W. Indies, long. 67° W, on 8.8.99 and by an ordinary path of hurricanes and depressions to the Mediterranean on 9.9.99 (M.O.).

4. From Vancouver Island 23.2.25 and Canadian border 24.2.25 round the world to Strait of Belle Isle on 23.3.25 (Mitchell).

Each of the four paths has its line specially marked and dated at intervals.

CHAPTER XI

HYPOTHESES AND REALITIES ABOUT ADVECTIVE AND DIVECTIVE REGIONS

Let the great world spin for ever down the ringing grooves of change.

The motion of air and the movement of depressions

OUR discussion of the dynamics of the atmosphere has led us to the opinion that for meteorological practice the study of trajectories is of more importance than the dynamics of revolving fluid. By way of frontispiece to further progress we give some examples of two kinds of trajectories, namely the trajectories of air in November and December, 1882 (fig. 73), taken from *The Life-history of Surface Air-currents*, and the trajectories of centres of cyclonic depressions (fig. 74) taken from various sources.

The first takes our thoughts back to the concluding chapter of vol. III in which we endeavoured to get a general notion of the flow of air from divective to advective regions with the idea of tracing its connexion with the normal distribution of pressure. The result was not exactly satisfying and the trajectories of air give some suggestions of the reason. The pressure-distributions are a guide to the instantaneous flow but not to the long treks. Speaking generally for northward moving air (equatorial air) centres of depressions are goals to be arrived at by passing from high pressure to low; for southward moving air (polar air) they aré marks to be passed and the transition is from low pressure to high. Here is one of the notable examples of the difference between $\partial p/\partial t$ and Dp/Dt for a parcel of moving air which is illustrated in the frontispiece of the *Life-history*.

The map of the trajectories of the centres of depressions is dynamically perhaps more complex. The localities and paths of centres of depressions in the year 1923 have been set out in maps of chap. VIII of vol. II, and those for January to April 1925 more elaborately by C. L. Mitchell in the *Monthly Weather Review* for January 1930; clearly they are characteristic of the region of conflict between equatorial and polar air. Years ago they would certainly have been regarded as suggesting the travel of an identifiable mass of revolving fluid. Now we have to think of the generally accepted view, already mentioned, that when the identity is most nearly evident the local distribution is dying out. And yet there is a sort of continuity of motion, if not of centres then of the general system in which a fresh centre is formed while an old centre is on its death-bed. The new centre must owe its development to a convectional impulse similar to that which caused its predecessors but in a different locality. So we may regard the travel which is denied to an individual centre as belonging to the convectional impulse which has a depression centre for its objective. That might be a general description of the sequence indicated by J. Bjerknes's diagram[1] of the alternate invasions of the cyclone belt by equatorial and polar air.

If the conflict between polar and equatorial air be indeed the expression of the reaction between the underworld and the overworld as we have suggested in chap. VIII of vol. III, then the travel which carries the centres of depressions may be that of some deformation of the isentropic surface which separates the two worlds.

Put succinctly we shall endeavour to show that, personifying the acknowledged difference, the aim of the equatorial air is by its motion to fill up or destroy depressions, that water-vapour intervenes and not merely frustrates that aim but actually secures the opposite—it is water-vapour that "rides in the whirlwind and directs the storm"; on the other hand, the aim of polar air is to create or enhance anticyclones, and so far as we can tell it succeeds, unless it loses speed.

ADVECTIVE REGIONS

Meteorological tradition—a change of front

When the original text of this volume was written the travel of cyclones and cyclonic depressions, as illustrated by fig. 74, was the aspect of the general problem of weather most present to the mind; cyclones or tropical revolving storms were accepted as revolving fluid travelling as vortices with vertical axes. There was continuity between them and the cyclonic depressions of middle latitudes by the travel of the vortices round the western extremity of the anticyclonic areas of the tropical seas.

The aim of dynamic meteorology at that time might have been stated as follows: to determine (1) the dynamical structure of the travelling entity of a cyclonic depression, (2) the physical causes of its maintenance or decay, (3) the laws of travel by which the direction and velocity of its motion could be anticipated.

The ambition of the meteorologists at the same time was to generalise on these principles the situation presented by the daily map and thus announce beforehand the motion of the cyclonic depressions across the map. Those responsible for the subject in Britain were still cognisant of the effort of *The Times* and *The New York Herald* to forecast the arrival of storms crossing the Atlantic in about four days. By extension of the area of observation the period of travel might be extended so that with a map of the northern hemisphere it might be possible to announce the behaviour of depressions for more than a day, perhaps for a week ahead. With that object telegraphic reports from Iceland and wireless reports from the Atlantic had been obtained. The investigation of the upper air was actively pursued with the co-operation of a number of volunteer workers in different parts of the country so that a representation in three dimensions seemed possible. We looked forward with some confidence to a time when a single efficient map of the northern hemi-

[1] J. Bjerknes and H. Solberg, *Geofysiske Publikationer*, vol. III, No. 1, Kristiania, 1922, p. 15.

sphere for one hour of the day might replace the three maps for 7 h, 13 h and 18 h of a limited area. Weather-maps were classified into types which might indicate a definite sequence[1]. The precedent of the tides, *longo intervallo* perhaps, might be followed. A general forecast for the whole region for a week might be issued to be punctuated locally from day to day.

It must be remembered that before broadcasting made the anticipations of the weather twelve hours ahead a matter of personal interest to millions, forecasting could have little practical application. Gale-warnings were transmitted by telegraph to some hundreds of sea-ports and notified by signal-cones, but they dealt only with the conditions of wind and naturally tended to lay stress on the distribution of pressure. Very few persons were aware of the prospect of rain, frost or fog before the event. The success or failure of each warning and forecast was however carefully scrutinised and in practice forecasting was the touchstone by which a meteorologist could test his own knowledge of the laws of weather, and an acknowledgment of a public demand, rather than the satisfaction of official and public need.

The difficulty that we sought to overcome was that we did not know exactly what travelled, nor what laws it followed in its travel. We knew that a spinning mass of air, provided it was suitably ended at bottom and top, might travel as a separate entity for a time and produce effects distinct from those of the current which presumably carried it, and we knew that some kind of identity could be traced for thousands of miles; but we did not know exactly how it might originate nor what process there might be for it to take advantage of local conditions to nourish itself during the journey, nor its relation to its environment. No other form of structure than revolving air was suggested as adequate for the task, but the records left by passing depressions were not at all conclusive for the hypothesis of rotation like a solid.

Consequently the demonstration that the irregularities of the movement of the surface-air indicated on an anemogram might be consistent with the existence of a layer of revolving fluid in the upper air at perhaps 500 metres was regarded as a definite step in the progress of our investigation.

The kinematics of air-masses

But now the attitude of meteorologists towards dynamics has changed. A modern meteorologist would probably claim (and we should agree) that the analysis conclusively disproved the existence of regular spin in the surface layer represented on the weather-map, and he might go on to say that revolving fluid in the upper air hangs on the thin hypothetical thread of motion under balanced forces; consequently the case for the travel of revolving fluid should be regarded as having broken down. What we tried to synthesise into a single living entity the modern view, led by the Norwegian school, analyses into air-masses labelled polar air, tropical or equatorial air, maritime

---

[1] W. J. van Bebber, *Lehrbuch der Meteorologie*, Stuttgart, 1890, chap. IX; E. Gold, "Aids to Forecasting: types of pressure distribution with notes, etc " *M.O. Geophysical Memoirs*, No. 16, 1920.

air or continental air as in fig. 135 of vol. III; the air-masses being separated by discontinuities into fronts, the phenomena of weather are the mutual action and reaction of the fronts. Air-masses become entities and if they derive their right to existence from any spin, it may be perhaps as a consequence of the earth's spin, but it is not a local spin.

So the cyclonic depression and the anticyclone become merely conventional names to indicate what in vol. III we have called advective and divective regions. Under those conventional names they are studied in minute detail with corresponding advantage to the forecasts.

The data which now go to make a day's maps are nearly four times as numerous as those which we thought it necessary to collect—7500 in 1930 instead of 2000 in 1914. Instead of reducing the triad of maps to one a day, a fourth map, for 1 h, has been added, and the forecaster has supplementary information for 10 h and 16 h. In fact the central offices are now doing what we thought would be natural local work when we had discovered the general laws of weather and announced the conclusions drawn from them.

The change of view originating in the exigencies of war came at a time when forecasters had acquired the advantage of wireless for the transmission of data and the broadcasting of inferences, and aircraft required details of weather which a close analysis of the behaviour of depressions could give. The general laws of atmospheric structure and the travel of depressions were of less practical importance than the accurate anticipation for a few hours ahead of wind, weather and visibility.

Discontinuities and "fronts" have become the working entities of the modern forecaster in place of revolving fluid. As soon as the stage of revolving fluid is arrived at occlusion of the equatorial air or warm front has occurred and death follows. "Now depressions when they reach the British Isles are nearly always occluded." A general account of the new method by E. Gold has been given in vol. II. The technical details do not come within the scope of this work; they have been set out officially for English readers in a new version of *The Weather Map*[1] by J. S. Dines.

We do not regard it as part of our present duty to discuss methods of forecasting; the finer details are far too much a personal matter for the forecaster who has the responsibility of translating to-day's map into to-morrow's weather. He may go upon certain recognised general principles but the judgment which he uses in wording an announcement to the wide world must be his own. A forecaster's heart knoweth its own bitterness, and a stranger intermeddleth not with its joy.

But the revised scheme of forecasting has moved the whole meteorological world which thinks of fronts whereas the older forecasters used to think of whole depressions. The detailed character of the forecasts makes them peculiarly acceptable to the spirit of this tachistocratic age, and the reader may expect some information about the principles upon which the revised method of forecasting relies.

[1] M.O. Publication, No. 225 i, 1930.

We must refer to the diagrams of vol. II, p. 387. "The normal birth, life and death of a cyclone are illustrated in fig. 3. Fig. 3 (*a*) shows a cold easterly current and a warm westerly current in juxtaposition. The cold air begins to bulge southwards, and the warm air northwards in (*b*). In (*c*) a cyclone has formed, and (*d*), (*e*), (*f*) show successive stages, until the whole of the warm air has been lifted above the cold air. The cyclone gradually dissipates, as indicated in (*g*) and (*h*)."

These proceedings are the result of the action and reaction of polar air, represented in the first diagram as forming a cold, dry east wind, and equatorial air represented as a relatively warm, moist west wind in juxtaposition. The opposite winds are separated by a straight line called the polar front which it is supposed can be identified upon investigation somewhere on every meridian of the hemisphere.

In the succeeding diagrams the polar front continues as a line of separation between polar and equatorial air, though it is bulged northwards by the movement of the warm air northward—an advance of the warm front—or by the cold air on the western side southward, turning east—the advance of the cold front. The portion of the cyclonic area covered by equatorial air bulging to the centre and nearly surrounded by polar air is called the warm sector.

The cold front is the dominant element of the partnership; it invades the warm front from the north and turning eastward continues its invasion until it has reached and overlapped the eastern portion of the polar front with the easterly wind north of it (somewhat diverted from its normal direction) and thereby isolates the portion of the equatorial air nearest the centre of the cyclone. It thus secures the "occlusion" of the cyclone. Access of equatorial air along the surface being denied, the cyclone perishes.

We may note that if a billion ton depression with a radius of 600 km were surrounded by dry air with an inward component of velocity of 5 m/sec we should expect it to be levelled in about 3 days. But here we must remark that the effective power of occlusion to annihilate depressions depends upon the surrounding air being dry. If part of it happened to be saturated by passing over warm water or being subject to some other influence to produce the same effect, the most intrepid forecaster would hardly be able to say offhand what would happen.

Meanwhile the cold front on the west "undercuts" the western margin of the warm sector, and with a wedge-shaped push causes comparatively sudden lifting of the face of the warm front, with all the consequences indicated in the trough of a V-shaped depression or a line-squall.

That line of contact with cold air invading warm air used to be called the squall-line of the trough of a depression but is now a principal cold front.

Pursuing these general principles in the study of the details of maps as of p. 401 of vol. III, we get minor differentiations, principal and attenuated warm fronts, principal and secondary cold fronts succeeding one another at irregular intervals in the manner suggested by the figure of breaking atmospheric waves on p. 27 of vol. III.

And besides occlusion we have areas of subsidence where air descends and displays exceptional dryness, and indeed in a northern climate we can hardly conceive of really dry air except as drawn directly from the upper regions where the maximum vapour-pressure is at most a small fraction of what is possible with surface-temperatures. So very dry air anywhere implies downward motion; but if our conception of motion along an isentropic surface from low pressure to high pressure under the influence of increased velocity is correct downward motion need not imply subsidence.

Behind each front is an air-mass the provenance of which can be traced and the mass identified as equatorial air from the region north-west of the permanent Atlantic anticyclone forming warm fronts, or polar air forming cold fronts coming from the American continent or from somewhere in the north-western or northern Atlantic and bending eastward. It may be modified by eastward travel over Atlantic water and become maritime air acting as a deputy warm front.

The cold side of the polar front may also be supplied by air from north-eastern or eastern Europe, and perhaps ultimately from Asia; but these supplies of cold air from the east are less noticeable in north-western Europe than those from the west, except on such occasions as the long frost of February 1929.

This description of the behaviour of fronts is amply supported by physical explanations of the weather incidental to the advances of the fronts. The general associations are represented in the diagram of fig. 2 of p. 386 of vol. II, which shows the air of the west-south-westerly wind of the warm front ascending over the cold air lying across the warm front at the surface as shown in the model of p. 389 of vol. II. In its progress it produces first detached clouds, then nimbus, then alto-stratus, then cirro-stratus, and finally cirrus: the sign to wise men in the east of an approaching depression.

The direction of the isobars in the warm sector generally between west and south-west is an indication of the direction of motion of the centre and for that reason the line of the polar front eastward from the centre was called the steering-line.

All this seems entirely reasonable and most valuable as providing a plan of detailed study of the surface-air and ideal language in which to express the behaviour of air. It began with *steering-line* and *squall-line*, both good, and *discontinuity*, only moderately good; and now *equatorial air, polar air, warm front—principal and attenuated, cold front—principal and secondary, occlusion, subsidence, refraction of isobars* make the language of the golden age of weather-study, which gave us weather-chart, forecast, cyclone, anticyclone, depression front and rear, secondary, satellite V-shape, trough, wedge and col, a back number. A modern meteorologist must now at least have learned to speak Norwegian.

The reader who wishes to appreciate the vitality and energy of the Norwegian school should read the application of the method to the analysis of the phenomena of six days weather ("Analyse der Wetterepoche, 9–14

Oktober 1923 ") by Tor Bergeron and G. Swoboda[1], with "Waves and whirls at a quasi-stationary boundary surface over Europe" as its subject. The analysis runs to a hundred pages and is prefaced by a table of thirty symbols of meteorological status.

The command of language is remarkable; every new chosen word conveys the right impression and no more; and it has done more for developing interest in the science than either harmonic analysis, correlation or integration.

Three points may engage our attention: first the diagram of origin, secondly the ascent of air in the warm front, and thirdly the convection at the squall-line.

The diagram of origin

With regard to the first, the diagram of origin, we find it easy to follow the process from the bulge of the warm sector, onwards, but the preliminary easterly wind of cold air in juxtaposition with a westerly wind of warm air is something we have never seen on any map and never expect to see. One may imagine its existence from Dove's expression of the meteorology of the northern hemisphere as the conflict of opposing winds, or from Helmholtz's scheme of discontinuity between isentropic surfaces for rings of air moving north or south respectively, which was the starting point of V. Bjerknes's theory of wave-motion in the formation of cyclones. And, in a way, the juxtaposition of the cold east wind of the underworld, north of the polar front, with the west wind of the overworld shown at the surface south of the polar front, may be claimed as a direct consequence of our analysis of surface-pressure into two parts, that of the underworld being opposite to that of the overworld if we take a horizontal section at 8 km or 4 km or even 2 km (vol. II, pp. 262–3 and article 13, p. 412).

Moreover, if we regard the polar air as proceeding southward from polar regions it ought to develop into an easterly current just as the trade-winds do at the equator. But if we trace back our depressions to their locus of origin over the Atlantic we find that the polar air instead of or after turning west turns east. Our diagram for the cradle of a cyclonic depression would be

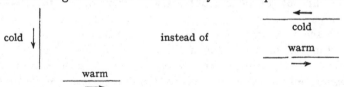

Depressions seem to originate in a rectangular kink in an isobar, not in a reversal of isobars. The reversal of isobars might indeed happen if the accounts of the underworld were not hopelessly complicated by surcharges of liability from the overworld. By all our laws of atmospheric motion, except one, there ought to be an easterly turn in the polar air, for a reason corresponding with that under which the equatorial air forms a west wind. That one super-dominant law is Law IX, the law of isobaric motion.

[1] *Veröff. d. Geophys. Inst. d. Univ. Leipzig*, Zweite Serie, Bd. III, Heft 2, Leipzig 1924, S. 63–172.

Here (Fig. 75) are two specimen cradles of depressions for 8 and 10 Sept. 1903.

The representation which we give of the cradle of a depression is in accord with Exner's suggestion of the weather of what we may call the depression belt being due to masses of cold air breaking away from the polar regions and invading the westerly current. And indeed a general explanation on those lines is an obvious necessity of the situation generally accepted.

The particular point to which we want to call attention is expressed in the question, why does the air flowing southward in the north-western area of the Atlantic turn left as it approaches temperate latitudes and join a westerly wind instead of turning right and providing an easterly wind as it ought to do in compliance with the law of conservation of angular momentum?

The answer which we offer may be of sufficient general interest to justify our setting it out. It is clear that the southward flowing air is prevented from obeying the law of conservation of momentum by the coercion exercised by the distribution of pressure which seems to be regulated by fluctuating centres of low pressure over the North Atlantic, "The large depression with its centre off the south of Iceland" of the broadcast reports.

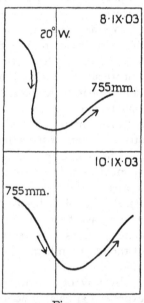

Fig. 75.

Tracing the isobar of 1004 mb over the Atlantic it will generally be found to include several low pressure centres one of them not far from Iceland. And in explanation of that generalised situation we would point out first that the air of the underworld, whatever its own proclivities may be, has to suffer the imposition of the distribution of pressure of the overworld above it, and this necessarily brings the southern fringe of the underworld into agreement with the established eastward motion of the overworld, and in fact makes a kind of fictitious displacement northward of the boundary-line of the two worlds beyond the limit prescribed by entropy. To a certain extent this behaviour coupled with the diurnal variation of entropy of the surface-air may account for the apparent dynamical chaos of the depression-belt which is perhaps even more conspicuous in the southern hemisphere than in the northern.

This in itself might perhaps carry the southward flowing air eastward but it is not likely that it would of itself completely override the influence of angular momentum throughout the southerly traverse. For that we want some additional influence that insists upon the development in winter of a centre of low pressure generally somewhere near Iceland.

For explanation we may appeal to the excess of temperature of the water over the air of the northern Atlantic especially in the western part, which is shown in the maps of fig. 16 of vol. II. The perpetual supply of warm moist air must ultimately lead to vigorous convection, and that again to the distribution of pressure which expresses itself as a persistent complex of cyclonic depressions over the North Atlantic. There is a similar complex, not perhaps so well-marked, in the North Pacific under the Aleutian Islands. That also is a source of travelling depressions which may begin like those of the Atlantic with a rectangular kink in the isobars of the quasi-permanent low. Thus the cyclone designer should not reckon on having a blank sheet on which to write his pattern—somebody has been there before him and has left traces of a primary "vortex" in the upper air and a secondary oceanic depression at the surface.

We may go a step farther and say that a right-angled kink in an isobar of the north-western Atlantic means an incipient cyclone which will be developed in the southern quadrant and occluded in the eastern quadrant. But we should like to draw a distinction between an isolated rectangular kink in an isobar, which means a centre, and a series of parallel kinks in successive isobars which imply rather action in a line and develop into a line-squall or squall in the trough line. These are exhibited as the refraction of isobars, an attractive part of the subject of the theory of fronts.

After all, the primary origin of cyclonic depressions can hardly be a matter of much practical importance because apparently our local maps are much more concerned with the final episodes of cyclonic history than with the initial stages. If it had been of serious practical importance no doubt the reality would have been substituted for an idea which is derived from theoretical considerations. The recent extension of the daily chart to include the whole of the northern hemisphere however may make it a matter of some scientific interest.

By way of parenthesis we may remark that the ideas which are here expressed are also practically embodied in a paper by the late Prof. Exner[1], "Ueber die Zirkulation zwischen Rossbreiten und Pol."

The ascent of the air of the warm front

In considering the relation of atmospheric movements to the entropy of the moving air we have assumed that the flow of air in the warm front over the cold air beyond the eastward prolongation of the polar front in a well-marked depression takes place primarily along an isentropic surface which divides the polar air of the underworld from the warmer air of the overworld above it.

The assumption has aroused incredulity in some quarters but, indeed, making some allowance for the chaos of the depression-belt, the actual conditions can scarcely be otherwise. The flow represented in the horizontal section of the ideal diagram of vol. II, p. 386, shows a series of trails crossing the warm sector from a direction which we may call south-west bending somewhat towards the north and impinging upon a mass of cold air shaded to

---

[1] *Meteor. Zeitschr.*, Heft 2, 1927, p. 46.

indicate rainfall, which implies the upward movement of air; and the vertical section shows two lines of air flowing from the cold front of the squall-line across the warm front and then up the slope of the polar (cold) air to heights ultimately marked by cirrus. The fact that the condensation is continuous implies that the departure from the original isentropic surface would be gradual and continuous too. The main part of the motion would always be along the isentropic slope with the current.

No doubt the formation of cirrus would take place several surfaces above the continuation of the original surface of the polar air, but in a very real sense that surface would be a guide to the thermal changes which would occur and which would cause the flow to lead away from the original surface of flow. At each stage the flow would be approximately along the isentropic surface then occupied and the slope of the flow would differ so little from the slope of the isentropic surfaces that the slope of those surfaces is the best guide we can suggest for the development of the physical processes experienced.

Near the ground the air will of course be somewhat turbulent and it will require a finite thickness of mixing air for its accommodation; but in so far as it is dry air, or as we have called it elsewhere gaseous air, it cannot penetrate *below* the isentropic surface representing its own condition. We have explained that Helmholtz discussed the question of the slope of the isentropic surfaces in the case of air which had acquired different entropies with different velocities of travel, and proved that when convection happened it would result in a mixture of portions of two consecutive layers and the convection would travel along the lower isentropic surface and tend to increase the difference of entropy between the surfaces in juxtaposition.

So we can only picture to ourselves the air of the warm front passing along an isentropic surface so long as it is gaseous; and, as soon as condensation begins, adapting itself to the next higher isentropic surface; but always with the understanding that it could not pass below the surface to find the place to which its entropy entitles it.

We take a mass of air of low entropy to be just as impenetrable as a mountain to air of higher entropy that has a correct appreciation of physical laws.

The convection at the squall-line

Conditions are quite different in the case of impact of the cold front upon the equatorial air at the squall-line. For that we can only refer to the picture of the isentropic surface (10·75 megalergs per unit of temperature) given as fig. 98 of vol. III, for the deep depression with a centre in the south-west of Ireland on 12 November, 1901, which, since its publication, has been represented on a model.

The isentropic surface shows a very steep slope in the quadrant south-east of the centre and it must be noted that that surface might act as a guide for air of the cold front to go downwards and/or for air of the warm sector to go upwards.

Air of the warm sector must certainly have gone up and have been replaced by air which, in some sense or other, came down. The effect is generally described as pushing a sharp wedge of cold air under the warm sector. Giblett (*J. Roy. Aero. Soc.* 1927) gave a wedge sloping 1 in 20 with a nose-point at the squall-cloud in the air at 500 m, four miles in front of the veer of wind.

The figure referred to gives only a single isentropic surface, that one not necessarily the one along which the warm air was forced up. The whole situation is complicated by the turbulence of the operative air as well as by condensation and by the restriction imposed by the action taking place at the ground where vertical motion is limited to the upward direction.

We cannot offer any adequate representation of the dynamical process which expresses the thermodynamic changes until we are able to get information about successive isentropic surfaces. For that we have to wait until we have access to pressures as well as temperatures and humidities at surface-level, and some means of ascertaining the lapse-rate of temperature in the layers just above the surface.

Cyclones and air-masses

An essential difference between the cyclone of the nineteenth century and the alternative air-masses of the twentieth as an expression of the characteristic behaviour of surface-air is suggested by the derivation of its name. Piddington derived the word as a name for tropical revolving storms from a Greek word, quite familiar to those who think of cycles, meaning the coil of a snake.

A snake has a pair of very well-marked sides with a quite inconspicuous but very insidious forepart that one hardly dares to call a front; whereas the air-masses have fronts so impressive that they might almost be called brazen and sides that, if they can be said to exist at all, pass unnoticed.

The winds of the cyclone are closely associated with the isobars; if the association were in perfect agreement with the law of isobaric motion the motion would be conservative in any horizontal plane, no gain or loss of mass would be required to maintain it; but the insidious foreparts of the coils of the snake have an incurvature which if it were at the same angle over successive isobars would have to be compensated by an ascent of air uniform over the whole area of the closed isobars (*The Air and its Ways*, p. 127).

Nobody can claim that that is an accurate representation of the convection of a cyclone, and indeed differences of entropy in different parts of the area would be a good reason for handing over the vertical motion to the care of the fronts, or if we have to make a suggestion we should continue the coils of the snakes upwards into spirals round a central core, with some pleasurable anticipation of the results of an investigation of the real facts. In that case there is nothing in principle to prevent the representation of successive horizontal sections by similar distributions of pressure, leaving a ring round the core for the upward motion of the air which converges at ground-level.

And here we come upon a real difficulty. The law of isobaric motion concerns itself only with horizontal motion; the phenomena of weather within the boundary of a closed cyclonic isobar are not adequately provided for by horizontal motion, and as regards the fronts any question about vertical motion is hampered by the fact that the earth's surface is a place where the laws of motion appropriate to the free air cannot find expression because, from the nature of the boundary, only upward motion is possible, downward motion, possible and indeed probable in the free air, is barred at the surface.

That circumstance is of no practical importance in forecasting, the inferences from the behaviour of fronts can be based on general ideas of what might happen; but it inhibits the possibility of having a horizontal section in the upper air similar to the surface section in the same sense as in the case of the cyclone. For real scientific progress a definite investigation of the structure above the surface is absolutely essential, and no investigation is really adequate which does not provide enough material to trace the isentropic surfaces.

Whether the investigation proceeds by the pv method or the $tt\phi$ method is of little importance, the same material is required for either. We look forward therefore to some effort to provide the necessary data. To ignore the doctrine of fronts is as futile as to ignore the doctrine of isobaric motion, and to leave out the limitation imposed by the doctrine of entropy is only to postpone the dawn of productive knowledge.

In any case in these days of maps covering vast areas some alternative for maps at sea-level is necessary; if p is correct v is not, neither is tt nor ϕ. Reduction of pressure to sea-level has done good service for nearly a hundred years in giving us an outline of the plan; when we come to real details something else is required.

THEORIES OF CYCLONIC MOTION

The methods which we have indicated are essentially empirical in that they depend on the recorded experience which we preserve in weather-maps, and have not found expression in algebraical formulae.

There have been however many efforts to express the behaviour of cyclonic depressions in algebraical terms. Ferrel's or F. H. Bigelow's summaries and extensions for one side of the Atlantic and Sprung's *Lehrbuch* and Oberbeck's memoirs for the other will be remembered by those who have any practice in theoretical meteorology.

At the present day in the home country there is a tendency to regard dynamical meteorology as a part of forecasting, and from that point of view the classical works which we have mentioned hardly call for detailed consideration. But there are two names which are frequently referred to in meteorological texts, these are Max Margules of Vienna who wrote on the energy of storms, and V. Bjerknes, formerly of Oslo, Leipzig and Bergen and now again of Oslo, whose theoretical considerations expressed in "The dynamics of the circular vortex" were the starting-point of the more empirical development of the Norwegian practice by J. Bjerknes.

We may therefore regard the contributions of Margules and the dynamics of V. Bjerknes as classics of meteorological literature contemplated in chap. II as belonging to the theory of advective regions, and in that sense we may now refer to them.

The underlying theory of discontinuity

The idea of discontinuity in the structure of the atmosphere as a basis of meteorological theory may be said to have begun with Helmholtz as we have already explained, and to have been pursued by Brillouin, exploited by Margules and used by V. Bjerknes as the basis of a theory which expresses the atmospheric vortex as a development of wave-motion in one of Helmholtz's surfaces of discontinuity.

On the theoretical side a good deal of stress has recently been laid on the importance of the contributions of Margules in which discontinuity is a notable feature.

Margules—the energy of storms

As a preliminary to his work on the energy of storms Margules[1] evaluated for certain special cases the energy represented by the distribution of pressure and compared it with that of the kinetic energy of the motion of the air. As a result of his investigations he showed that in the cases considered the kinetic energy is many times the equivalent of the potential energy.

For a cyclone of radius r in an isothermal atmosphere in which the pressure ranges proportionally to distance from 745 mm at the centre to 760 mm in the undisturbed region the potential energy A estimated as $\int dm \int p\,dv$ is given as $A = \pi r^2 \times 6 \cdot 3$ kg-cal/m^2, or for a radius of 555 km, $6 \cdot 1 \times 10^{12}$ kg-cal, sufficient to raise the temperature of the whole cyclone (under constant pressure) by $\cdot 0026°$ C.

The variation of pressure being linear the energy per unit of area is approximately proportional to the square of the range of pressure between centre and margin.

For a stationary whirl of frictionless air in which the particles describe circles round the vertical axis and in which there is no variation of velocity with height so that the whole mass shares the kinetic energy, the ratio of the kinetic energy to the potential energy is $4p_0/(p_0 - p)$ where p_0 is the undisturbed pressure and p the pressure at the centre, i.e. 200 times for a centre 745 mm, 100 times for a centre 730 mm in an environment of 760 mm.

For a whirl on a revolving horizontal plane surface with the usual equation of strophic balance and a pressure distribution increasing outwards according to the square of the radius over a range of 30 mm and radius 1080 km in latitude 45°, the ratio of kinetic to potential energy is 20 for a rotating earth and 76 for a non-rotating earth.

In these calculations we must note that Margules takes the velocity as that relative to the earth. Helmholtz in somewhat similar circumstances took it as the motion relative to axes fixed in space. The effect on the numerical expression of the relation of the kinetic energy of motion to the potential energy of pressure would be considerable.

Moreover, if we understand him rightly he regards the kinetic energy of wind and the potential energy of lifted mass, expressed by pressure-difference, as alternatives. It is, however, obvious that the two forms of energy belong to the same side of the energy-equation, not opposite sides; they rise

[1] 'Über den Arbeitswert einer Luftdruckverteilung,' *Denksch. Wien. Akad. d. Wiss.*, Bd. 73, 1901, tr. C. Abbe.

and fall together under the same originating influence or cause, whatever that may be.

Margules points out that, on the theory which used to be accepted, an atmospheric cycle might be formed for air ascending to a height of 6000 m in a low pressure area considered warm (288tt at 740 mm at the surface) and descending in a high pressure area considered cold (770 mm at 273tt); he works out the efficiency of the cycle at ·21 for a circulation of dry air under adiabatic conditions and ·126 for saturated air. We could thus obtain a transfer from high to low at the ground on account of the pressure-difference and from the cyclone to the anticyclone in the upper air on account of the reversal of the pressure-gradient due to the difference of lapse-rate of pressure in the two areas consequent on the difference of temperature. Margules himself notes that the conditions of temperature postulated, namely cold high pressure and warm low pressure, are not in accordance with the results of the investigation of the upper air. The hypothetical circulation is not necessary, as we shall show that the rotation of the earth itself provides for a flow from low pressure to high in the upper air unless the pressure-gradient is sufficient to keep the flow along the isobars.

In his main paper Margules[1] sought to discover the conditions in which a sufficient supply of geopotential energy could be stored to maintain a storm. He conceived the source of energy to be in the juxtaposition of two masses of air of different temperatures. He dealt only with adiabatic motion and assumed that the heat produced by friction was withdrawn. The position of equilibrium of the masses was calculated from the fact that in the final stage entropy must increase with height. His general enunciation of the problem is as follows:

A mass of air in a closed system is at the beginning at rest and has a given initial internal distribution of temperature and pressure. It is set in motion by its tendency toward a condition of stable equilibrium. If there were no friction the individual portions of the mass would oscillate about their positions of equilibrium. In the presence of friction the final condition is attained by the gradual consumption of kinetic energy. We seek the maximum values of $\delta\overline{K} + (\overline{R})$ [i.e. of the increase of kinetic energy of the system + the kinetic energy lost by the whole system through friction] which we designate as the available kinetic energy of the system. Since the initial values of $\overline{P}$ [the potential energy of the system due to its position and the action of gravity] and $\overline{I}$ [the internal energy of the system] are known our problem is to compute their values for the final stage.

Margules considers first cases of discontinuity in the horizontal in which masses of air of unequal temperature at the same level are separated by a sharp boundary. These conditions are typical of the gusts of wind which are accompanied by rapid increase of atmospheric pressure and rapid fall of temperature—the modern line-squall. The horizontal discontinuity may be one of entropy or of pressure—both cases are dealt with but the former gives rise to much larger values of the kinetic energy than the latter.

[1] 'Über die Energie der Stürme,' *Jahrb. k.k. Zentralanst. f. Met. und Geodyn.*, Wien, 1903, tr. C. Abbe.

Horizontal discontinuity. Let the mass of air in the lower part of a closed system [very long in comparison with its height] be initially divided by a screen into two parts. Let the cold air be in the left hand chamber but the warm air in the right hand. Each mass of air to be at rest in either stable or neutral equilibrium; the whole mass of air in the enclosure above these two chambers takes no part in the successive processes; we can suppose it replaced by a movable piston. What amount of kinetic energy becomes available when we remove the screen and let the masses move adiabatically?

Provided that the pressure on the upper surface of the two masses remains unchanged there would be developed kinetic energy corresponding with a mean velocity of 13 m/sec of the total mass for a difference of temperature of 10tt extending to a height of 2000 metres, or 23 m/sec if the discontinuity extended to 6000 m. From which it follows that with chambers 2000 m high and a difference of temperature of 10tt storm-velocity could not be attained by more than one-fourth of the air, and even a still smaller fraction if the loss due to friction be allowed for.

If on the other hand the initial state is such that the entropy is constant in a horizontal layer, but the pressure and consequently the temperature of one mass is greater than that of the other, equilibrium is attained by horizontal motion only. If initially each of the two masses is in stable equilibrium then for a difference of pressure of 10 mm (which implies a difference of temperature of 1tt) the available kinetic energy is that appropriate to a velocity of only 1·5 m/sec for the whole mass. The figure is therefore considerably smaller than in those systems that have horizontal differences of entropy. Margules concludes: "It seems now to have been abundantly demonstrated that the available kinetic energy of such a system is not dependent materially on the horizontal differences of pressure but on the distribution of entropy and the buoyancy dependent thereon."

"A great velocity of a mass of air over a broad area under the influence of a horizontal pressure-gradient can only arise when this gradient is maintained by some outside source of energy; otherwise it would disappear before any portion of the mass of air had attained the velocity of a moderate wind. Dry air possesses such a store of energy when horizontal differences of entropy of ordinary amount exist at any level; and not only when there is a sharp boundary between warm and cold air but also when there is a steady continuous horizontal gradient of entropy."

Margules considers further the kinetic energy for a system in which masses of air of unequal entropy are superposed in unstable equilibrium; but he points out that though such a condition would give sufficient energy for the development of storms there is little or no evidence that such a condition could exist for any length of time over extensive areas.

Vertical instability. If two isentropic columns of air each 2000 m high are superposed one above the other with a difference of temperature of 3tt at the surface of separation such that the column of lower entropy is at the top, the energy developed by the overturning of the masses to a position of equilibrium would be sufficient to give an average velocity of 14·85 m/sec to the mass. The height of the column in the new position would be 4·14 metres greater than in the original position.

Margules considers the suggestion that the source of energy of a storm may be found in the latent heat of condensation but concludes that in the circumstances postulated "the latent heat of condensation contributes nothing to the energy of the storm."

He summarises the conclusions of his work as follows:

The kinetic energy of a mass of air is derived from its internal energy and from the work done by the force of gravity. In the case of a continuous distribution of density the importance of gravity in the production of great velocities can be concealed, whence we derive the very common belief that the horizontal gradient of pressure produces the storm. But it is now demonstrated that, even when the distribution of pressure at the base is as observed in storms, still the horizontal movements of the masses have a potential energy that is only a small fraction of the observed kinetic energy. So far as I can see the source of storms is to be sought only in the potential energy of position. A system in which the masses are disturbed vertically from equilibrium can contain the necessary potential energy. Hence, therefore, the storm-winds develop by reason of the velocity due to descent and that due to buoyancy, notwithstanding the fact that these evade attention because of the large horizontal and small vertical dimensions of the storm area. The horizontal distribution of pressure appears as a translation of the driving power of the storm; by means of it a portion of the mass can attain greater velocity than by simple ascent in the coldest or descent in the warmest portion of the storm area. Here we come into the presence of problems that cannot be solved by a simple consideration of the energy alone.

We will not follow the details of the computations, we only remark that the whole process of physical reasoning is hampered by the limitation of height and the consequent implied limitation of thermal convection of moist air for which there is no justification.

We notice that the whole tendency of Margules' discussion is towards regarding entropy as one of the controlling factors of the process. It is understood from the beginning that in the condition of equilibrium entropy must increase with height and when the reverse condition obtains kinetic energy will be developed in the process of readjustment.

Margules comes very near to the position set out in what we have called the law of isentropic motion that the motion is guided by the isentropic surfaces. If the motion is horizontal when the isentropes are horizontal it follows conversely that when the isentropes are not horizontal neither is the motion, and later we shall suggest that in the formation of anticyclones when the air is moved from low pressure to high pressure it is guided by the isentropic surfaces. The shapes of the isentropic surfaces are indicated in fig. 64 and the relations of temperature and entropy in fig. 65 of vol. II.

Bjerknes—The dynamics of the circular vortex

V. Bjerknes[1] approaches the subject from a different point of view. He develops the distribution of pressure in a cyclonic depression as the horizontal projection of wave-motion (fig. 13 of vol. III) in a surface of discontinuity inclined to the horizontal at a slope of about 1 in 100; and on the surface of discontinuity, which is in fact an isentropic surface, many of the phenomena of the cyclonic depression are staged.

The general method of the paper is to examine the dynamical conditions of a circular vortex on the lines of *Dynamic Meteorology and Hydrography*, with pressure and density as the variables, first in a homogeneous medium when the conditions are barotropic, secondly when there are two media each homogeneous in itself with a surface of discontinuity separating them. The

---

[1] 'On the dynamics of the circular vortex with applications to the atmosphere and atmospheric vortex and wave-motions,' *Geofysiske Publikationer*, vol. II, No. 4, Kristiania, 1921.

conditions are then barotropic in each medium but baroclinic at or across the surface of discontinuity.

For liquids, homogeneous or barotropic means equality of density, ρ is the same always and everywhere; for a gas, equality of potential temperature or entropy, pv^γ is the same always and everywhere.

Incidentally the author points out that the classical researches on vortex-motion assume one or other of those conditions from which it follows that "circulation" is conserved and vortices can neither be created nor destroyed. The limitation is imposed in order that the effective forces may be regarded as belonging to a conservative system. Some thirty years ago Bjerknes explained that the limitation was not necessary, but with one exception all the text-books of hydrodynamics retained it. The incident is quoted as an example of the way in which assumptions get rivetted on to physical science as expressed by algebraical equations.

The lesson has been repeated recently by S. Sakakibara[1] in "A note on the generation of vortices and its independence of viscosity."

There are seven chapters in Bjerknes' treatment. Chap. I expounds the dynamics of barotropic and baroclinic fluids and forms an excellent introduction to dynamics for students of meteorology; chap. II gives "various examples of preceding principles" and includes the theory of wave-motion at a surface of discontinuity with a note on the transition from extreme displacement in gravitational wave-motion to complete circular motion; chap. III expounds the dynamics of a two-dimensional circular vortex (1) under constant vertical force and (2) under a central force. The vortex is assumed to obey one or other of a selection of laws of variation of velocity with distance from the centre. A special combination of an interior and exterior law is chosen as an appropriate type though the choice is arbitrary.

In these three chapters no account is taken of the rotation of the earth, but that is introduced into those which follow.

Chapter IV, on relative motion, describes a simple cyclone and anticyclone of unusual form.

Chapter V treats the earth's atmosphere as a circular vortex and locates certain surfaces of discontinuity the chief of which is the tropopause. There are two within the troposphere, one between the trades and anti-trades and the other the polar front, which we have indicated as the surface of separation of the overworld from the underworld.

These account for low pressure and easterly winds at the equator, then westerly winds; then going northward to the underworld a second zone of easterly winds with probably a final zone of westerly winds nearest the pole.

We do not know if the sliding surfaces of discontinuity go continuously round the earth, but it is more probable that they exist only over the oceans and end with the more or less marked borders near the coast of the continents, which may account for the parabolic motion of tropical cyclones.

[1] *Proc. Phys.-Math. Soc. of Japan*, 3rd ser., vol. XII, No. 2, Tokyo, February 1930.

Chapter VI treats of the disturbances of the surface of discontinuity. The disturbances are assumed to be wave-motion and the chapter gives the wave-theory of cyclones and anticyclones.

Chapter VII summarises the results and deduces a zonal view of the general circulation of the atmosphere.

The conclusions are traversed by Ryd[1], who deals especially with the consideration of the transformations of energy which are included in the cyclonic process.

"Dr Ryd approaches the subject from the synthetic side; he imagines a certain pressure-system to hand and predicts, so to speak, the observations. He takes up the subject where it was left in the *Life-history of Surface Air-currents* but he assumes a travelling vortex not at the surface but in the free air, perfect at 5·5 km. He deduces surface-phenomena which are a fair representation of what are now called the phenomena of the polar front. His treatment is not very realistic because he leaves water-vapour out of account and the atmosphere without water-vapour would be different from that which we know."

The energy of the circulation

Thus the question of the energy of the circulation and its relation to the distribution of pressure comes up again. The kinetic energy can be computed in so far as we know the mass which is moving and the velocity with which it moves. The only residual difficulty is to decide whether the velocity is to be estimated as motion relative to the earth's centre or the earth's surface. Certainly it would be the motion relative to the centre if we are relying upon the conservation of momentum with reference to the polar axis for the production of the observed velocity.

The question of potential energy is different. At first sight it might appear that the excavation of a vast mass from the area of a depression and the piling of it to form high pressure in the surrounding region would create a store of energy depending on the differences of pressure at sea-level or other horizontal surface. We have seen on p. 294 that Margules assigns to the potential energy of pressure-difference a very small value compared with the energy of motion, and relies upon entropy as the agency at work in that case. We should like to reinforce that conclusion and to lay stress upon the reasons for regarding the pressure-distribution not as in itself a store of energy but merely as a static index of the energy of atmospheric motion.

Let us think of the distribution of pressure as the deformation of an isobaric surface, originally horizontal, by the removal of air from beneath the area of the depression and its storage by lifting the region surrounding it. We may think of a section across the centre showing the isobaric curve of the depression with an anti-cyclonic hump on either side. Then if we suppose the surface to resume its horizontal position the energy derived from the descent of the surface at the humps will be pv, where p is the pressure-index of the isobaric surface and v the volume of the humps. On the other hand energy will have to be spent in lifting to the horizontal the pressure-surface over the depression, again pv'. These are equal if v is equal to v', and the balance of energy is zero except in so far as the volume of the displaced air is altered during its displacement. So we count nothing for the energy of pressure-distribution in itself, only for the change of entropy associated with it, thermodynamic energy not simply dynamic.

Readers may be incredulous of this result but may be reassured if they consider

[1] 'Meteorological Problems, I. Travelling Cyclones,' *Det Danske Met. Inst., Meddelelser*, Nr. 5, Kjøbenhavn, 1923.

the dynamical effects of the motion of the air. We may find an analogy in the rotation of a pendulum-bob controlled by a string. The energy of the system will be that of the motion of the bob; though the string will exert the force necessary to control the motion of the bob its energy will be only the negligible amount represented by the elastic deformation of its structure. So with the atmosphere, the energy at any time is expressed by the motion, with some allowance for the thermodynamic alteration of its structure.

We exclude the surface-layer from the conditions here described. Something more is required to deal with the energy effects of turbulence.

Discontinuity in practice

On the practical side as exhibited in the weather-map discontinuity may be said to have originated with Dove. It was investigated in detail by R. G. K. Lempfert to explain the sudden lull shown on an anemogram at Pendennis Castle on 24 February 1903, and described in *The Life-history of Surface Air-currents*. It appeared in diagrammatic form in *Forecasting Weather*, and beginning with 'The structure of the atmosphere when rain is falling[1]' has been developed in the publications and practice of the Norwegian school under the guidance of J. Bjerknes. It has become a cardinal principle of the meteorology of the northern hemisphere at least of the eastern side of the Atlantic. On the western side the travel of cyclonic depressions is still being traced over the hemisphere[2]. It appears as a common characteristic of the isobars of modern maps, but the run of lines on a weather-map is very much at the discretion of the artist. In 1903 the contemporary artist missed an undeniable discontinuity, but "the whirligig of time has brought in its revenges." An interesting example of accurate plotting of the details of isobars by R. Corless is given in *Forecasting Weather*, 1923, fig. 144.

About discontinuity as a basis of the development of a cyclonic depression there is this difficulty, that we find it most clearly indicated in the central regions of a well-formed travelling depression. There is a general principle in dynamics that the particular condition which constitutes a cause tends to exhaust itself and disappear while the cause is producing its effect. We should expect therefore, working backwards chronologically from the developed form of a depression approximately circular, to find an initial condition with a discontinuity more conspicuous before the cyclone began than in the finished product; but as a matter of fact the discontinuity is not often noticeable, in the regions that are going to be affected, before the cyclonic depression appears there[3].

So with Margules an observed distribution of potential and kinetic energy can of course be regarded as equivalent to a discontinuous separation of air

[1] V. Bjerknes, *Q. J. Roy. Meteor. Soc.*, vol. XLVI, 1920, p. 119. For the later developments the reader may be referred to contributions by J. Bjerknes, E. G. Calwagen, T. Bergeron, H. Solberg, and others in the *Geofysiske Publikationer* of the Norske Videnskaps-Akademi at Oslo.

[2] C. L. Mitchell, 'Cyclones and anticyclones of the northern hemisphere, January to April inclusive, 1925,' *Monthly Weather Review*, Washington, vol. LVIII, 1930, p. 1.

[3] See, however, Bergeron and Swoboda, *loc. cit.* on p. 288. We note also that S. Fujiwhara in a paper in the *Geophysical Magazine* (vol. II, No. 2, Tokyo, 1929) gives examples of cyclones over Japan which formed on a line of discontinuity.

of different entropy; that is purely a matter of calculation; but that there ever actually was such a separation before the energy was developed is another proposition which is very difficult to verify, and it would be easier to postulate the development of a discontinuity by the operation of kinetic energy than the reverse.

The interpretation of the anemogram which was laid down in chap. 1 and which still remains as the fundamental problem of dynamical meteorology is of too general a character to cover the work of a forecaster now as it was intended to do a dozen years ago. And we have reviewed the subject of the advective and divective regions of the weather-map as one which may be treated independently of the idea of revolving fluid, so far independently indeed that the principle of discontinuity becomes the fundamental idea and the text of what we have to say.

And yet we would not have the reader unmindful of the fact that the "two views" of advective regions as revolving fluid on the one hand and the reactions across discontinuity on the other are not so far dissociated as might appear at first sight, for an illuminating paper by T. Kobayasi[1] shows that the travel eastward of the pressure-distribution of a column of revolving fluid over a surface-field of temperature with gradient northward will press the isothermal lines into discontinuity and curl them up under the more central region of low pressure. V. H. Ryd comes to a somewhat similar conclusion in the paper to which we have referred.

As a matter of fact the conclusion to be drawn from the evidence of turbulence and the consequent failure to find any indication of revolving fluid in the anemogram is that the surface layer with its solid or liquid substratum is a most unsuitable specimen of the atmosphere for the study of its structure and its laws.

One of the incidental disadvantages of choosing the ground-level as the basis of dynamical ideas is that it is liable to give rise to automatic depreciation of the importance for all the rest of the atmosphere of the relation between pressure-gradient and wind. We have already noticed that Hesselberg and Friedmann in the Skipper's Guide (p. 50) regard the geostrophic relation as something which may be skipped in the dynamical equations. We have moreover quoted a passage (p. 210) which indicates that what we treat as an inexorable consequence of the earth's rotation, V. Bjerknes and others regard as fortuitous coincidence. We should gather that this has been the case in the theory of fronts by the lines which are drawn to represent isobars on the sea-level maps. They present angularities which are inexpressible in dynamical terms without allowance for the complications of convection represented picturesquely in the commotion of the squall-line. We find it difficult to believe that the complications inevitable in such cases are not in fact to some extent smoothed out in the air above those lines, and we look forward to the time when material is available for the consideration of that question.

[1] 'On the mechanism of cyclones and anticyclones,' *Q. J. Roy. Meteor. Soc.*, vol. XLIX, 1923, p. 177.

DIVECTIVE REGIONS

This leads us to the question of the origin of divective or anticyclonic areas, which has not received the same attention as that of cyclonic depressions, and yet the high pressures have greater controlling influence than the low, though they have not such striking incidents of weather.

Margules says about them:

In an area of high pressure the columns of air not only have a temperature that is high for the season, but also one that is higher than anywhere in the surrounding region of lower pressure. In the lower portion the air flows steadily away in the direction of the gradient; while the anticyclone as a whole remains stationary and often for a week or more. We therefore must necessarily assume that in the upper layer there is an inflow (toward the anticyclone).

Under this assumption the upper inflow can only take place *against* the gradient. The differences of pressure do not disappear with elevation, but become relatively larger in the upper level provided the whole column in the area of high pressure is warmer than the surrounding air. If we assume a circulation to exist under these conditions then we cannot assume any heat to be converted into work. The pressural forces do the work below the lower layer, but for the inflow in the upper layer work must be expended, and more than we gain in the lower layer. Hence this system or cycle cannot be considered as in any manner similar to that hitherto considered but must maintain itself by drawing directly from a store of kinetic energy that feeds the upper inflow. Like all other movements on the earth, this kinetic energy can only have its ultimate source in heat; but in order to get an idea of the whole process we must consider the conditions prevailing over a very much larger region and for that purpose must devise some scheme that shall include the conversion of heat into work.

(Max Margules, 'The mechanical equivalent of any given distribution of atmospheric pressure, etc.' tr. C. Abbe, *The mechanics of the earth's atmosphere*, 3rd collection, Washington, 1910, p. 528.)

Anticyclones and the strophic balance

So far as we are aware the mode of creation and development of anticyclones has been hitherto one of the unsolved problems of meteorology. Regarded from the point of view of the strophic balance their formation becomes so much a part of the ordinary sequence of events in the atmosphere that it seems almost incredible that it has been overlooked for so many years. It must be contained implicitly in many equations but we have seen no expression of the implication.

The principle of the strophic balance sets out that in a current of air when there is no acceleration and only negligible curvature the deviating force of the earth's rotation, $2\omega V\rho \sin \phi$ for the velocity V, is balanced by the pressure-gradient.

It follows that if the velocity is *too small* for the gradient there will be a margin of deviating force directing the flow from high pressure to low pressure. That we are quite familiar with at the earth's surface where the flow is retarded by friction.

Equally it follows that if the velocity is *too large* for the gradient there is a margin of deviating force directing the flow from low pressure to high pressure.

If we think of a case in which a flow is originated in an air-field without gradient, as, for instance, in the nocturnal katabatic valley-wind leading to an open plain, it is evident that there will be a force across the flow from left to right of the moving air; and if the area available for supply and delivery be circumscribed, the air will be removed from the left-hand side and piled up on the right-hand side until the proper gradient is attained. In other words the formation of a "high" will have been begun on the one side and of a low on the other.

Thus the initial stages of both high and low are simultaneous expressions of the effect of the earth's spin on air-currents so arranged as to circumscribe the areas of supply and delivery. The first stage in circumscription may be a linear distribution, two adjacent currents in opposite directions; casual circumstances would soon indicate centres of advection or divection and consequent closed systems.

It may be remarked that this process is three dimensional. The diversion of the current, to heap on the one side air taken from the other, occurs at every level of the air-current with velocity in excess of the gradient and may extend over a number of kilometres of vertical height. The reverse process of flow from high to low in consequence of friction is confined to the surface layers, or with gradually diminishing effect above the surface in what we have called the foot of the structure.

We have still to inquire how flow can be increased. A natural suggestion is the horizontal accompaniment of convection, but a more effective arrangement is displacement along the earth's surface from south to north for westerly currents and north to south for easterly currents. Other things being equal a south-westerly current of air will automatically increase its westerly component as it travels, and *vice-versa* with a north-easter. Other things, as we have seen, include the distribution of pressure.

The area of supply or delivery is circumscribed by being surrounded by the current which is the subject of investigation. Thus if there be a current in geostrophic equilibrium separating a high pressure from a low pressure the effect of an increase of strength of current will be an increase of the high pressure by the delivery to it of air obtained from the low pressure.

Thus we get a curious picture of the relation of high to low as incidental to the disturbance of the strophic balance. When velocity is increasing there is flow in the upper air from low to high compensated partly by adjustment of pressure-gradient and partly by compensating flow *along the surface* from high to low in consequence of the friction. Surely for a meteorologist one of the most beautiful exhibitions of nature's ingenuity is to be found in an air-flow forming high pressure by quietly raising the pressure and lifting the stratosphere, with consequent reduction of temperature and all the rest of the effects, by simply passing air along the isentropic surfaces in the neighbourhood, until the necessary task is done, while at the surface we get the initiation of the divective effect of high pressure and the advective effect of a low pressure without the assumption of discontinuity or the preliminary of wave-motion or thermal convection.

But the advective effect of the low pressure brings together air-masses of different origin and composition and the physical processes that are there involved may take direction of subsequent events. According to the note on p. 299 the pressure-distribution has in itself no energy and the only energy required for the development of high pressure is that due to the alteration of the physical state of the air.

For the gradual growth of high pressure in the upper air we may refer to the models of the distribution of pressure and temperature included in fig. 60 of vol. II. The increase of pressure by the spreading of cold air at the level of the tropopause is clearly indicated, but the scale of the photographs is too small to give the figure for the winds at the several levels.

One set of observations of pilot-balloons for 29 July, 1908, is represented in fig. 41 and shows a north wind gradually increasing with height to a maximum at 12 kilometres, and in that connexion we must note that even if the law of isobaric motion suggests a means of creating appropriate pressure-differences we are left with the problem of a strong north wind at a high level, as indicated in the case cited, and also in fig. 51, chap. VIII, still unsolved.

THE DISPLACEMENT OF MASS BY VELOCITY

For the reasons explained in the previous section we may now look upon the transfer of air from the left to the right of the advance in a current in the upper air under the influence of the rotation of the earth as a valid cause of anticyclones. Others may certainly be caused by the accumulation of cold air on the earth's surface due to radiation and the katabatic effects which are incidental thereto. We may therefore examine cyclonic depressions and high pressures from the point of view of the defect or excess of air over the area covered by them.

An estimate of the amount of air removed in the development of a small cyclonic depression over northern Europe in July-August 1917 was given in *The Air and its Ways* as 70,000 million tons; a number of computations are given in the heading to chap. VIII of vol. II. The subject has been pursued by M. Loris-Melikof and A. Sinjagin of Leningrad in a paper entitled 'Numerische Charakteristiken der Zyklone und Antizyklone und ihre synoptische Interpretation[1].'

With the normal pressure as their datum line and the area enclosed by the corresponding isobar as the limit of influence they have computed the amount of air lifted to produce certain cyclonic depressions and the amount of air accumulated in the formation of certain anticyclones. The amounts are of the order of 10^{11} to 10^{14} tons, and the authors go so far as to suggest a special unit of 10^{11} tons for the expression of a cyclone or anticyclone and to call that unit a "shaw."

Adhering for the present to the ton as unit these computations give 1·08 billion tons for a depression on the chart of Europe at 7 h of 17 October, 1925, being 1·4 per cent of the total mass of air normally on its area, and 1·11

---

[1] *Gerlands Beiträge zur Geophysik*, Bd. 24, 1929, S. 121–67.

billion tons or 1 per cent of the normal total on 26 August, 1926; and for cyclones over the Philippines ·29 billion tons, 2·3 per cent on 24 November, 1912, and ·30 billion tons or the same percentage on the next day.

For the total deficit of cyclones in the northern hemisphere estimated from normal charts 19 billions, for the southern hemisphere 25 billions, and for the whole earth 45 billions.

Curves are given for the variation of defect with the time, for a cyclone on 1 to 6 September, 1925, with a maximum of 1·55 billions, and on 3 to 7 January, 1927, with a maximum of 1·35 billions.

For the excess of mass they find for European anticyclones: 26 September, 1913, 1·1 billions; 8 February, 1926, 2·71 billions; and for the Siberian maximum in winter, 23 November, 1907, 6·82 billions. It may be remembered that in chap. VI of vol. II we noted 10 billion tons as expressing the transfer of air from the southern to the northern hemisphere between July and January and *vice versa* between January and July.

The exposition of these computations is accompanied by a description of a number of characteristics of cyclones for which the reader may be referred to the original memoir. Here we mention only the determination of the centre of gravity of the deficit and its travel across the map, which is said to be more regular or less irregular than the travel of the centre of minimum pressure.

The initial stages of a cyclonic depression

From what has been written in the preceding section it may be understood that the creation of a cyclonic depression is an undertaking in the art of removing air which altogether transcends the capacity of human agencies.

There are two natural agencies for the achievement—or, as perhaps we may refer to it for theoretical purposes—the action. One of the two is the unbalanced component of the acceleration due to the earth's rotation, $2\omega V \sin \phi - bb/\rho$, which is operative whenever the velocity is out of proportion to the gradient. The other is the effect of penetrative convection removing air by physical means from lower to higher levels.

If we start from the hypothetical picture of Bjerknes's theory of the origin of a cyclone, namely an east wind opposite to a west wind some distance to the south, and remembering that the effect of unbalanced velocity is to produce an acceleration from left to right of the air's path, it is evident that the motion westward in the north and eastward in the south would produce a line of low pressure between the two. We may look upon that line of low pressure as the beginning of cyclonic depressions.

The development of the circular from the linear form

For the further progress in the development of isolated cyclonic depressions we have first to remark that the linear distribution described in the preceding section is unstable. It is subject to change by any local intensification of the flow and for the same distribution of mass circular forms of motion maintain gradients with less action of excavation as represented by velocity than the linear form.

If instead of the original picture of two opposing currents separated by a line of discontinuity of motion, temperature and humidity, we take the one suggested on p. 288 of a right-angled kink in the run of an isobar in the south-west quadrant of an existing depression, then perhaps we may suppose the delivery of the cold air from the north into the region of the westerly current to be possible only if the westerly current can be increased in velocity, and that increase will imply the development of gradient from south to north by the excavation of air from the north side and its delivery to the south, thus some beginning of a depression may be made which may be developed by the convection arising from the juxtaposition of two masses of air of different temperatures.

The excavation might be completed by horizontal velocity alone if it were possible to develop the velocity from the kinetic energy of the air; but in the absence of any effective suggestion for developing the necessary velocity apart from the convergence of air towards the axis, on the analogy of the vortex, we must turn our attention to the influence of convection.

Displacement by convection

We have here to bear in mind Aitken's experiment on the formation of secondaries by dropping sand through the current as it travels. Regarding the undisturbed circulation of the upper air as a mass revolving round the polar axis we must suppose rain falling from a travelling cloud to produce a like effect as a secondary in the circumpolar circulation.

And the like would occur with upward convection as illustrated by cumulus clouds or other forms of convection.

Vigorous convection would carry air away by what we have called eviction, the air so carried up might be distributed over the surrounding regions of higher pressure if the velocities of the currents were suitable.

Convection within the area of a cyclonic depression may be due to two independent causes. First the convergence of air towards the centre caused by the motion across the isobars which is incidental to the reduction of velocity in the lowest layers by the turbulence due to the surface friction; and secondly the condensation of water in the free air as it passes upwards along an isentropic surface. Either or both of these may be illustrated in the case of a well-marked depression by the rainfall of the trough-line in the one case and of the advancing warm front in the other. There is no reason to suppose any general convergence of air in the layers above those affected by the surface. It is not easy to trace the course of the convected air. One may regard the upward flow along an isentropic surface as expressing itself in cumulus cloud which in the early stages may be isolated and subsequently cover the sky, and the convection in the trough with its squalls may be similarly localised; but in the case of the tropical revolving storm which has a calm and clear centre it seems reasonable to suppose that the air converging towards the centre passes upward in a spiral path round the central region and by the effect of its horizontal velocity may enhance the pressure-difference.

The normal position for convection at the surface is on the line of the polar front at or near the centre of a depression. There the surface of separation of the overworld from the underworld cuts the earth's surface with a slope upward of 1 in 100 to the north. Hence of the volume of air above a depression 1000 kilometres in diameter, not much more than one tenth is likely to be polar. A core of revolving fluid if formed at all will be in the equatorial air.

The disposal of the débris

If we allow that the formation of a cyclonic depression involves the removal of some hundreds of thousands of millions of tons of air (some "shaws," as our Russian friends suggest, would certainly be economical of words) from one environment to another outside its basic area, it is obviously important to discover if possible how the process of removal is arranged to deliver the air into its new environment.

Wherever the air is dumped the pressure which it would produce must necessarily be taken into account; so it is not unfair to argue that corresponding with the depression there must be high pressure within near range of the depression.

Also the only form of disposal that suggests itself is delivery into some passing current in the upper air which acts the part of scavenger, not necessarily the same current for all the "shaws" nor even current at the same height; there may be suitable distribution in the vertical as well as in azimuth. It is not unnatural to suppose that these scavenging currents belong to the primary cyclonic vortex of the hemisphere.

We have suggested that the process of excavation is twofold; part of the air is removed by the operation of the cross-acceleration left to right in a current in which the velocity is in excess of that corresponding with the existing gradient, another part by the travel of the warm front along the isentropic surface to the cirrus level, and another part by convection in spirals round the core of a depression. Some convection also takes place at the surface of contact between the cold front and the warm sector in the squall-line, but that may be carried by the increased westerly current of the upper regions of the warm front.

With regard to the three first-mentioned we must recall the relations between the temperature of high pressure and low pressure in the upper air obtained from the records of balloon-ascents by W. H. Dines in England and by Patterson in Canada.

The results show that the horizontal section at 9 km conforms with the adiabatic relation of dry air between the temperature of high and low, and this would correspond with horizontal motion of air at that level. Below that level the temperature of the high pressure is relatively higher and would correspond with the dip of an isentropic surface leading from low to high; and above that level the temperatures are consistent with the diagram which shows a slight upward slope of the isentropic surfaces from low to high.

20-2

Hence we may regard the currents at the 9 km level, not inappropriate as an estimate of the cirrus level, as being the limit of the delivery of convected air from the surface or the supervening layers and the loci of increased pressure corresponding with the weight of air removed. Other limits might be suggested for colder air as indicated by the lower levels at which the adiabatic lines of saturated air become practically horizontal in the tephigrams of chap. VI of vol. III.

These considerations would confirm the suggestion of the westerly currents of the primary atmospheric vortex as the scavenging currents, and the increase of velocity with height appropriate to the Egnell-Clayton law would indicate increased efficiency of the upper layers in that capacity.

The explanation is necessarily somewhat vague because once more the results of actual observations are wanted and, indeed, a technique of balloons applicable to the special circumstances; but it is fair to claim that a mechanism is available for the disposal of the air which passes up over the cold front, and also for that which we have supposed to work upwards in spirals round the core of a cyclone.

It would be very agreeable if we could assume that the convected air of the outer ring of the spirals in upward movement under the influence of its water-vapour acquired horizontal velocity greater than that necessary to balance the gradient, and so entitled itself to acceleration across the current to the higher pressure outside.

We should then have completed the ideal of the model exhibited in fig. 220 of vol. II, and have provided not only for the convection appropriate to the rainfall but also for the disposal of the refuse air of that particular depression, and incidentally for a distribution of pressure at about 9 km. corresponding with the distribution at the surface by the formation of appropriate anticyclonic regions.

The picture is an attractive one but very hazy in respect of the question of maintaining the idea of the distribution of pressure as a vertical structure in view of the variation with height of horizontal motion, which echoes the Egnell-Clayton law. We have to imagine the persistence of a cyclone even if the upper air of the current which carries it is travelling with twice the speed of its lower part. However, Exner (*Dyn. Met.* 1925, p. 368) indicates the position of the upper centre as north-west of the lower centre and we may remember with advantage a term $2\omega w \cos \phi$ in the equation of acceleration in the direction of motion on p. 46, which would appear as an acceleration opposed to the motion and might help to keep the columnar structure upright so long as the vertical velocity w is vigorous. We see from p. 50 that when w is large the order of magnitude for middle latitudes is the same as the geostrophic acceleration due to horizontal velocity. Another term uw/r would help in the same way but is less important.

The whole subject of the formation of cyclones and anticyclones and their relation to the theory of fronts is treated by Fr. Ahlborn in *Beiträge zur Physik der freien Atmosphäre*, Bd. XII, S. 63.

RETURN TO TRADITION

Thus after consideration of the various aspects of the structure and behaviour of the atmosphere with regard to advective and divective regions
the casual anticyclone becomes a natural result of the automatic redistribution of mass, a static index of the velocity of the moving air, and it may
be withdrawn from the position of prominence which it used to occupy in the
dynamics of the atmosphere. The behaviour of the cyclonic depression as
a physical entity still remains the primary expression of atmospheric energy.

Dynamical meteorology may still regard the structure, origin and travel of
depressions as constituting its main problem with weather-maps of the world
and forecasts of longer range as its ideal. And the problem is likely to demand
increasing attention with the attempt to establish trans-oceanic routes for
airships. We may conclude our survey with a note of the situation from
that point of view, though what we have to say can be little more than a
recapitulation of what has been included in chap. VIII of vol. II.

Relation of spin to environment

If we imagine a layer of fluid between two smooth horizontal planes AB,
$A'B'$ and suppose the velocity of motion of the fluid so arranged as to
be proportional to the distance from a centre C, pressure will be arranged so
as to show a minimum at C and uniformity in successive circles, increasing
with the distance from the centre, and so far as the proportionality of velocity
to distance extends the fluid will be rotating like a solid.

In the circumstances described the rotating air will be freed from the frictional effect of relative motion and subject only to the friction of the upper
and lower bounding planes and to the effect of viscosity or turbulence between
the external ring of the quasi-solid and the fluid which surrounds it and which
provides the uniform pressure within which the fluid can rotate.

Across that external boundary there will be diffusion of momentum forming what Rankine calls a simple vortex, each ring having equal moment of
momentum, which will ultimately wear down the motion, but otherwise the
decay will be very slow, and if we could devise upper and lower plane or curved
boundaries of the fluid layer which would ease the relative motion there we
could regard the revolving fluid as a kind of live entity which would be self-
supporting until its quasi-solid rotation was worn away by the friction at its edge.

Thus it would appear that wherever in the atmosphere there is found air
which is rotating like a solid it may constitute itself a separate entity which can
travel along like a vortex in a stream, preserving its identity until its motion is
worn down by friction with the fluid external to itself.

Bearing this view in mind we can look upon the "vital" part of revolving
fluid as the central column which, like that of fig. 70, has quasi-solid rotation.
The outside may be regarded as developed by the relative motion between the
quasi-solid and its environment.

This view is rather different from that suggested by Rayleigh which regards an atmospheric vortex as the result of the convergence caused by the annihilation of fluid in the central region, but it may be that we have here two separate actions in different levels which combine to produce the final result. At the surface in consequence of the friction we have the convergence which would superpose the "simple vortex" on a rotational system, but above the surface there is no provision for convergence and the distribution may perhaps be attributed to the diffusion of motion from a central core with quasi-solid rotation.

In this case the surface of the earth forms a boundary which is not exactly smooth, and to preserve the identity of the quasi-solid rotation we must provide a boundary in the upper air, keeping as far as possible the law of solid rotation. In a lecture before the Royal Institution[1] it is explained that a lid of that kind might be provided by the stratosphere which provides also a natural limit for the convection of saturated air.

The fact remains that the rotation of air like a solid constitutes it an entity which it takes time to destroy. A question to be considered is how quasi-solid rotation can be produced.

On a very large scale the normal rotation of the upper atmosphere round the polar axis is an example of quasi-solid rotation[2]. On an intermediate scale it may apparently be produced by convection within a local area of a larger spin.

On a small scale it can be visibly produced by a strong current of air passing a street corner; on a still smaller scale in the vortex ring which arises from the turbulence of the motion of an impulsive air-current passing through a circular opening, and every impulsive current impinging on a solid obstacle will presumably have somewhat similar effects as shown by the incomplete eddies of fig. 34.

An experiment might show that quasi-solid rotation about a vertical can be given to a mass of air with two plane horizontal boundaries, or with the floor for one boundary and a gradually diminishing rotation for the other, pointing the column like a sharpened pencil. Successful experiments of this kind are sometimes carried out spontaneously by the accidental rings of tobacco-smoke.

The structure of a developed cyclone

We have already explained on various occasions that the isolation of a mass of air as a separate entity seems to make the rotation of a portion of air like a solid a necessary postulate, with the understanding that the solid rotation may be confined to a portion of the interior and must certainly be surrounded by rings of air of smaller velocity. The solid interior itself may spread itself gradually over a larger area until the rotation becomes inappreciable.

The theories such as those of Oberbeck and Rayleigh, which provide a

---

[1] 'Illusions of the upper air,' *Proc. Roy. Inst.*, 1916.
[2] Vol. II, p. 372; *Q. J. Roy. Meteor. Soc.*, vol. L, 1924, p. 69.

structure of revolving fluid, are for the most part two-dimensional, possibly applicable to one horizontal section though not necessarily true for all.

We have to consider whether the approximately circular isobars with discontinuities between polar and equatorial air at the surface can be associated with a single layer of revolving fluid, or something in the nature of spiral motion equivalent to it, in some upper level perhaps at half a kilometre or a kilometre above the surface; and if so whether the system of revolving fluid or spiral motion can be extended to the stratosphere or terminated in some lower stratum of inversion.

At present the observations of the upper air are not so frequent and regular or so well-placed as to make the preparation of a map of upper levels as effective as one for so-called sea-level; and the means of representing the co-ordination of the information that exists is not yet organised. Observations which throw light on structure are sometimes available, for example M. A. Giblett[1] cites an occasion at Cranwell, 19 October, 1923, in which a pair of observations before and after the passage of a line-squall showed "polar air" up to 10,000 ft. at four hours' distance (estimated at 200 miles) behind the surface-line of front; he cites a similar case at Baldonnell in 1921.

In the absence of co-ordinated observations of the upper regions of cyclonic depressions we cannot give an effective account of the structure which has to be explained. A great deal of attention has been paid to the subject[2], particularly in the United States, both as regards tropical revolving storms and the cyclonic depressions of middle latitudes. An excellent account is given in McAdie's *Aerography* and in Humphreys' *Physics of the air* based to a considerable extent on the work of F. H. Bigelow, who made out a scheme of flow at different levels from the observations of clouds over a cyclonic depression. The conclusions are rather vague because they are aggregated as mean values and the individual cases are too irregular for the application of mean values. We know for example that easterly winds do occur in the various levels up to the stratosphere, they have to be accounted for but they make no show in mean values. There is also a considerable literature of the subject in the *Beiträge zur Physik der freien Atmosphäre* and in the *Meteorologische Zeitschrift* upon which Exner relies for his account of the behaviour of cyclones and anticyclones.

Personal observation suggests that the movements of clouds when a centre of depression is passing near by are very diverse and irregular, and it is rather astonishing that with the expansion of the means of observation an investigation of atmospheric structure in a depression is not a part of organised meteorological enterprise. One can understand that a tropical revolving storm by its very intensity defies investigation; but the cyclonic depressions of the temperate latitudes are not so formidable.

[1] *Nature*, vol. CXII, 1923, p. 863.
[2] F. H. Bigelow, *Report on international cloud observations*, Washington, 1900; A. Peppler, 'Windgeschwindigkeiten und Drehungen in Cyklonen und Anticyklonen,' *Beitr. Phys. fr. Atmosph.*, Bd. 4, 1912, p. 91; A. D. Udden, 'A statistical study of surface and upper air conditions in cyclones and anticyclones passing over Davenport, Iowa,' *Monthly Weather Review, Washington*, vol. LI, 1923, pp. 55–68.

For a tropical revolving storm with its calm centre we can picture to ourselves a revolving mass of some kilometres diameter extending to the ring of maximum velocity as in fig. 70 with motion diminishing thence to the limit of the barometric depression. At the surface there will be the convergence due to the friction. The air which flows in may be convected in spirals upwards round the revolving core and perhaps distributed over the high pressure in the upper air on account of the excess velocity of the circulation there.

We can set no satisfactory limit to the height of the column. It must be considerable because, as Kobayasi[1] has described, it can pass over mountains and develop a circulation at the surface on the other side, and it would not be surprising to find the cyclonic circulation extending up to the stratosphere.

With the cyclones of temperate latitudes the case is different. All that we are sure about there is that there may be the equivalent of revolving fluid at the height of say 500 metres. If that condition can be represented by a revolving disc or comparatively thin layer of revolving fluid we can say with confidence that below it the motion is confused by friction and that the convergence caused thereby brings together air of different conditions with convection as the result. Nevertheless in many notable cases the distribution of pressure at the surface which represents the weight of the whole superincumbent atmosphere can hardly be said to differ from the equivalent of that of revolving fluid.

Also from the investigations of J. Bjerknes we may conclude that the right front, beyond any solid core, is to be regarded as air flowing upwards along the isentropic surface of discontinuity and consequently is not the same kind of air as that on the left unless the depression is occluded, in which case the motion decays.

All the cases of vortex-motion that have been displayed in experimental work have narrow revolving cores and the convection which disposes of the air-flow towards the centre is in spiral whirls of no great thickness surrounding a central area, the inner portion of a sheet of air as represented in our model.

It may be that the solid core of a natural vortex which forms the nucleus of the spin is of small dimensions, the convected air may pass round it as a sheet into the westerly current in which the depression is formed as represented in fig. 220 of vol. II. The same kind of motion is suggested by Ahlborn in the paper cited above.

The travel of cyclonic depressions

In accordance with the conditions indicated in previous sections the travel of an ordinary depression means the removal of a billion tons of air from positions successively in front of the centre, and the filling up of a like amount in the rear. The simplest way of visualising that result is undoubtedly to regard the depression as formed and carried along in what may be called

---

[1] 'On a cyclone which crossed the Corean peninsula and the variations of its polar front.' *Q. J. Roy. Meteor. Soc.*, vol. XLVIII, 1922, p. 169.

a solid current[1], the defect may then be transported without internal change. The alternative is to imagine the action of some kind of air-screw that carves air from in front and delivers it in the rear, in which case the depression may travel through still air as a torpedo or a submarine travels through still water. For the atmosphere action of this kind could only be by the operation of convection with a travelling locus of activity such as might be conceived for a stream of tropical air impinging on a cross stream of polar air with a continuous displacement of its own line of flow.

The alternative is much more difficult to visualise than the process of transport in a permanent current and the best example of a permanent current of suitable extent and velocity in temperate latitudes is the west to east circulation in the upper air.

According to Helmholtz's ideas of the formation of the tropical belts of high pressure the circulation would be from east to west in the region south of the belt of high pressure and from west to east in the region north of it; and, so far as we know, that is a correct representation of the actual conditions throughout the thickness of the atmosphere in the intertropical regions; in the temperate regions, intermittently on the surface, from the tropical zone of high pressure to the common path of cyclonic depressions, and in the upper air of the regions still farther north.

We conclude that the best method of dealing with the travel of depressions is to trace its relation to the upper circumpolar cap, a canopy which, as we have seen, according to the isobars shows a normal rotation of about 10° per day at the level of 4 km in July and a rotation of about double that amount at the same level in January. At higher levels in the summer the normal rotation is in inverse proportion to the density of the air. These speeds are of the same order of magnitude as those of the travel of depressions. We have shown in vol. II (p. 355) that the mean rate of travel over the British Isles ranges from a minimum of 8 m/sec in May to a maximum of 12 m/sec in December, 10° to 14° per day with an extreme range from zero to 37 m/sec (44° per day) over the same region. Humphreys[2], citing results of Bowie and Weightman, gives 10·9 m/sec in the summer and 15·6 m/sec in the winter as the corresponding values for the United States, and quotes Hann's figures of 7·8 m/sec and 12·4 m/sec for Japan. Over the Bering Sea he gives a maximum of 10·3 m/sec in the summer, and a minimum of 8·5 m/sec in winter and spring.

We have no very definite information as to the level at which the normal rotation begins, but in accordance with the ideas developed by Helmholtz it is not unreasonable to think the boundary surface not higher than the first isentropic surface which completely surrounds the globe. It may indeed be indicated by the isentropic surface which, according to the scheme set out

[1] The view that cyclones are carried along in the general circulation is adopted by W. J. Humphreys, *Physics of the air*, 1929, p. 173, and by S. Fujiwhara, 'On the behaviour of lines of discontinuity, cyclones and typhoons in the vicinity of Japan,' *Geophysical Magazine*, vol. II, No. 2, Tokyo, 1929, p. 129.

[2] *Physics of the air*, 2nd edition, 1929, p. 174.

in chap. VIII of vol. III, separates the overworld, to which the circumpolar rotation from west to east naturally belongs, from the underworld in which the circulation of air is more definitely related to local conditions of pressure.

The evidence of the persistence of the west to east circulation exists in those long stretches of the oceans of the northern hemisphere where there is a general westerly wind which is understood to have a tropical origin and which supplies the warm sectors of the cyclonic depressions of the temperate zone and forms the southern side of the discontinuity which is characteristic of those depressions.

The discontinuity separates the westerly or south-westerly current from the easterly, north-easterly, north-westerly currents of polar air of the under-world, which is carried so far south in the end as to replace the westerly wind fed by south-westerly air of tropical origin and "occlude" the depression, or cut off the tropical supply so far as the surface is concerned.

In that case the Norwegian school has shown that the tropical air may be found at some height above the surface travelling upwards over the polar air to form a succession of cloud and rain; and the lowest boundary of the cap of rotating air may be regarded as the isentropic surface which reaches the earth's surface farther south and passes upward over the polar air at an inclination of about 1 in 100 as calculated by Helmholtz.

The relation of cyclones to this may be indicated by the accepted principle that the travel of a depression is along the direction of the surface wind in the warm front, which might clearly be explained if the depression were equivalent to a mass of revolving fluid in the rotating cap, between the bounding isentropic surface and the stratosphere, of which the wind of the warm front would form part.

In support of this conception we may note first that if the lowest layer of the revolving cap were not more than 1 km up then the pressure-distribution at the surface would be mainly that of the revolving mass, which may extend upwards over more than half the equivalent thickness of the atmosphere between 1 km and 10 km, about 900 mb to 300 mb. The corresponding pressure-distribution would be transmitted to the surface. The kilometre from the surface is concerned only with about one-tenth of the atmosphere and the disturbance of pressure by that layer would not generally amount to as much as ten millibars.

It should be noted that the equivalent thickness of the revolving canopy diminishes with increase of latitude, because the isentropic boundary slopes upward with latitude and its isobaric surfaces presumably downward. The extension of the isentropic surface beyond the range that is already explored, as exhibited in the table under fig. 97 of vol. III, is a matter of considerable interest. It would give some clue to the way in which the entropy of the air can be reduced otherwise than by contact with a cooled surface.

Secondly, we are aware that observations of pilot-balloons which have been carried beyond the limit of the stratosphere show easterly winds as well as westerly winds and in those cases apparently the relation of easterly wind to westerly wind, which is quite common in depressions at the surface, must occur also throughout the troposphere. The occasions of these observations cannot be regarded as normal because with a depression on the surface the vision of pilot-balloons is generally prevented by clouds, but the existence of easterly winds in the upper air cannot be disregarded.

Zistler's[1] investigation of winds in the stratosphere shows northerly winds second only in frequency to the preponderating westerlies.

The travel of secondaries

It is natural to suppose that if a secondary is formed in part of the current of a cyclonic depression of considerable dimensions it will be carried along in the current in which it is formed. The direction of motion should therefore be along the isobars, but as the isobars are themselves in process of change the isobar must not be regarded as the trajectory of the air-current or of its load.

J. S. Dines writes:

The movement of these systems is generally one of rotation round the parent system as a centre in a counter-clockwise direction, to which is added the movement of the parent depression itself. The speed at which they travel seems to be governed to some extent by the steepness of the pressure-gradient associated with the main depression in the region in which they are formed. If the pressure-gradient is steep they travel more quickly than if it is slight. There appears in some cases to be an even more intimate association between their speed and that of the winds prevailing at high levels in front of their advance. When clouds of cirrus type which form at a height of some 5 miles above the earth's surface are moving rapidly from the west in front of a secondary, it may be taken as an indication that the secondary will travel more quickly than if the cirrus is only moving at a slow speed.   (*The Weather Map*, 1930.)

Stationary cyclones

Within the subject of the travel of depressions we must include those cases in which the depression does not travel, and in the absence of definite observations it is not reasonable to explain their behaviour by assuming that the rotation of the upper layers has been suspended. Examples of such stationary cyclones are not infrequent within the margin of the complex oceanic depression over the Atlantic, with a centre to the south of Iceland frequently referred to in forecasts. Other special cases in other localities that may be noticed are the depression of July to August, 1917, which originated over the North Sea and decayed after 3 days without travelling, and a depression, again over the North Sea, south-east of Spurn Head, which remained stationary, 21 to 24 July, 1930.

And here we must perhaps look to the effect of convective rainfall producing a depression not in the area of rain itself but not far away. The depression of 28 July to 6 August, 1917, was preceded by heavy rainfall over northern France and the eastern part of the English Channel; and the depression of July 1930 may similarly be related to the so-called cloud-bursts of north-eastern England on 20 to 23 July which caused disasters in the neighbourhood of Whitby.

[1] *Beitr. Phys. fr. Atmosph.*, Bd. XIV, 1928, p. 65.

We may perhaps also cite the case of the small depressions represented in fig. 72, the second of which might be revolving fluid with a maximum velocity of 16 m/sec, and which was attended by hardly any wind over the surface and no rain in the immediate neighbourhood but followed heavy thunderstorms over the English Channel Islands and south-east England.

We cannot say whether the same suggestion is applicable to the case of the Icelandic depression; we have no observations of rainfall with which to relate it; but we have seen that the area so favoured as a depression-centre is to the north-east of that part of the Atlantic Ocean where the difference of temperature between warm sea and cold air, conducive to convection, is most pronounced. If convection produces or intensifies depressions persistent convection would naturally produce a persistent depression. This leads us to another aspect of the question.

Loci of origin

We have a number of examples of travelling revolving storms and travelling depressions, and we may seek the loci of origin by tracing the paths back to their beginning. The evidence is fairly clear as regards revolving storms; they are formed near the equator possibly at the junction of northern and southern air, an intensification of the doldrums, and travel westward until they are carried northward and eastward. Generally speaking they become merged in depressions travelling eastward, but some supply of energy is probably necessary to keep them in being. Corresponding experiences are recorded for the typhoons of the Far East[1] and the cyclones of the South Indian Ocean[2].

We have already explained that well-marked depressions which are being occluded in the region of the British Isles may be traced back to kinks in the isobars of the south-west quadrant of the oceanic secondary.

Some years ago Willis Moore quoted the Aleutian low as a source from which depressions were calved. A. J. Henry[3] does not fix upon any special locality. We should certainly regard the north-west Atlantic and the Mediterranean as loci of origin of energy, if not for the commencement of depressions for the reinforcement of them.

C. L. Mitchell, in the paper cited on p. 300, gives an ample account of the behaviour of cyclones and anticyclones of the northern hemisphere, January to April inclusive, 1925, which includes four maps, one for each month, of the geographical positions of beginnings and endings of primary and secondary cyclones, and other four for the beginnings and endings of primary and secondary anticyclones, as well as a map of the around-the-world cyclone, 23 February to 23 March, 1925, which we have shown in fig. 74, and very much other valuable information which cannot be summarised in a few paragraphs. We extract the following:

[1] J. Algué, *Cyclones of the Far East*, Manila, 1904.

[2] R. H. Scott, *Cyclone tracks in the South Indian Ocean, from information compiled by Dr Meldrum*. M.O. Official Publication No. 90, London, 1891.

[3] 'Weather-forecasting from synoptic charts,' *U.S. Dept. Agric., Misc. Pub.* No. 71, 1930.

There are two Arctic areas from which nearly all outbursts of polar air take place. The first area is that part of the Arctic Ocean north and north-east and occasionally north-west of Alaska, extending, roughly, from longitude 120° to 180° W. The second is the area north and north-east of Scandinavia, and this area extends approximately from Spitzbergen to Nova Zembla. By far the greater number of anticyclones, which may be designated polar anticyclones, move out and spread southward and south-eastward from the first area. During the first four months of 1925 there were 27 from this area and only 4 from the second (Spitzbergen to Nova Zembla). It is because of the numerous polar anticyclones, together with the large number of cyclones from the Pacific Ocean, that Canada and the northern and middle portions of the United States east of the Rocky Mountains experience more frequent material fluctuations of temperature than any other area of the northern hemisphere.

The question that interests us at this stage is the relation of the polar front, which is the basis of the Norwegian system, to the boundary of the underworld and overworld indicated by the local values of entropy that we have postulated in chap. VI of vol. II.

With reference to this question we note that, apart from the cases of occluded centres which are not infrequent on our maps, the polar front is taken to be the line which joins the centres of successive cyclones. It has polar air on its northern side and equatorial air on its southern side. On that account we should expect it to coincide with the isentropic boundary of the underworld because a discontinuity of temperature without any discontinuity of pressure must mean a discontinuity of entropy, and the region between the line of centres of depressions and the tropical high pressure is a region of westerly winds associated with the overworld.

It is as difficult to trace a line of discontinuity of temperature round the northern world as it is to trace a line of discontinuity of entropy, and the reasons for this difficulty are perhaps more easily apparent in the examination of the conditions of entropy than in those of temperature.

Common experience, confirmed by Mitchell's maps, finds cyclones in origin, in full vigour, or in a debilitated condition over a very large belt of latitude from 40° northward, and the distribution is very much affected by the intrusion of the oceans between the great land-areas. The same may be said of the conditions of entropy which at the surface, particularly over the level surfaces of the oceans, are controlled mainly by temperature. Over the land-areas however, which are certainly important from the point of view of forecasting, the conditions depending on local pressure as well as temperature are affected not only by the diurnal variation of radiation, solar and terrestrial, but also by the interference of clouds and other conditions of weather with those fundamental agencies.

Over a continent a vast area may be attached to polar air in the early morning and to equatorial air in the afternoon, to polar air to-day and to equatorial air to-morrow.

This state of things may obliterate the local identification of the polar front and imposes a serious burden upon the dynamical forces. How it affects the inferences to be drawn from the conditions of entropy we have not yet had opportunity to examine.

Some time in the future it may be possible to investigate further the relation between these two aspects of the general meteorological problem.

We have visualised a primary permanent circulation in the upper air round what we have briefly designated the pole, and a complicated quasi-permanent oceanic secondary related to the south of Iceland from which we have drawn certain inferences with regard to the formation, development and travel of cyclonic depressions farther south. We must not be understood as regarding either of these circulations as having unalterable centres or structures. Accepting a very proper warning by W. H. Hobbs to be rigorously objective in such matters we remark that even the solid earth in its rotation round what we call the pole has its periods of precession and nutation, and we need not be surprised if the movements of the fluid air are at least four thousand times as skittish without being suspected of wilful disregard of law.

We must still ask observers to give us the actual conditions prevailing from time to time in the upper air of any locality and draw from them conclusions as to the fluctuations of the aerial pole.

CHAPTER XII

RETROSPECTIVE AND PROSPECTIVE

We look before and after
And pine for what is not.
To a skylark. P. B. Shelley.

LOOKING BACK—SOME COMMENTS ON SCIENTIFIC PROGRESS

WE have now completed our survey of the structure of the atmosphere and its changes, of the means by which they have been ascertained and expressed, and of the physical and dynamical principles by which it is sought to explain them.

We have referred to the structure and laws of travel of cyclonic depressions as the principal object of solicitude of meteorologists twelve years ago and as the object with which this Manual was originally planned.

Whatever the intention may have been the result stands exposed in the pages of its four volumes; and it is fitting that, in taking leave, we make a brief retrospect of what has been achieved and express the hope that even if the work has failed to attain its original object its readers may still have been encouraged to think the study of weather worth pursuing, not only for its practical utility but also for its engaging intricacy and its scientific interest.

The first volume was devoted to the history of the science, the instruments devised for its observations and the methods of expressing its accumulating information. It led up to the evolution of the weather-map and ended with a list of contributions to the theory of the phenomena and a note of regret for the subordinate character of the rôle of theory in the play of actual meteorological practice. We may suggest some reasons for that.

Comparative Meteorology

The second volume displays in pictorial or statistical form a summary of the information about the weather that has been collected from all parts of the world. The most notable feature is the established position of the month as time-unit, in spite of its inequalities and its maladjustment to solstices or equinoxes; and, next to that, the restriction of observations to the more regularly inhabited or civilised parts of the planet's surface.

That there should be no effective information about rainfall over the sea, and still more that, in spite of a succession of international conferences beginning with 1853, apart from a few squares cared for by the Netherlands, there should be no regular international system of periodical publication of the available data for the oceans, must always surprise those who wish to regard meteorology as a geophysical science. There are, of course, difficulties; but it is certainly the ambition and generally the duty of science to overcome them.

The lack of observations from the polar regions is equally obvious: there was an international effort for the polar year of 1882–3 which is to be repeated

(319)

in 1932–3 as a jubilee commemoration; but, with the interests of the whole world involved, and means of communication transcending the wildest dreams of the pioneers of the weather-map, the supply of knowledge of the current weather of the polar regions might fairly be regarded as a duty to humanity and not as an occasional luxury of geophysical students. That two millions of able-bodied persons should have to be maintained in idleness in one country, five millions in another, while science is enfeebled for want of workers, seems to indicate something defective in social or political organisation. Perhaps the League of Nations may see to it.

A revision of the articles

Our second volume included a final chapter entitled "The Earth's Atmosphere" which contained a series of articles that might be the foundation of meteorological theory.

We shall conclude this final chapter by citing the additions which we regard as justified by the subsequent volumes.   In the meantime the reader may support us in the opinion that too little attention was paid to Helmholtz's separation of the atmosphere into thermodynamic shells by surfaces of equal potential temperature or equal entropy. Nor was attention adequately drawn to the close reciprocal relation between the distribution of pressure and the distribution of velocity in the upper air under the influence of the earth's rotation that follows from Helmholtz's conceptions.

The physical processes of weather

A further comment on the original series of articles is to note that entropy has been found to be a more useful and convenient term than potential temperature or mega-temperature for indicating the separation of the atmosphere into thermodynamic shells (vol. III, chap. VI).   So long as the purpose is merely to indicate the geography of a separating surface either term is sufficient; but when we enter upon numerical values and propose to use them in computation then it is entropy which gives energy when combined with temperature. Mega-temperature requires to have its logarithm taken before it can be so used.

The doctrine of entropy in meteorological science occupies a considerable space in vol. III to which we now refer. We may fairly claim that the tephigram is an eloquent expression of the possible developments of energy by the action of its environment on dry or saturated air.

The purpose of vol. III is to describe and illustrate the physical processes of the atmosphere which are operative as transformations of energy in the phenomena of weather. It is energy which is the real subject of the volume, and processes are examined from the point of view of the part they play in energy-transformations. The reader is invited to satisfy himself as to the importance or otherwise of any form of energy that the atmosphere may exhibit. The forms which are recognised as important are thermal, gravitational, electrical and kinetic.

It is needless to say that any display of atmospheric energy is the consequence of variation in the balance of solar and terrestrial radiation. One of the features of the chapters on those subjects is our lack of knowledge of the processes of expression of radiation in terms of the temperature and entropy of air, guarding as they do all the avenues of transformation of solar energy into other forms of energy in the atmosphere.

The chapters on thermodynamics begin with the energy of radiation expressed in kilowatt-hours as an engineer's unit, and pass on to temperature, water-vapour and the theory of heat. They are carried forward to include the use of entropy in the representation of atmospheric processes and the expression of energy as the product of temperature and change of entropy.

The suggestion is made that maps should be drawn on isentropic surfaces which might be done if the surface-pressure and temperature and the lapse-rate were known. At a recent meeting of the International Union for Geodesy and Geophysics an appeal was made for the publication of material that would make the construction of isentropic surfaces easier. A reply was suggested to the effect that, as reduction to sea-level was a recognised process, it was easy to recalculate the surface-pressure from the published values of sea-level pressures, temperature and height; but the idea of a weary student with a weather-map of the northern hemisphere in front of him trying to recover by a separate calculation for each station within millions of square miles observations which are already in somebody's possession, and wondering whether the formula used at Verkhoiansk is the same in January as in July, or even if the same formula is used for Denver and for Calgary, is sufficient to indicate that the need for a real as distinguished from an imaginary map has not yet become obvious to the scientific world. Perhaps the time may come when isobars on maps may be regarded as a sufficient indication of sea-level pressure which need not be repeated in columns of figures as well. Some of the implications of the reference of phenomena to an isentropic instead of an isogeodynamic surface are given in the summary which closes this chapter.

Volume III concludes with an endeavour to represent the general circulation at the surface in terms of convection as indicated by rainfall, or as the relation of advective and divective regions. From the analysis we infer that rainfall is a summer phenomenon over the great land-areas and a winter phenomenon over oceanic islands and exposed coasts, presumably therefore over the sea. Some coastal areas combine both systems; but an inland sea, like the Mediterranean, associates itself with the oceans.

Terminology

The chapter raises the question of meteorological terminology and its shortcomings. It is an aspect of the subject that deserves consideration.

Scientific terminology should carry definite and unmistakable meanings. Who can estimate the loss which the science has suffered by calling loss of air in a particular locality depression, or who can enumerate the meanings that a reader is expected to attach to the words gradient, or temperature, or inversion? These things may be of small importance for the limited society of experts who understand the stenography of the science; but they cannot fail to be repellent to the vast number of persons who are interested in weather and have intelligence enough to pursue the study of the science if the initial steps in the path were judiciously paved instead of being unintentionally obstructed.

As an example of the value and power of well-chosen terminology, on the other hand, let us think of the achievement of the Norwegian meteorologists who have provided a technical language for the behaviour of the component parts of cyclonic depressions that within ten years has spread over the whole world and has made it impossible even to think of "cyclones" of the temperate zone.

If, with no disrespect, we may call the special aspect of the science with which they deal the physical geography of the turbulent layer, the treatment of the subject is ideal; we can offer unstinted admiration for the language used in its expression.

Units

Nor should we fail to mention the lack of uniformity among the nations of the world in the matter of units and the irony of the explanation that uniformity is impossible because the laity of the nations are themselves interested in the subject; that they cannot understand anything with which they are not already familiar or have not learned at school, however serviceable it may be from the point of view of their ultimate comprehension of a science and its meaning. What a debt of gratitude chemistry must feel to the late Sir Henry Roscoe for an introductory chapter on units in an elementary text-book used in schools, which induced the laity of the whole chemical world to use the same units as the leaders in their science.

Meteorological Calculus. Pressure and Wind

In this fourth volume we return to those theoretical aspects of the subject which can be treated by the customary methods of dynamics of fluids or solids. We have realised therein the efficiency of the strophic balance as a means of maintaining the proper adjustment of pressure and velocity; also the curious facility afforded by the earth's rotation for the travel of a vortex as a vertical column.

Therein also we may find some explanation of the lack of application of meteorological theory. In many cases, to facilitate computation, assumptions are made with the understanding that the result will be a sufficiently accurate picture of nature. But so long as the effect of the assumptions is not realised the conclusions lack vigour. We may give some examples.

Weather-mapping

We can hardly suppose it possible to get effective theoretical results when the elements are not rigorously co-related. Temperature and wind are surface phenomena and to relate them to pressure on some other plane than the actual surface is to invite failure in any theoretical endeavour.

There can be no justification for the tacit assumption that the natural motion of air is horizontal. It is true that the horizontal scale of the phenomena is much greater than the vertical scale. The whole thickness of the atmosphere might be represented by that of a very few sheets of the paper on which the map is printed. But large distances are included in the weather-maps and the vertical consequences are by no means negligible. Atmospheric motion must be considered with due regard for isentropic surfaces which will guide the air upwards or downwards. It is not unreasonable to regard the sequence of cloud and rainfall in a cyclonic depression expressed in the Norwegian scheme as originating in the motion of the air from near the surface upwards along an isentropic surface. The map of an isentropic surface in fig. 98 of Vol. III suggests a rise of 3 km in 900 km or 1 in 300, and to assume that isentropic surfaces are horizontal is to lose the key of the thermodynamical situation.

The other implication with regard to weather-maps for the current study of weather is that the horizontal section is only one of many possible sections of a three-dimensional subject. It requires to be supplemented by a vertical section in order that the dynamical conditions may be exposed. Whether it would be possible to give a vertical section of isobaric and isosteric surfaces or of isentropic and isothermal surfaces as a supplement to a weather-map is ultimately a question of meteorological organisation; but a single enthusiast might find a provisional answer. In days gone by, with the aid of Sir Arthur Schuster, twenty-four balloons were sent up from Manchester or Glossop Moor, one each hour on two separate days. It would not be much more costly or more difficult to send up twenty-four balloons simultaneously from different stations. Aeroplanes might take the stations where balloons might be lost. So, with good luck, it might be possible to determine the structure of some typical barometric distributions.

In this enterprise the universities ought to share and bring themselves once more into co-operation with the investigation of the atmosphere.

In the new language, the meteorology for which we have suggested Maxwell wrote his *Theory of Heat* has been occluded by the bi-polar air of his own laboratory. It waits with weary expectation for some pious benefactor to develop a secondary with its library, observatory and laboratory, that will remove the reproach common to universities of the British Empire of ignoring that branch of physical science which is most closely in touch with every phase of human life and endeavour.

Until a vertical section is possible it is to be feared that dynamical meteorology must be largely guess-work, as are indeed the sample sections given on p. 226 of this volume, and all the other sections with which its author is acquainted.

Mean values once more

In this connexion we may revert again to the question of mean values and remind the reader of the many occasions in the course of the work in which it has been apparent that a mean value hides much more than it shows. Mean values have undoubtedly been useful in the past for statistical or geographical purposes; but it is more than doubtful whether any good can come out of them for dynamical purposes.

It may indeed become possible to understand the structure of this or that individual cyclone; one can hardly conceive the possibility of developing the dynamics of the "mean cyclone"; hence our chance of employing the vitality of meteorological theory lies most probably in the possibility of organising detailed synchronous observations of the structure of individual cyclones. We cannot too often remind ourselves of Maxwell's suggestion that the determination of forces from observations of motion is a more trustworthy process than the computation of the motion from assumed values of the forces.

Beyond the weather-map

No-one can at present describe the reality of the structure of a single example of what has been assumed for a hundred years to be a vortex with a vertical axis. There are still many gaps in our knowledge of its structure and circulation which must be filled before our material is, in the proper sense, ready for the mathematician. In so far as it is used before it is ready in that sense, the mathematician who is asked to deal with the material is in a less favourable position than Newton was when he gave his attention to the dynamics of the solar system. We have therefore still to protect our observers from the suggestion that from the scientific standpoint their work is done and the subject may be left to those whom a hundred years ago Forbes classed as "philosophers who, in pursuit of other objects, step aside for a moment from their systematic studies and bestow upon the science of meteorology some permanent mark of their casual notice of a subject which they never intend to prosecute...."

Real progress in the science requires the continuous co-operation of the faculties of the observer and the mathematician, not merely the occasional assistance of the one or the other as separate specialists. Taking the widest possible view of the reasoning which we have put forward, the science of meteorology is already in debt to mathematics and the obligation will certainly increase as time goes on.

As this volume passes under his notice the reader who appreciates the limitations under which meteorology labours in the attempt to solve its problems by algebraical equations may have been reminded of old Kaspar's dilemma about Blenheim:

'And every body praised the Duke
Who this great fight did win.'
'But what good came of it at last?'
Quoth little Peterkin:—
'Why that I cannot tell,' said he,
'But 'twas a famous victory.'

For the author the situation is an intimate one because, for 50 years, after following Helmholtz's guidance in the study of hydrodynamics at Berlin, it became his duty, and for 20 years his responsibility, to bring together whatever might be useful in practice. He may therefore be allowed to set out for any reader who sympathises with little Peterkin his own expression of the "good that came of it at last."

(1) Accepting the record of an anemograph as the statement of the problem of the dynamics of the atmosphere we have learned from the reasoning of mathematical formulae: that the motion of the air at the surface is a composite picture of the motion of the air in every layer above the surface into which the atmosphere is or may be divided, suitably adjusted (as to density) for combination, but marred beyond recognition by the turbulence at the surface.

The only hope of a true picture lies in well-directed co-operative effort in the investigation of the upper air.

(2) The idea of pressure-distribution as the continuous automatic adjustment of pressure to wind in the upper air on the analogy of the banking of an aeroplane in its flight is a direct consequence of the equation originally given by Coriolis, and derivable from the Newtonian equations of motion.

(3) The uprightness of a rotating column of air in its advance may be inferred from the cosine term of the same equations.

(4) Following the mathematician's advice to analyse the motion of the air into u, v, w, that is west-east motion, south-north motion, and up-down motion, we may by inquiry assure ourselves that, when the motion is of a general character and not merely local spin, the west-east motion is derived from the earth's spin, south-north motion from the distribution of pressure, and up-down motion from the distribution of entropy.

(5) The meteorological "molecule" is the eddy of an elementary spinning mass of air to the accretion, combination or development of which must be ascribed all the varieties of motion from the shift or gust at the surface to the circumpolar rotation of the upper air of middle latitudes.

We are not aware that anybody has yet expressed meteorological gratitude for these conclusions, but if Peterkin will regard them as the results of famous victories, as he ought, he will not deny to the dukes who won the fights their meed of praise.

A FAREWELL VIEW OF THE ATMOSPHERIC CIRCULATION

Our retrospect shows that the process of exploring and expressing the available information about the atmosphere leads, as might be expected, to some modification of view in those parts of the subject which are imperfectly formulated. We conclude our exposition therefore with a brief summary of the ideas about the circulation of the atmosphere which seem best in accord with the facts and processes set out in the course of the work. We can best explain the view taken by exhibiting overleaf a snapshot (taken in our own camera) of the circulation at 7 h on 30 December, 1930, as represented in the Meteorological Office map of that date (fig. 76) together with a diagram (fig. 77) picturing the "top hamper" on that occasion.

SYNTHESIS OF THE GENERAL CIRCULATION OF THE NORTHERN HEMISPHERE

Features indicated are:

Orographic relief: contours of sea-level, 200 m and 2000 m.
The regions of intertropical easterly winds.
The prevailing westerlies and the associated cyclonic areas.

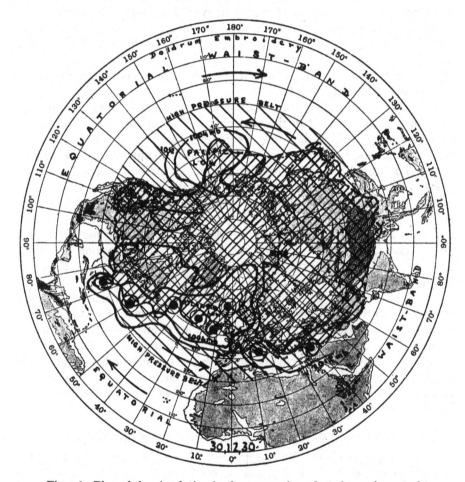

Fig. 76. Plan of the circulation in the upper air and at the surface at 7h
on 30 December, 1930.

The stronger continuous lines are isobars and carry numbers. The closed curves are
the isobars of 1004 mb which define the Atlantic "low" and the Pacific "low."
The isobar of 1012 mb is also drawn and includes with the Atlantic low the
Mediterranean extension on the east and a West Indian extension on the west.

⊙ The small circles with black centres indicate cyclonic depressions, regions of "fronts"
at the time of the map, 7h in Europe, 1h in America. The diameter of a small
circle is about 750 km.

SYNTHESIS OF THE GENERAL CIRCULATION OF THE NORTHERN HEMISPHERE

The features indicated are:

1. **Equatorial waist-band**, a region of general easterly winds at the surface and in the upper air.
2. A **polar canopy** with cyclonic circulation round a variable centre in the polar regions, forming the overworld. Westerly winds at the surface between 30° and the irregular line of the polar front form part of the canopy.
3. The **underworld** of polar air separated from the overworld by a surface which cuts the earth's surface in the polar front.
4. The **oceanic lows**, two sectors of the underworld affected by thermal convection of moist air.
5. **Cyclonic depressions** centred on the polar front.
6. **Convection** in the canopy leading from the front to the stratosphere.

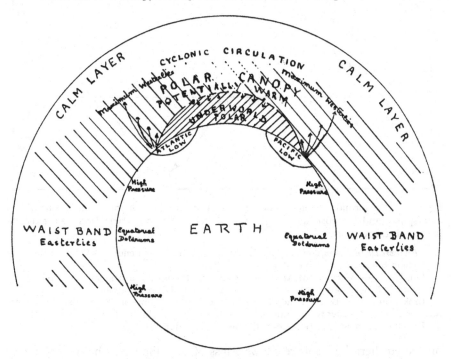

Fig. 77. Vertical section of the atmosphere along the meridians of 10° W
and 160° W at 7 h on 30 December, 1930.

Shows the region of the **polar canopy**, north of the equatorial waist-band with W to E rotation, and westerly winds at the surface between the southern margin and the polar front.

Shows the region of the **underworld**, below the polar canopy. The boundary at the surface is the isotherm of 40° F and indicates at the same time approximately the irregular southern limit of the polar front. The underworld is a region of polar air with anticyclonic tendency over-ruled intermittently on its southern margin by the pressure-distribution of the canopy.

The relation of the hemispheres

We have to think of the distribution of the atmosphere over the whole globe and the changes therein which are brought about by the motion of the air represented in the general circulation with all its local developments of advection and divection. All these changes are universally acknowledged to be ultimately due to the variation, periodic or otherwise, in the solarisation of the water, ice and land of the rotating globe in its annual path round the sun.

We have no simple formula for the details of the effect of solar radiation but as a general result we would not forget the close relation between radiative energy and air-mass on the northern hemisphere briefly referred to in chap. VI of vol. II. Exhibited in diagrammatic form in fig. 78 the recorded mass of air

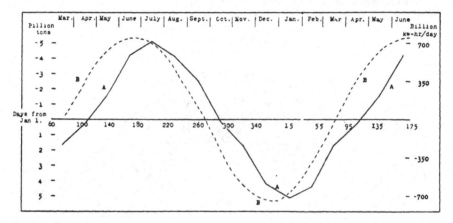

Fig. 78. The normal seasonal curve of variation of the mass of air in the northern hemisphere and the corresponding curve of the energy of solarisation. An exercise in graphical integration.

"An *increase* of 750 billion kilowatt-hours per day in the solar energy is followed by a *loss* of 5 billion tons of air 27 days later."

Full line ——— A (reversed): the mass of air over the northern hemisphere in billions of tons obtained by the graphical integration of the monthly pressure-maps.

Broken line ----- B: solar radiation over the northern hemisphere in billion kilowatt-hours per day according to the number of days from the beginning of the year.

Both curves show deviations from the mean values for the year.

on the northern hemisphere shows a loss of 10 billion tons between January and July with a sequence closely following the curve of variation of the receipt of energy from the sun.

This change in the total mass implies a transfer of appropriate amount across the equator from all the latitudes of the hemisphere, and consequently motion from north to south, which is not by any means apparent in the maps that we have put before the reader. We certainly show a steady flow of air towards the equator in the permanent trade-winds part of which has to come from the north as polar air, but we show a large part of it returned to higher latitudes round the tropical high pressures of the oceans to assume the rôle of equatorial air, and none crossing the equator to correspond with the diagram. Indeed, during the summer months when the delivery of air to the southern hemisphere is approaching its completion, there is an immense delivery of air going on from the southern hemisphere over the equator in the Indian Ocean to maintain the summer monsoon of India (vol. II, fig. 161), and conversely there is a great delivery in the other direction during the winter months to maintain the NW monsoon of Australia.

We must therefore regard the delivery of the 10 billion tons as taking place by some air-route, but at what level we cannot say. The maps of the distribution of pressure in the upper levels show a west to east circulation for the maintenance of which some drift towards the pole is necessary, with corresponding drainage from the lower levels.

But the 10 billion tons is a very small fraction of the whole mass of air, about one-fifth of one per cent; and the delivery spread over six months could hardly be represented on the wind-charts and might easily be lost sight of in the doldrum junction of the north and south.

Without undue neglect therefore we may regard the transfer between the hemispheres as a small incident in the atmospheric fortunes compared with the maintenance of the west to east circulation that extends over half the hemisphere in the upper air with a fringe at the surface covering a belt between 35° and 60° on the oceans.

Air in cold storage over the continents .

The complexity of the situation is strikingly illustrated by the fact that a great deal of the air imported from the southern hemisphere between July and January, if not the whole or even more, is stored in the cold winter anticyclones of Siberia and America. The weight of the additional air of the Siberian anticyclone on 23 November, 1907, is estimated by Loris-Melikoff and Sinjagin at 6·82 billion tons. The change of load from summer to winter over the whole Eurasian continent would be much greater. We note from the pressure-maps of vol. II that whereas the pressure over the American continent, with an alignment of its orographic features N and S, ranges only from 1012 mb in June to 1020 in the winter months, that over Asia, north of the Himalaya, ranges from 1005 in June, July to 1035 in December, January, and south of the mountains from 996 in July to 1015 in the five months November to March. The Atlantic Ocean supplies some air for the great winter load, the pressure in the "low" ranges from 1010 mb in the summer (1015 in May) to 1000 in December, January whereas the tropical high maintains a steady figure with only a rise of 5 mb in June, July. It is not surprising therefore that the distribution of pressure over the oceans shows the greatest gradients in winter—the deepest excavation—when the supply of air for the hemisphere is at its maximum.

An interesting point about this form of cold storage which amounts to an imprisonment of the air on the continents is the maintenance at sea-level of the gradient of pressure by a circulation of the right magnitude and in the right direction round the store. It is in accordance with the principle of the strophic balance on a very large scale and the occasions when the guard fails to restrain the cold air, in consequence of lack of velocity or from some other cause, mean a good deal in the matter of the weather of the eastern Atlantic, the Mediterranean and the Baltic.

Harmony and syncopation of the atmospheric sequence

The relation of the hemispheres as thus expressed may be regarded as an annual atmospheric tide which is full in January, ebbs southward from January to July when it is at its lowest in the northern hemisphere, and flows northward from July to January. This regular ebb and flow has its full tide and low tide within a month of the solstices. In fig. 78 we have related it to the variation in the receipt of solar radiation in the northern hemisphere which has its maximum and minimum at the solstices; but the lowest of the air-tide corresponds with the highest of the radiation. In the diagram we have drawn a parallel between the departure of the air-tide from its mean value and departure of solar radiation from its mean, reversing the curve in the latter case to show the parallelism. Thus we have a new basis of comparison expressed by the same curves, namely the rate of loss of air compared with the rate of increase of solar energy.

If we plot the rates of increase and decrease instead of the actual states of the tides we get the maxima and minima about the equinoxes and the states of no change at the solstices. Thus we have two types of curve which can be associated with the cycle of the seasons; one represents a static condition of solar radiation and air-tide, with its solstitial maxima, and the other represents a dynamic condition, a rate of change which has its maximum and minimum at the equinoxes.

If the circulation is really general we ought to find examples of seasonal changes in the meteorological elements associated with the tide of solar energy or air-load which may suggest some more intimate relation between the different parts of the circulation than is commonly recognised.

Temperature is obviously one of the static characters with a syncopation that is different in different countries. Its relation to the solar radiation is a subject in itself. Turning over the pages of the three earlier volumes we find data of seasonal sequence for a number of other elements.

Two of them are what we have called solstitial, one is the latitude of the northern limit of vigorous convection of the equatorial rain-belt displayed in the maps of chapter x of vol. III. The other is rainfall on the western coasts of the great oceans which may be connected with the westerly current of equatorial air as a sort of inductive current due to the drift of equatorial air poleward. The example which we have selected is the rainfall of Valentia in SW Ireland; many more are exhibited in the "new view of the rainfall of the globe" in vol. III, p. 406. And other rainfall diagrams syncopated by half a period from that of the west coasts can be found for many continental stations.

As examples of dynamic effects with maxima and minima about the equinoxes we have four examples besides the ebb and flow of air between the hemispheres. They are the strength of the NE trade which harmonises with the ebb of the air; and the strength of the SE trade which harmonises with the flow of the tide in the opposite northern hemisphere. Between these is the latitude of the belt of doldrums of the Atlantic which harmonises with the NE trade but is syncopated by half a period from the SE.

Finally the rainfall of England S which, as seen in fig. 183 of vol. II, harmonises well with the SE trade of the southern hemisphere and has its maximum in October.

That is in all probability a composite curve made up of the combination of west coast rainfall and continental rainfall with an accumulated accentuation of the summer influence.

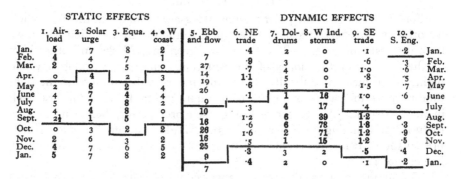

Figures in black type are above the mean value.

Data are taken from the following sources:

1. Excess of air-load over the N hemisphere in billions of tons (vol. II, p. 213, vol. IV, p. 328).
2. Solar urge over the N hemisphere in hundred billion kw-hr/day from the mean (vol. IV, p. 328).
3. Northern limit of convective rainfall over Africa in degrees N of 7° (vol. III, figs. 137–48).
4. W coast rainfall Valentia in cm. above 12 (vol. II, p. 182, vol. III, p. 406).
5. Ebb and flow of air in "shaws" computed from the figures in 1.
6. NE trade in m/sec from 4·9 (vol. II, p. 277).
7. N limit of the doldrums in degrees N of 6° (vol. II, p. 242).
8. W Indian hurricanes, 1887–1923 (vol. II, p. 364).
9. SE trade, 1892–1903, in m/sec from 8·2 (vol. II, p. 286).
10. Rainfall S England, 1866–1900, in inches from 2·5 (vol. II, p. 286).

Data for all these variations are given in the table. They have all been rounded off to avoid a multiplicity of figures and they are reduced to express differences from the mean for the year to facilitate comparison.

There is undoubtedly some dynamical relation between these various elements which expresses itself in the difference of phase; but we must look for an opportunity to pursue that line of thought in the future and take the preliminary question of the causes of motion.

The initial stage of atmospheric movement is the transformation of radiant energy by thermal change which increases or diminishes the entropy of the air and brings the energy within the operation of the second law of thermodynamics; and in the course of its operations causes winds.

PROCESSES WHICH CAUSE WINDS

The earth's rotation

If we concentrate our attention on the normal west to east circulation of the upper air we must, with Helmholtz, regard the velocity as due to the contraction of rings of air along the earth's surface towards the polar axis under the influence of the earth's rotation. The transfer of a belt of air, or an arc of a belt, from 50° N to 60° N would develop a west to east velocity of 240 km per hour. Other plans of developing velocity in the upper air are transient and local compared with the universal aptitude of the earth's rotation. But the continuous movement northward requires room to be found for the air in the higher latitudes, and must take time enough for a good deal of the incidental energy to be lost in turbulence.

Slope effect of radiation

For the drainage in the polar regions necessary to maintain the circulation we bring in the effect of a slope losing heat to the sky by radiation which we have discussed on p. 227. The slopes of Greenland, Alaska and other high northern lands, or the slopes of the Antarctic continent, working day and night, would make no difficulty about inducing a few billion tons at the right time to make its way over the sea to maintain the trade-wind and lose itself in the doldrums. The danger of losing itself prematurely lies in its turning eastward and finding itself over water that would take off the chill of self-preservation. It might become maritime polar air, which behaves very much like equatorial air.

The amount of air available in this way could be calculated, but at present no actual observations of slope-effect are at our disposal and calculation would involve some assumptions which we cannot verify.

Another possibility for the development of cold descending air which would have a chance of bearing the rôle we have given to slope-effect is the cooling by radiation from water-drops or water-vapour in the air; that is also a subject which invites attention before any effective statement of its influence is possible.

On account of the defect of the necessary entropy cold air moving southward cannot leave the surface, it will turn westward unless otherwise guided by pressure.

The trade-winds express this, they are always colder than the sea-surface. At the equator they can go west or go up.

The drainage of the polar regions causes a flow of air northward and consequent west to east circulation which may extend until it adjoins the east to west current; between the two, high pressure must be formed some of which leaks at bottom but is maintained above the surface.

Upward convection of wet air

In any recital of the influences available we must not forget the most impressive of all the atmospheric forces—the upward convection of saturated air, the primary cause of thunderstorms for the energy of which in various circumstances the tephigram provides an estimate. That form of convection certainly provides for a rapid passage of the water-laden air to the upper layers of the atmosphere, stopping short perhaps only at the tropopause. But its influence hardly extends to the relation between the hemispheres: it is generally restricted to developing the kinetic energy of local circulations which are incidental to the local removal of part of the air of a vertical column.

We draw a distinction between this form of convection and the upward movement of air *along* an isentropic surface which we suppose to take place in the eastern portion of the warm sector of a depression.

Convection is often attributed to warm air apart from any consideration of the water-vapour which it contains; but the regard which has to be paid to the distribution of entropy as set out in chap. VI of vol. III makes it clear that the result of local warming of air is a pool of air in isentropic or convective equilibrium, without any power to penetrate the upper layers unless provided with a suitable supply of water.

According to our view downward convection of cold air combined with displacement of the upper layers, which would be persistent even "if the earth went dry," is responsible for the primary circulation of the atmosphere from the polar regions to the equator at the surface and from the equatorial regions to the pole in the upper air; the upward convection of wet air for the development of the secondary circulations in the primary, and the descent of cold air, again, for certain striking incidents in the behaviour of the secondaries. These aspects we will now recall.

THE EVIDENCE OF THE MOTION

We pass on from the consideration of the most potent influences to the evidence of the motion which they produce.

Starting from the anemometric records as the statement of the problem we can easily recognise four classes of wind. If we discount as belonging to the fourth class the most obvious irregularities, which are due to turbulence, we may pass on to recognise first the steady current like that of fig. 3 which always shows a close relationship to the earth's spin, second the wind of a well-formed cyclonic depression with its curved isobars and its apparent local spin (fig. 8), and third the intense but temporary squall such as that of the Kew record (fig. 5) which has no direct relation with the earth's spin.

The primary system

Limiting ourselves primarily to the winds of the first or strophic type we may assert that the identification of any separate mass of air is a question of spin. The stream of air which can be identified as a separate current and traced from one part of the world to another as a moving air-mass is conditioned by the earth's spin. By the earth's spin we explain the westerly winds over the surface of the oceans north of the tropical belt of high pressure and the easterly winds of the same oceans south of the same belt, and in the upper air we find the continued expression of the same action in the circulation of the upper air of the northern hemisphere. It is indicated by the distribution of pressure at the level of four kilometres which shows an oval-shaped circulation round the pole stronger in January than in July. The comparative quiet of the conditions of July must be secured by the removal of air from the levels below 4 km because the weight of air at that level is enhanced in the summer. It is apparently the underworld which is warmed and not the overworld. The warmest specimen of the overworld in England is claimed, by the *Daily Weather Report*, for October, the coldest for April.

According to the diagrams of fig. 180 of vol. II the layer at 8000 metres makes a complete revolution in 23 hours in July, at 4000 metres in $23\frac{1}{4}$ hours, at 2000 metres in $23\frac{1}{2}$ hours; in January in $22\frac{1}{2}$ hours at 4000 metres; while over the equatorial zone at 4000 metres the revolution takes 24 hours $12\frac{1}{2}$ minutes and at 2000 metres 24 hours 10 minutes.

That surely constitutes the primary cyclonic circulation of the northern hemisphere, the great circular vortex associated with a continual drift of air from the equator towards the pole in the regions above the surface, an immense vortex which is always present but varies in its general intensity and in the location and strength of its secondaries.

It is this drift polewards which we regard as the agency for the conveyance of warmth established by Dr Simpson's work on radiation (vol. III, p. 169).

It must be remarked that the isobars from which the motion of the primary cyclone is inferred, being oval, would not indicate the steady revolution of the upper air as described in fig. 180 of vol. II except as a mean result for all the different longitudes; but here again we want more definite observations. The isobars are drawn for the average of all the months of July and January during which observations have been made. We want at least a sketch of the motion at one epoch. With the multitude of observations by pilot-balloons now available it might be possible if some provision could be made for soundings from the sea-surface to obtain a sketch free in part at least from orographic disturbance.

The primary circulation which we have cited is only part of a general scheme which includes a corresponding primary circulation from west to east in the upper air of the southern hemisphere and a circulation from east to west throughout the atmosphere above the intertropical zone. The whole zone between 30° N and 30° S covers one-half of the earth's surface and the circulation over it is in fact itself not quite homogeneous. There is a northern belt fed by the north-east trades and a southern belt fed by the south-east trades; these may be of very different qualities when one comes from a summer hemisphere, the other from winter. They have a junction through the doldrums which exhibit the kind of weather natural to the junction of air from two different sources. Here let us remember the tropical hurricanes of the North Atlantic, associated with the strength of the SE trade and the northing of the doldrums.

We find the doldrums junction over the oceans; over the land is a belt of convection marked by rain (vol. III, chap. X) which is on the southern side of the equator in the southern summer but moves to the northern side for the northern summer, and extends then indeed over the equator to include the monsoon area of the Indian Ocean.

These three circulations in the upper air, one round the north pole, the other round the south pole and the third a dual one in the intertropical zone, can all be referred directly to the effect of the earth's rotation upon air moving north or south. They constitute the primary general circulation of the atmosphere. The polar circulations are true perennial cyclones with pressure-gradients balancing the motion.

The west to east circulation round the pole is separated from the east to west circulation of the intertropical belt in each hemisphere by a zone of high pressure over the sea and a succession of deserts over the land. The position of the high pressure at 30° N or S is deduced by Helmholtz as a consequence of the displacement of the rings of a quiescent atmosphere northward or southward under the earth's spin.

The underworld

In writing of these circulations we have referred them all to the upper air. If we carry our thought downward to the surface we find some interesting results. In the equatorial belt, as we have seen, the motion is still easterly over the oceans except in the monsoon region; over the land it is irregular. And for a certain distance northward on the north side of the high-pressure belt we have the west to east circulation of the primary cyclone in being at the surface but not infrequently interrupted by incursions of cold air from the north. At the surface the air is derived in part at least from that which has come round the western end of the high pressure and is called by the Norwegians equatorial air.

The belt of west to east circulation at the surface forms part of the lowest layer of what in chap. VIII of vol. III we have called the overworld, and as we proceed farther north between the latitudes 40° to 60° we find the line of separation of the overworld from the underworld. Farther north still, beyond the line of separation is the underworld itself with its own surface circulation formed of air which has come southward from the polar regions, or westward from the cold continent—the polar air of Norwegian meteorology.

The normal existence of these two systems is quite clearly shown by the analysis of the normal surface-pressure into the distribution at 4 km and a nearly equal and opposite distribution for the layer between 4 km and the surface which follows closely the lines of surface-temperature. We find similar opposition between the distribution at 8 km and that which represents the layer beneath the level of 8 km.

We can find similar indications of opposite equivalent circulations by a section at 2 km but the surface of separation between the primary cyclone and the lower air is not itself a horizontal surface. According to our view the surface of separation is an isentropic surface which is subject to variations in shape and extent due to changes in the temperature and pressure of the air at the junction.

In this connexion we may remind ourselves of the uniformity of density at 8 km shown in the mean values with the consequent conclusion that, normally, below 8 km the density of air increases with latitude and above that level, on the contrary, decreases. The increase with latitude will be greater at lower levels.

There is a corresponding uniformity of temperature at about 12 km interrupted by the tropopause about latitude 50° and a similar reversal above as compared with below of the temperature-gradient in the true sense of that word. Below that level there is a gradient northward and above it a gradient southward.

This kind of reversal seems to be characteristic of the distribution of all the atmospheric elements. P. R. Krishna Rao[1] suggests uniformity of temperature of 220tt at 25 km, and beyond that we have to allow for increase of temperature up to 300tt in the ozone layer.

The air of the overworld at 4 km carries the isobars of a cyclonic circulation, round the pole, which extends to the surface where the west to east circulation is experienced. It is also to be found, by rising above the surface, in that part of the hemisphere which is within the underworld, north of the line of separation. And thus we form an idea of the overworld as a revolving canopy with the region of westerly wind north of the high-pressure belt as the margin at sea-level on the earth's surface. The isentropic surface which forms the boundary leaves the earth at a very small angle from the line of separation, about 1 in 400 to 1 in 100, and passes over the pole at the level of several—perhaps six—kilometres. It is carried beyond 3 km in fig. 98 of vol. III.

North of the line of separation, the discontinuity of Norwegian meteorology, the isentropic surface in the free air separates the overworld above from the underworld beneath, and as the underworld is supplied by air descending the slopes of the polar regions on its way southward, or by air which has lost heat on account of radiation from its water-vapour, we have a region of polar air in which the tendency is for air to turn towards the west subject to a régime quite different from that of the overworld.

There must be a certain amount of flow into the underworld from above and consequently a flow out of the underworld across the boundary of the overworld. This crossing expresses itself as cyclonic depressions and the result of the influence of the colder air is that the underworld obtains access to the persistent trade-wind.

It is this persistent rotation of the overworld which we have regarded as accounting for the travel of depressions. In so far as they travel they must be in some way the expression of spin. What form the expression takes can only be determined ultimately by observation, for which at present there is no adequate provision.

---

[1] 'Distribution of temperature in the lower stratosphere,' *India Meteorological Department*, Scientific Notes, vol. 1, No. 10, Calcutta, 1930.

Secondaries

Over this underworld the circumpolar circulation of the overworld proceeds. The marginal line at the earth's surface between the two circulations marks the position of the surface discontinuity, the line of centres of cyclonic depressions, the polar front of the Norwegian system.

In that margin are many irregularities due to the variations of temperature which control the distribution of entropy by night or day. The juxtaposition of equatorial and polar air causes local circulations in the form of large cyclonic depressions of the Atlantic and north-western Europe which are in reality secondaries of the great primary of the northern hemisphere; and these depressions again are identified by spin, the spin about the local axis which may be incomplete at the surface.

The great secondaries, usually counted as primaries on a map of western Europe, express themselves as the deportation of an amount of air of the order of a billion tons, and in like manner the formation of a European anticyclone represents the dumping of a billion tons. So likewise the formation of what we have called the primary cyclonic circulation of the hemisphere must represent the removal of several billions of tons and its maintenance the continued removal of any natural inflow that would deteriorate the circulation. It seems probable that this is the only complete and persistent cyclonic circulation of any considerable size in the hemisphere.

The question of how the removal of such vast masses can be accomplished is answered partly by the immediate and automatic operation of the tendency to compensate for any departure from the strophic balance; any increase of wind brings into operation a force transferring air across the current up the gradient and this operates throughout the whole depth of the current. But undoubtedly penetrative convection of dry or saturated air comes in also to develop a gradient towards the centre in a way that is not yet explicit.

For the understanding of the rôle of convection in the formation of secondaries let us remember that since the surface separating the overworld from the underworld is isentropic the air of the underworld cannot cross the boundary and penetrate the overworld unless it is supplied with heat as by conduction, or by radiation, or by the condensation of its water-vapour. We can allow that the air of the overworld is free to move along the isentropic surface of separation and attain a level of some kilometres, ultimately developing cloud and possibly rain as described in the Norwegian life-history of the warm front. Thereby it may obtain the energy necessary for penetration, but as between the overworld and the underworld we have to recognise the juxtaposition of warm front and cold front at the earth's surface.

Upon the situation represented by the fluctuations of the boundary of the underworld at the surface we have to impose the differences of temperature and wind at the line of discontinuity and the possibility of convection from the juxtaposition of these different kinds of air.

If circumstances are favourable the convection may penetrate through the overworld at some part of the front and develop something analogous to the traditional motion of a cyclonic depression as long as the juxtaposition of different kinds of air is able to maintain the convection. Thus convection may form a secondary to the circulation in which the convection takes place. And on the largest scale, when the secondary is developed in the persistent cyclone, the juxtaposition may be continued long enough to form what is recognised in ordinary map-work as a great depression covering hundreds of thousands of square miles and corresponding with the displacement of a billion tons of air.

The possibility of a vertical column which preserves its attitude in spite of the differences in the motion of the air at different levels seems to be provided by a term in the general equations of motion which is proportional to the vertical velocity and the cosine of the latitude.

A vigorous secondary or large depression of this character has a trough-line where the polar air moving from west or north-west is displacing the equatorial air, leaving only a protuberance northward of equatorial air as an encroachment of the warm front of the overworld upon the underworld. This appears on the map as an advance southward and eastward of the cold front, and when the last of the equatorial air has been surrounded and occluded the play of convection is practically over; the underworld has advanced southward and the air belonging to it is thrown into a steep ridge by the arrest of its westerly motion in consequence of its translation southward.

The site for the display of convection and development of secondaries is thereby pushed southward to the new discontinuity and the decay of the original centre is followed by the development of a new one in the new position of the discontinuity.

At the new centre convection may begin, how far up it will extend will depend upon various circumstances—the condition of the upper air in respect of entropy, the difference of condition on the two sides of the discontinuity, and the duration or maintenance of the difference. Hence the development of the secondary is a speculative problem.

There is no lower limit to the volume of fluid in which a convective current will produce a spin if judiciously applied; for experimental purposes a few cubic inches are sufficient, so that in any air-current which is itself or by its environment dependent on the earth's spin the juxtaposition of air of different densities is likely to be attended with the analogy of a tropical revolving storm. So we are tempted to recall L. F. Richardson's transformation of a well-known couplet:

Big whirls have little whirls that feed on their velocity
And little whirls have lesser whirls and so on to viscosity.

But viscosity hardly seems an appropriate limit for spin: it may follow the same statistical law but it is not of the same physical construction; atomicity is perhaps the ultimate limit of spin with the solar system as a large-scale example in the other direction.

All these secondary spins are carried in the current in which they are formed, and, if we regard them all as successively derived from the original permanent cyclone of the northern hemisphere, that will carry the whole series with it on its way. So far as maps are concerned the circulations are modified in their appearance by the motion, as they are also by a translation northward or southward. The most vigorous spin will hardly be able to show any motion eastward if it is displaced southward by as much as one or two degrees of latitude. It must be a wise forecaster that recognises all the spins on his map.

Occlusion

In working with a modern weather-map we hear a good deal in Norwegian about depressions being occluded. In our own language it would be the impounding by the underworld of a bit of the overworld.

If that were the only kind of result of the perpetual conflict between the two worlds it is evident that in course of time the underworld could capture the whole of the surface display of the overworld. At the surface the whole of equatorial air would be occluded by the polar air.

This obviously does not happen. The conflict is repeated time after time in the same locality and that can only occur if the overworld gets its own back by capturing air from the underworld, or perhaps the process may be better expressed as allowing the polar air to advance southward. Indeed the cut-off of the margin of the polar air must be of much greater order of magnitude than the occlusion of small portions of equatorial air if, as we think, the flow of the trade-winds depends upon the supply of air from the polar regions.

To develop the implications of this view requires the examination of the line of discontinuity in the region between the well-developed depressions. We cannot enter into that subject here and now, but we should like to introduce the aspect of the processes of depressions which is obtained by representing them on an isentropic surface instead of on a geodynamic surface.

Katabatic winds in the isentropic surface

Let us pass therefore from the consideration of what we may call a normal air-current controlled by the earth's spin to the spasmodic wind of the mountain-slope, the trough-line squall or thunder-squall—boisterous winds which pay practically no attention to gradient.

The examples which are most open to inspection are the katabatic winds of mountain slopes, the blizzards of the mountainous north or south. We are not prepared to deny that if allowed to go on long enough the downward flowing air, in consequence of the uncompensated centrifugal force, would gradually build up an anticyclone on the right of its flow with a corresponding depression on its left; but there is not time for that. The phenomena of katabatic winds deserve more attention than hitherto they have received. We commend, therefore, to the generous reader the further investigation of slope-effect.

We pass on to the consideration of squalls which are experienced in special types of weather without any visible guiding slope. From all that is known about them we may conclude nevertheless that they also are katabatic and represent a cataract of cold air descending the forward slope of a cushion which the cold air itself has formed. The example in our statement of the problem of dynamical meteorology is that of the severe squall of very short duration on 1 June, 1908, fig. 5, which was very destructive of trees in the celebrated avenue of Bushy Park. With some assumptions the sacrifice of geopotential of the air during descent in such a squall as that of Kew could be calculated from the wind-velocity; it would be several hundred geodynamic metres.

Other examples are to be found in the records of line-squalls and the troughs of cyclonic depressions, and the point which we wish to make here is that their natural indication is a steep slope on the isentropic surface in the immediate neighbourhood of the cyclonic centre or the localised atmospheric disturbance.

Corrugations in the isentropic surface

The explanation which we should give is that the polar air advances east by pushing its lower layers under the air of the warm front and forms the kind of cataract suggested in *Forecasting Weather* as representing the conditions at the trough with a surface-wind enhanced by downward convection. The surface-wind often takes the form of a violent squall, and is in every way like the katabatic wind which forms a blizzard; except that the slope down which it moves is that of an air-cushion instead of a mountain side.

From the point of view of the isentropic surface the advance of polar air under the overworld may thus be represented as the pushing forward of spurs of polar air from a plateau in the north, like the fingers of the hand from the palm, or the ridges and furrows of Norwegian fjords, and forming corrugations in the isentropic surface. From the principle already enunciated it follows that the mere fact of their moving southward tends to arrest any eastward motion that they may have and gives the corrugations a steep face on the eastern side.

Thus in the language of the isentropic surface a succession of trough-lines would imply the passing of a succession of corrugations each with a steep slope on the eastern side. The dynamical conditions during the continuance of the squall might be analogous to those of the inner part of a small secondary where we might expect vigorous convection of wet air.

If we use the isentropic surface with its successive corrugations as the surface upon which to draw the dynamical or thermodynamical lines we may take fig. 98 of vol. III as a specimen representing the vigorous cyclonic depression of 11 November, 1901, a notable secondary in the perennial northern cyclone.

Judging by the single example we find the centre of the depression to be a point near the edge of the steep slope of the corrugation of the isentropic surface of discontinuity pushing out from a plateau over the far north of Scotland and terminating in a tongue off the south of Ireland.

On this surface the isobars, which are also isotherms and isosteres, are not circular or even closed; they also are fingers pointing southward with sharply curved ends just passing round the centre. A finger or thumb of the opposite kind is pushed out at the foot of the slope by the warm front on which the lowest thermodynamic line has also a sharp curvature the opposite way round.

This is a novel feature of cyclonic depressions; they have generally been regarded as symmetrical with regard to the centre, but clearly that is not so in respect of entropy. There may be symmetry in a west to east section but the south to north section makes no suggestion of that kind.

The relative motion

The diagram of fig. 79 is intended to represent the boundary line of the underworld of the northern hemisphere which is very deeply indented on account of the distribution of land into two continents with intervening oceans.

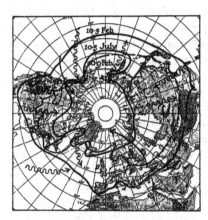

Fig. 79. Approximate boundary of the underworld on 5 February and 25 July, 1930, being the line of separation between polar air and equatorial air indicated by the isentropic line of 10·5 million c,g,s units; with the lines of outflow of polar air from and inflow of equatorial air to the polar regions suggested by F. M. Exner for 23 February, 1914, plain lines, and 2 March, 1914, wavy lines.

The line of 10·0 million c,g,s units on .5 February, 1930, is given in broken line for the contrast between land and sea.

We may expect that on a circumpolar chart an isentropic surface will mark a line of discontinuity between the overworld and the underworld, with corrugations like those in fig. 79 for 10·0 c, g, s in February and 10·5 in July. On that surface with that boundary the air of the overworld in its revolution round the pole will be passing. The underworld of polar air will be affected by the distribution of pressure of the overworld but will have an independent scheme of motion. The upper air of the overworld will be passing over the corrugations of the underworld and the indentation of the surface of separation will appear as waves in the barograms of those stations over which the dual system passes.

Where the slope of the passing isentropic surface is steep, secondaries may be established.

The southern hemisphere

In the southern hemisphere the arrangement is simpler beyond latitude 40° because there is a free ocean surface between that latitude and the Antarctic land and ice about latitude 70°, except for the projection of South America to latitude 50°.

The discontinuity boundary will be an indented line somewhere between 40° and 60°. The concentration of the high land round the south pole gives unity to the shape of the upper surface of the underworld; the corrugations should be defined by the intermittent downward streams of cold air.

The revolution of the overworld, the primary cyclone of the southern hemisphere, over the underworld is not so much disturbed by large secondaries and the alternations of pressure at the surface will be more regular. Whether these correspond with the waves of pressure described by Dr Simpson in his account of the meteorology of the Antarctic regions may perhaps be determined by subsequent observations. We have already indicated in vol. II, p. 372, a sequence of pressure-changes passing round the hemisphere in 38 days.

The elusive vortex

A horizontal section at sea-level will cut the separating surface and show sea-level isobars belonging in part to the one layer and in part to the other, an unhomogeneous arrangement to which attention is called in maps of the Norwegian type. But the section may be called an unnatural one as compared with the natural one which keeps to the dividing isentropic surface and shows the isobars as thereon represented in fig. 98, vol. III. There apparently the depression may be a region of sharp curvature but it has no centre. The isobars show no circular form, but as they are also isotherms and lines of equal density they give the simplest representation of the structure. The "un-naturalness" of a horizontal section is apparent in the sudden transition from warm to cold, north of cyclonic centres, in the trajectories of air which are shown in *The Life-history of Surface Air-currents*.

We have still much to learn about maps on isentropic surfaces before we can use them with the facility to which we have become accustomed with sea-level maps, and it will indeed be a surprising result if the column of revolving fluid with circular isobars which used to represent the cyclonic depression of middle latitudes—a cyclic form which has been hunted assiduously for more than a hundred years—turns out to be a mirage after all, a picturesque physical integration of the distribution of pressure at all levels.

Stranger things than that have happened in the history of science, and we leave the future in the hands of the reader, confident that there is yet much to learn and much to enjoy in the learning.

OUR MYSTERIOUS ATMOSPHERE

At the end of vol. II we cited a series of articles, forty-five in number, as "pegs upon which to hang or centres round which may be grouped, the conclusions of observation, experiment or reasoning as they occur."

As the result of the experience derived from the writing of the subsequent volumes we supplement the series of forty-five articles with the following:

ARTICLE 46. *The causes of weather.* The operative agencies which are responsible for the changes in the atmosphere connoted by "weather" are:

1. Radiation acting through entropy. Only that part of the energy of solar or terrestrial radiation which is "entropised" by absorption, and thereby becomes subject to the second law of thermodynamics as well as the first law, is available for causing motion in the atmosphere. In that respect there is an essential difference between the energy of radiation and that of gravitation.

2. Gravitation acting through geopotential. It is geopotential which controls the slope-effect and the downward convection of cooled air.

3. Water-vapour acting through latency. It is the latent heat of the vapour of water which supplies the driving-power of the vigorous upward motion of penetrative convection.

4. The earth's rotation with gravity acting through conservation of angular momentum, (i) geostrophic or (ii) cyclostrophic.

(i) Geostrophic—it is the displacement of a body of air polewards or equatorwards that is the chief cause of west or east motion; and

(ii) Cyclostrophic—the high velocities of hurricanes and tornadoes are related to the convergence of rotating rings of air.

5. Friction as a modifying influence, acting through turbulence. It is turbulence which is the chief factor in the restraint of winds within normal limits.

ARTICLE 47. *The rôle of entropy in relation to air-motion.* It is inadvisable to assume that the motion of air represented by the wind is horizontal because the instrument used to measure the wind records only the horizontal component.

Apart from penetrative convection due to the excess or defect of the entropy of a parcel of air above or below that of its environment, the motion of air is confined to an isentropic surface which may be horizontal or vertical or at any inclination to the horizon. It is a mistake to assume an isentropic surface to be horizontal.

ARTICLE 48. *Weather-mapping.* The effective representation of the facts of weather requires the computation of the form of successive isentropic surfaces upon which the distribution of pressure, temperature and density is represented by a single family of lines.

The addition of the real wind-velocity at a number of points of observation, if it were possible, would complete the effectiveness of the picture.

ARTICLE 49. *Pressure-gradient* (1) *the rôle of the earth's rotation.* Apart from the influence of penetrative convection the gradient of pressure at any given level is controlled by the earth's rotation acting through the uncompensated component of horizontal velocity.

The force $bb - 2\omega V\rho \sin \phi$ is always in operation at any level for re-establishment at that level of the condition $bb - 2\omega V\rho \sin \phi = 0$, and air is transported up or down the gradient to secure the adjustment.

ARTICLE 50. *Pressure-gradient* (2) *the rôle of water-vapour.* Water-vapour is the agency by which the entropy of air is increased locally to such an extent as to produce penetrative convection of air that may carry hundreds of thousands of millions of tons of air from the lowest levels to the highest attainable by convection in the circumstances indicated on a tephigram of the condition of the environment.

The ultimate limit of penetrative convection due to water-vapour is the tropopause.

ARTICLE 51. *Discontinuity.* In order to produce penetrative convection there must necessarily be discontinuity between the ascending air and its environment, which may be provided by the juxtaposition of masses of air of different life-history and consequently different composition.

ARTICLE 52. *The dumping of refuse air.* Penetrative convection implies the delivery of hundreds of millions of tons of air into the upper layers of the atmosphere which has to be carried away by the currents of the upper layers. The method of distribution of the air to the layers which carry it away is not yet determined by observation.

ARTICLE 53. *Revolving fluid.* The motion of the air at some one level in a cyclonic depression will approximate to that of a disc of revolving air, and, with the assistance of the earth's rotation, even a vertical column of revolving air extending from the ground through the canopy to the stratosphere, is possible but not inevitable.

The development requires the extension of convection of saturated air to the tropopause and its maintenance requires a discontinuity between the air which is to be convected (whether from the surface or elsewhere) and its environment.

Every cyclonic depression with closed isobars represents an endeavour, not always successful, to develop a revolving column in the overworld.

ARTICLE 54. *The result of the operations.* The atmospheric motion as produced and controlled by the operating agencies may be generalised as follows:

In the upper air:

1. The equatorial waist-band between the tropics with doldrum embroidery.
2. The northern and southern primary cyclones from lat. 35° north or south.

At sea-level

3. The quasi-permanent oceanic secondaries of fig. 76—representing the excavation of 2 billion tons.
4. The casual tertiaries (ordinary cyclonic depressions that travel), 1 billion tons.
5. The occasional quaternaries, tropical revolving storms, sub-tropical tornadoes and other small revolving systems, less than 100 million tons, and sometimes less than 1 million tons.
6. Eddies due to friction at the surface.

ARTICLE 55. *The energy of horizontal atmospheric motion.* By an easy formula if we take as unit of energy N, 162,500 kilowatt-hours, the energy of horizontal motion of the air in a slab at any level 100 m thick lying on a square between a pair of isobars without curvature 5 mb apart is $N \csc^2 \phi / \rho$.

Air-currents horizontal or inclined, not centres, are the working agents of the atmospheric polity.

ARTICLE 56. *A venturesome proposition.* Upper cloud-layers mark more probably an isentropic surface at which condensation of water-vapour has begun than the effect of turbulence in a horizontal surface between two layers with relative motion.

ARTICLE x. *An unsolved problem.* Motion of the atmosphere for which an explanation is asked is:

(1) A north wind or a south wind of 40 m/sec at the level of 8000 metres.

The motions of the atmosphere are controlled by the changes of entropy caused by solar and terrestrial radiation but what we call heat remains one of the mysteries of the universe.

It will indeed be interesting if the representation of the dynamical universe as the result of deformations of a space-time continuum should have its humble counterpart in the representation of the dynamics of the atmosphere as the result of deformations of an entropy-temperature continuum.

Two Voices are there; one is of the Sea,
One of the Mountains; each a mighty voice.

INDEX

POSTSCRIPT

ONE of the conclusions from the perusal of this volume seems to be that there is no fundamental principle for the sequence of wind and weather at all comparable with that of gravitation for the motion of the heavenly bodies and no picture of atmospheric motion comparable with Kepler's ellipses.

Mathematical computation is not by any means lacking but the best estimates of the prospects for to-morrow are based upon practical experience in the interpretation of weather-charts with an abundance of detail of past and present weather, not directly on numerical calculation. Even in the writings of comparatively successful forecasters one may find vague dissatisfaction with the position of the subject which one does not expect from successful exponents of a science and does not find in the other sciences.

The subject is perhaps unfortunate in having to submit the anticipations of its theory to the unavoidable check of practical experience of weather. No doubt wireless publicity has some disadvantages from the point of view of the advancement of science. A daily issue of the reasons for yesterday's happening might be a more direct avenue for the improvement of natural knowledge than a daily anticipation of to-morrow's weather. Scientific progress is mostly a matter of careful elaboration behind the scenes.

Sooner or later meteorologists must find an explanation of the origin of the general sense of dissatisfaction with, and disinclination for that part of the subject which is generally regarded as a first charge on the resources of the science. The author may be forgiven for indicating the contribution to the general theory of wind and weather which the reading of the proof-sheets of this volume has suggested and which was expressed in letters to *Nature* on June 27 and August 8.

For reasons which are set out tentatively in Chapters XI and XII it would appear that the lack of a successful theory may be due to the dominance of the idea of the depression and the anticyclone as the *controlling* agents of atmospheric changes, with an associated theory which explains the wind as *derived* from the distribution of pressure—such dominance relegates to a position of insignificance the main currents of the atmosphere which lie between the centres of depression and the anticyclones—the most notable examples of geostrophic wind-carrying energy which requires millions of kilowatt-hours for its expression.

Instead of looking to the centres of high pressure and low pressure as controlling powers Chapter XI proposes to regard them as created by the distribution of the geostrophic currents hitherto disregarded.

This paradoxical view may perhaps be more easily tolerated if we recall other examples of a certain habit of nature perverting her apparent intentions and outwitting the unwary meteorologist. The natural object of the air-flow towards the centre of a depression would be, *prima facie*, to fill it up; but the air finds its store of water-vapour used to maintain and intensify what it set out to destroy (Vol. IV, p. 286). *La lune mange les nuages* because loss of heat by radiation is made by natural descent to give a rise of temperature (Vol. III, p. 346) and, *per contra, le soleil ne mange pas les nuages* when added warmth lifts the cloud and consequently means more water (Vol. III, p. 189). In every department of physics except meteorology pressure is a typical store of potential energy; but in meteorology, as there can be no "buying" of air but only "borrowing," the net result is nugatory (Vol. IV, p. 299).

If we cannot rely upon the distribution of pressure other means must be found for tracing the winds to their ultimate cause which cannot be other than solar and terrestrial radiation, gravity and the earth's rotation. The expression of their effect is kinetic energy. And if we inquire what that energy amounts to, relying on the geostrophic relation for the horizontal component, we can refer to the maps of the distribution of pressure at various levels with the understanding that there may be also a vertical component, and indeed the three components, W to E, S to N and vertical are all of independent meteorological significance.

The horizontal energy is indicated by the distribution of pressure. We have given (p. 345) $162500 \operatorname{cosec}^2\phi/\rho$ kilowatt-hours as the energy of a slab of air 100 metres thick contained within a square between consecutive isobars with $5mb$ interval.

Local radiation (solar and terrestrial) will alter the entropy of any parcel of air and consequently change the position to which the parcel is entitled—thence a change of level of the air with its acquired energy. Loss of heat by radiation from air or ground must be generally characteristic of the polar as compared with the subtropical regions (figs. 76–7) and the vertical displacement of moving air thus involved implies cross winds at all levels, leading to surface-winds with the energy of S to N components as well as that of W to E.

The normal values of these components may be inferred from the oval shapes of the circumpolar isobars of Vol. II, figs. 164 to 172. Where those isobars cut the lines of latitude there must be superposed N to S components of energy for isobars leading *from* the pole, and S to N components for those leading *towards* the pole. The flow is not necessarily confined to the position indicated in the normal, which does however indicate a region of maximum intensity or maximum frequency.

Signs of superposed vertical displacement are already obvious. Clouds about sunset often dissolve into thin air as they travel and rain is often intensified after long solarisation. The horizontal energy goes with the descending or ascending air.

Herein would seem to lie a foundation for a workable hypothesis of a theory of winds on the basis of energy which may be developed when our knowledge of solar and terrestrial radiation can be successfully "broken in" to harness. In a very real sense the credit of the science of meteorology seems to rest upon deciphering the signs of the local peculiarities of the distribution of effective radiation.

MANUAL OF METEOROLOGY
VOLUMES I–IV
1926–1931

The drama of the atmosphere
HISTORICAL, GRAPHICAL, PHYSICAL AND MATHEMATICAL

Dramatis personae
AN INTEGRATION OF THE CONTENTS OF THE FOUR VOLUMES

CAMBRIDGE
AT THE UNIVERSITY PRESS
MCMXXXI

MANUAL OF METEOROLOGY

THE DRAMA OF THE ATMOSPHERE
DRAMATIS PERSONAE

Contents of the Manual, Vols. I–IV

⊙

● rainfall, ✱ snow

⌐ wind, ⌐ gale, *B* Beaufort number, T Thunder

V wind-velocity, *P* thrust per unit area

⌒ cloud-amount, *q* vapour-pressure

⌒ dew-point, ⌐ relative humidity

mb pressure in millibars

ρ density, *E* entropy

tt temperature

z height

≡ fog

☿

SUMMARY OF CONTENTS OF THE FOUR VOLUMES

A. INTRODUCTORY

B. EARTH, SEA AND SKY

C. *THE ATMOSPHERE AS WE FIND IT*

D. *A THEATRE OF ENERGY-TRANSFORMATIONS*

SUMMARY OF CONTENTS OF THE FOUR VOLUMES

E. EPIGRAPH

Bibliographies

Forms

Printed in the United States
by Bookmasters

Printed in the United States
By Bookmasters